Probability Theory
and Applications

D. REIDEL PUBLISHING COMPANY

A Member of the
Kluwer Academic Publishers Group
DORDRECHT/BOSTON/LANCASTER

Probability Theory and Applications

ENDERS A. ROBINSON

INTERNATIONAL HUMAN
RESOURCES DEVELOPMENT
CORPORATION
Boston

Library of Congress Cataloging in Publication Data

Robinson, Enders A.
    Probability theory and applications.

    Bibliography: p. 411
    Includes index.
    1. Probabilities.   I. Title.
QA273.R553   1985   519.2   84–29678

ISBN-13: 978-94-010-8877-0   e-ISBN-13: 978-94-009-5386-4
DOI: 10.1007/978-94-009-5386-4

Published by D. Reidel Publishing Company
P.O. Box 17, 3300 AA Dordrecht, Holland in co-publication with IHRDC

Sold and distributed in North America by IHRDC

In all other countries, sold and distributed by Kluwer Academic Publishers Group, P.O. Box 322, 3300 AH Dordrecht, Holland

Contents

v

Preface

Probability theory and its applications represent a discipline of fundamental importance to nearly all people working in the high-technology world that surrounds us. There is increasing awareness that we should ask not "Is it so?" but rather "What is the probability that it is so?" As a result, most colleges and universities require a course in mathematical probability to be given as part of the undergraduate training of all scientists, engineers, and mathematicians.

This book is a text for a first course in the mathematical theory of probability for undergraduate students who have the prerequisite of at least two, and better three, semesters of calculus. In particular, the student must have a good working knowledge of power series expansions and integration. Moreover, it would be helpful if the student has had some previous exposure to elementary probability theory, either in an elementary statistics course or a finite mathematics course in high school or college. If these prerequisites are met, then a good part of the material in this book can be covered in a semester (15-week) course that meets three hours a week.

The purpose of a course in probability is to provide the student with an understanding of randomness and give some practice in the use of mathematical techniques to analyze and interpret the effects of randomness in experiments and natural phenomena. In addition, a major aim of a first course in probability theory is to give many examples and applications relating to a wide variety of situations.

The chapters of this book make up a list of topics that are both important and reasonably possible to offer in a first course. There is less disagreement in the topics to be covered in a probability course than, say, in a statistics course. In any case, instructors should feel free to experiment with the order and emphasis given to these topics to find the arrangement most suitable for the interests of their classes. A typical arrangement might be as follows.

Chapter 1 on the algebra of sets and the axioms of probability is covered in the first week (week no. 1). Chapter 2 on the theorems of probability, combinatorial analysis, conditional probability and independence, and Bayes' theorem is covered in the second week (week no. 2). (If the students have not had an elementary introduction to probability previously, then an extra week should be spent on Chapter 2, and as compensation one week should be removed from the end of the course.)

The next three weeks can be spent on Chapter 3, namely, week no. 3 on random variables and probability mass functions, week no. 4 on distribution functions and probability density functions, and week no. 5 on random sampling and change of variable. The change-of-variable section is the critical section in the book. Although this section is strictly nothing more than the technique of transforming variables in calculus, some of the students will master this section and others will not. The ones that do are the A students, and the rest of the semester is a coast downhill for them. The other students will still have to work hard.

One week (week no. 6) is enough for Chapter 4 if only the first two sections (applications of mathematical expectation and the Chebyshev inequality) are covered, and the third section (applications to operations research) is omitted.

Three weeks (no. 7, 8, 9) can be spent on Chapter 5, which covers multivariate distributions. Weeks no. 7 and 8 would be spent on the sections on joint, marginal, and conditional distributions, and week no. 9 on the sections on expectation, covariance and correlation, regression curves, and the law of large numbers. The last section, change of several variables, would be omitted.

One week (week no. 10) is enough for Chapter 6 if the first two sections (probability generating function, gamma functions and beta functions) are omitted, and only the last two sections (moment generating function, applications) are covered.

The ten weeks up to now have prepared the student with the mathematical techniques required to handle the material in Chapters 7 and 8. These two chapters, which cover the basic distributions of probability theory, will provide the student with the concepts needed to apply his knowledge to real-world problems. Chapters 7 and 8 represent the fruits of the probability course, and the sections are organized so as to bring out the interrelationships among the various probabilistic models.

If its last section (negative binomial distribution) is omitted, and its first section (review of combinatorial methods) is gone over rapidly, then Chapter 7 can be covered in two weeks (weeks no. 11 and 12). The sections covered would be on uniform distribution, hypergeometric distribution, the Bernoulli process and binomial distribution, the Poisson process and Poisson distribution, and geometric and Pascal distributions. The final three weeks (no. 13, 14, and 15) would be spent on Chapter 8, with the final three sections (gamma distribution; beta distribution; and chi-square, F, and t distributions) omitted. The sections covered in Chapter 8 would be on uniform distribution, the Cauchy distribution, exponential and Erlang distributions, sums of random variables, normal distribution, central limit theorem, reproduction properties of the normal distribution, and bivariate normal distribution. Time does not allow for the final chapter of the book, Chapter 9 on regression and correlation, to be included in a one-semester course.

Exercises for the student are given at the end of virtually all sections, and they should be assigned and graded at least on a weekly basis in order to monitor the progress of the students in an intensive course of this kind. Because the material is difficult and involves a lot of homework on the part of the student, it is important that the classes be small and a close student–teacher relationship be maintained. However, the payoff is large. The student will not only develop a better understanding of calculus with its attendant skills, but will also be well prepared for subsequent courses in stochastic processes, time-series analysis, and mathematical statistics, as well as for courses in specific professional areas. Additional exercises appear in Appendix A.

A word might be said about the use of computers in a probability course. In the standard fifteen-week course with three lectures per week, it seems that the student has enough to do to master the calculus techniques for studying probability as given in this book. However, if an additional two-hour per week computer laboratory session is added to the course, then the computer techniques acquired would be invaluable. A companion volume, *Computer Laboratory Manual on Probability*, is in preparation to go with this volume for such an expanded course.

The University of Tulsa is at the center of a high-technology area with major industrial enterprises and research laboratories in petroleum, chemicals, aerospace, computers, electronics, and precision machinery. The College of Engineering and Applied Sciences at the university attracts many serious students and professional engineers who expect and want difficult courses such as this one. They also expect the attendant computer laboratory experience in order to translate theory into practice. The origins of this book go back to the time when the author taught this course in the Mathematics Department of the Massachusetts Institute of Technology with Professors George P. Wadsworth and Joseph G. Bryan. I learned much in working with such distinguished mathematicians, especially regression analysis, and to them I own my sincere thanks and deep appreciation. The value of this course can be judged by the fact that many of the students who took this course years ago have achieved recognition in probability theory and statistics, and hold high positions in universities, industry, and government. More importantly, several have stated that their first introduction to probability theory led them into endeavors that gave them much personal enjoyment and a sense of fulfillment.

I want to thank Terry Saunders for her excellent and caring work in conveying the various manuscripts to such beautiful form on the word processor.

1

Basic Probability Concepts

*In each of seven houses there are
seven cats. Each cat kills seven
mice. Each mouse would have eaten
seven hekat of grain. How much
grain is saved by the cats?*

AHMOSE THE SCRIBE (ca 1650 B.C.)

Introduction to Probability

Algebra of Sets

Axioms of Probability

INTRODUCTION TO PROBABILITY

Probability theory is the branch of mathematics that is concerned with random events. An event that unavoidably occurs for every realization of a given set of conditions is called a certain or sure event. If an event definitely cannot occur upon realization of the set of conditions it is called impossible. If, when the set of conditions is realized, the event may or may not occur, it is called *random*. Probability theory, or the mathematical treatment of random events, has both intrinsic interest and practical interest. There are many successful applications within the physical, social, and biological sciences, in engineering, and in business and commerce.

At first sight, it might seem difficult or impossible to make any valid statements about such random events, but experience has shown that meaningful assertions can be obtained in a large class of situations. These situations encompass the random phenomena that exhibit a *statistical regularity*. For example, in tossing a coin, we can make no definite statement as to whether it lands heads or tails in any single toss, but for a large number of tosses the proportion of heads seems to fluctuate around some fixed number p between 0 and 1. Unless the coin is severely unbalanced, the ratio p is very near $1/2$. We can think of the probability that the coin will land heads in a single toss as being given by the number p.

In general, the probability p of a random event can be interpreted as meaning that if the experiment is repeated a large number of times, the event would be observed about $100p$ percent of the time. This interpretation is called the *relative frequency interpretation of probability*. Such a way of thinking about probability is very natural in many applications of probability to the real world, especially in the experimental sciences where an experiment can be repeated a large number of times under identical conditions. However, in other applications the relative frequency interpretation may seem quite artificial, especially in cases where an uncertain event can in principle happen at most one time, and cannot be repeated. In such cases, other interpretations of probability are required. In mathematics, the actual interpretation of probability is irrelevant, just as in geometry the interpretation of points, lines, and planes is irrelevant. In this book, we use the relative frequency interpretation of probabilities only as an intuitive motivation for the definitions and theorems.

Probabilities are always associated with the occurrence or the nonoccurrence of events, such as the event that three defectives are found in a sample of 100 electric motors, the event that a car will last at least 150,000 miles before it must be scrapped, the event that a switchboard has 10 incoming calls during a certain period of time, and so forth. Thus, in connection with probabilities, we always refer to an individual outcome or to a set of outcomes of an experiment as an event.

In statistics it is customary to refer to any process of observation as an experiment. Thus, an experiment may consist of the simple process of noting whether an assembly is broken or intact; it may consist of determining what proportion of a sample of people prefers Brand X to Brand Y; or it may consist of the very complicated process of finding a new oil field. The results of any such observations, whether they be simple "yes" or "no" answers, instrument readings, or whatever, are called the possible outcomes of the respective experiments. The possible outcomes of an experiment are so defined that, in any given trial of the experiment, (1) part of a possible outcome cannot occur (*elemental property*); (2) one of them does occur (*exhaustive property*); and (3) only one of them does occur (*exclusive property.*) That is, the possible outcomes are elemental, exhaustive, and exclusive. The totality of the possible outcomes of an experiment is called the *sample space*, or *universe*, or *population*, of the experiment and this universal set will be denoted by the letter U. The basic object in the mathematical model is the *element* or *sample point*, which is just a formal name for possible outcome. The collection of all sample points thus makes up the sample space. A sample space with at most a countable number of sample points is called a discrete sample space; a sample space with a continuum of sample points is a continuous sample space.

An event is defined as a subset of an appropriate sample space. To illustrate, suppose that an assembly consists of two subassemblies, with the first containing 4 components and the second containing 3. If we are concerned only with the total number of defective components in each subassembly (not with what particular components have failed), then there can be 0, 1, 2, 3, or 4 defectives in the first subassembly, and 0, 1, 2, or 3 defectives in the second subassembly. Thus the number of elements in the sample space is $5 \cdot 4 = 20$, which can be represented by the discrete points (k,j) where $k = 0, 1, 2, 3, 4$ and $j = 0, 1, 2, 3$.

It is often convenient to classify sample spaces according to the number of elements they contain. Thus, a sample space with a finite number of elements is referred to as a finite sample space. To consider an example where a finite sample space does not suffice, suppose that a person inspecting manufactured items is interested in the number of items that must be inspected before the first broken one is observed. It might be the first, the second, . . ., the hundredth, . . ., perhaps a thousand or more would have to be inspected before a broken one is found. In this case, where the size of the number of interest is not known, it is appropriate to consider the sample space as the whole set of natural numbers. Thus, the number of elements in this sample space is countably infinite. To go one step further, if the length of an item is being measured, the sample space would consist of points on a continuous scale (a certain interval on the line of real numbers). The elements of this sample space cannot be counted since they cannot be put into one-to-one correspondence with the natural numbers. We say a sample space is *discrete* if it has finitely many or a countable infinity of elements. However, if the elements (points) of a sample space constitute a continuum, such as all the points on a line, or all the points on a line segment, or all the points in a plane, the sample space is said to be *continuous*.

At times it can be quite difficult to determine the number of elements in a finite sample space by direct enumeration. To illustrate a method that can simplify things, let us consider the following problem. Suppose that three tests are carried out on a machine. Let a_1, a_2, a_3 represent the three levels of performance for the first test, let b_1 and b_2 represent the two levels for the second test, and c_1, c_2, c_3 represent the three levels for the third test. The possible outcomes of the three tests may then be visualized by means of the "tree" diagram of Figure 1.1. Following (from left to right) a given path along the tree we obtain a given outcome of the test, that is, a particular element of the sample space of the experiment. Evidently, the tree has 18 paths and thus the sample space of the experiment has 18 elements. We could also have determined the number of elements of this sample space by noting that there are three "a branches," that each a branch forks into two "b branches," and that each b branch forks into three "c branches." Thus, there are $3 \cdot 2 \cdot 3 = 18$ combinations of branches, or paths. This result is generalized by the following theorem:

THEOREM 1.1. If sets A_1, A_2, $\cdots$, A_k contain respectively n_1, n_2, $\cdots$, n_k elements, there are $n_1 \cdot n_2 \cdot \cdots \cdot n_k$ ways of selecting first an element from A_1, then an element from A_2, $\cdots$, and finally an element from A_k.

This theorem can be verified by constructing a tree diagram similar to that of Figure 1.1.

EXERCISES

1. An experiment consists of rolling a die and then flipping a coin if and only if the die came up odd. Draw a tree diagram and count the number of possible outcomes.

2. A tire manufacturer wants to test four different tread designs on three different kinds of road surfaces and at five different speeds. How many different test runs are required?

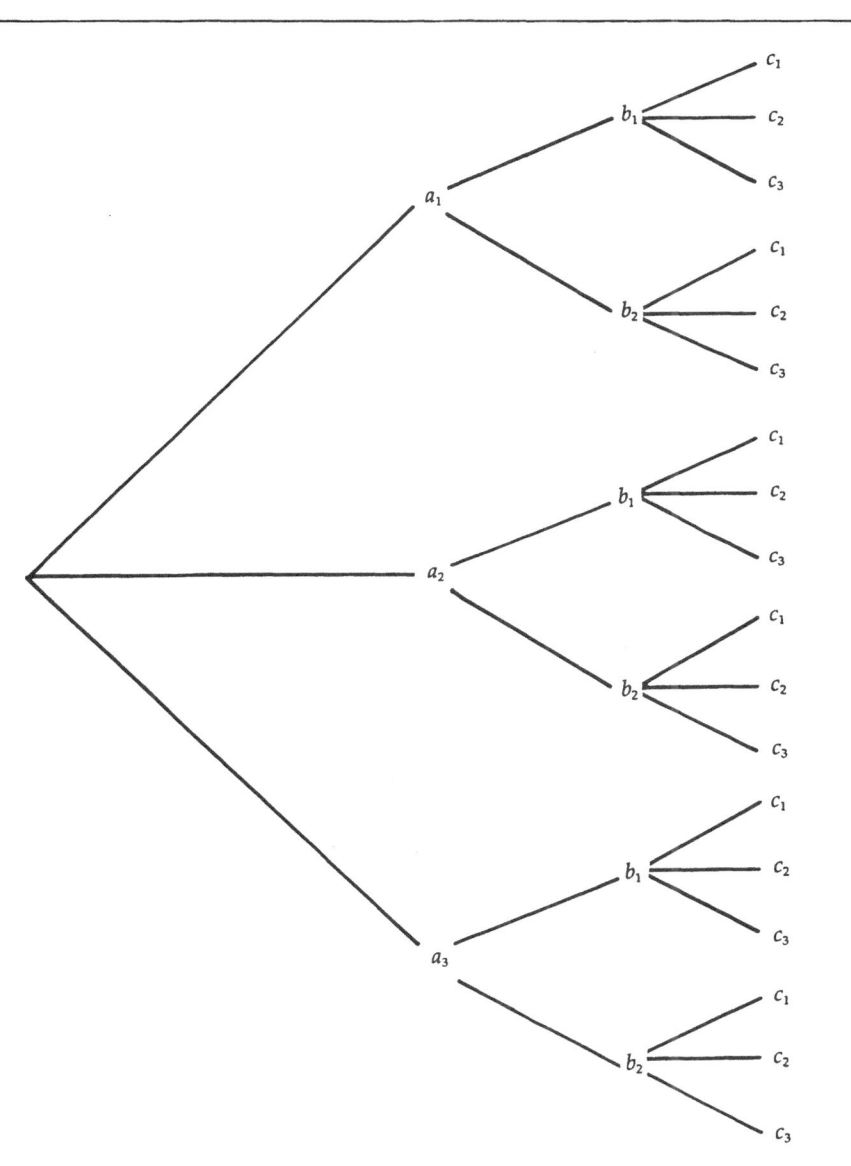

FIGURE 1.1 Tree diagram.

3. Construct a tree diagram to determine the number of ways a coin can be flipped four times in succession such that throughout the series of flips the number of tails is always greater than or equal to the number of heads.

4. In each of the following experiments decide whether it would be appropriate to use a sample space that is finite, countably infinite, or continuous:
   (a) One of ten people is to be chosen president of a company.
   (b) An experiment is conducted to measure the coefficient of expansion of certain steel beams.
   (c) Measurements of radiation intensity are made with a Geiger counter.
   (d) A policeman measures the alcohol content of a driver's blood.
   (e) A coin is flipped until the first tails appears.
   (f) A traffic survey is made to estimate the number of cars in the city with defective headlights.

5. An experiment has four different paints to apply to both sides of a sheet of steel. In how many ways is it possible to coat both sides of the sheet of steel if
   (a) both sides must be coated with the same material,
   (b) the two sides can but need not be coated with the same material,
   (c) both sides cannot be coated with the same material.

ALGEBRA OF SETS

An explicit language for the development of probability theory is provided by the algebra of sets. The words used in probability theory often reflect its involvement with the real world. Thus, in a probability setting, "point" becomes "sample point" and "space" becomes "sample space." A set of sample points is also given a name having a special empirical flavor: *event*. An event is simply a set of sample points—any set of sample points. In particular, the whole sample space or universe U is an event, and is sometimes called the *sure* or *certain event*. In probability theory, the null or empty set ϕ is called the *impossible event*.

Because events are sets, the notions and notations of elementary set theory are appropriate to their discussion. Thus, if A, B are events, then

A', the complement of A,
$A \cup B$, the union of A and B, and
$A \cap B$, the intersection of A and B

are all events.

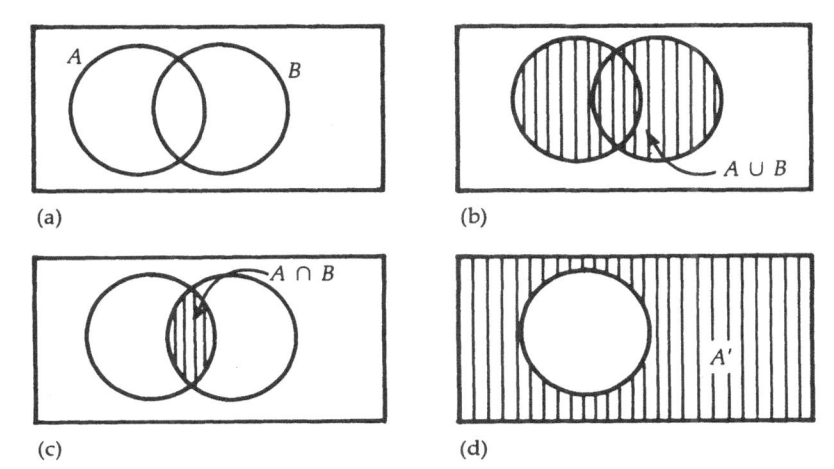

FIGURE 1.2 Venn diagrams.

Event A' is defined as the collection of all points in the sample space U that are not included in the event A. The null set ϕ contains no points and is the complement of the sample space. Event $A \cup B$ is the collection of all points that are either in A or in B or in both. Event $A \cap B$ is the collection of all points that are contained both in A and B. Diagrammatically, if A and B are represented by the indicated regions in Figure 1.2(a), then $A \cup B$, $A \cap B$, and A' are represented by the shaded regions in Figures 1.2(b), (c), and (d), respectively. Such diagrams are called Venn diagrams.

To illustrate these concepts, let A be the event "red on roulette" and let B be the event "even number on roulette." Then the union $A \cup B$ is the event that either red or even-number comes up. The intersection $A \cap B$ is the event "red even-number." The event A' occurs if a red does not come up.

In the ordinary Venn diagrams the sample space is represented by a rectangle, while subsets or events are represented by circles, parts of circles, or various combinations. For instance, Figure 1.3(a) depicts the relation "A is a *subset* of B," that is, $A \subset B$, while Figure 1.3(b) depicts the relation that "A and B are *mutually exclusive*," that is, $A \cap B = \phi$. Another term for mutually exclusive is *disjoint*. That is, if all the elements of one set are in common with elements of a second set, then we say that the first set is a subset of the second set. If two sets have no elements in common, then we say that the two sets are mutually exclusive (or disjoint). In symbols, sets A and B are mutually exclusive if and only if $A \cap B = \phi$.

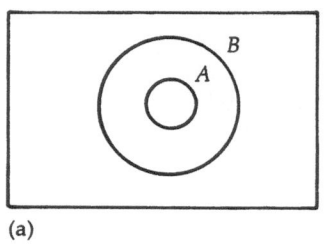 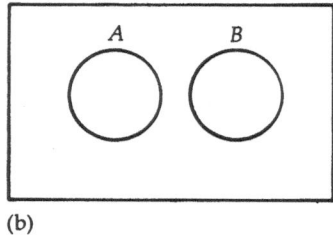

(a)   (b)

FIGURE 1.3 (a) $A \subset B$, (b) $A \cap B = \phi$.

The set operations we have introduced, together with appropriate axioms, lead to the algebra of sets, or Boolean algebra. In our work, we justify whatever results are needed with aid of Venn diagrams. As an example, let us demonstrate that

$$(A \cup B)' = A' \cap B',$$

which expresses the fact that the complement of the union of two sets equals the intersection of their respective complements. To begin with, note that the shaded region of Figure 1.4(a) represents the set $(A \cup B)'$. The crosshatched region of Figure 1.4(b) was obtained by shading the region representing A' with lines going in one direction and that representing B' with lines going in another direction. Thus, the cross-hatched region represents the intersection of A' and B', and it can be seen that it is identical with the shaded region of Figure 1.4(a).

EXERCISES

1. From the following equality find the random event X:

$$(X \cup A)' \cup (X \cup A')' = B$$

ANSWER. $X = B'$.

2. Prove that $(A' \cap B) \cup (A \cap B') = (A \cap B)'$.

3. Use Venn diagrams to verify that
   (a) $(A \cap B)' = A' \cup B'$,
   (b) $A \cup (A \cap B) = A$,
   (c) $(A \cap B) \cup (A \cap B') = A$,
   (d) $A \cup B = (A \cap B) \cup (A \cap B') \cup (A' \cap B)$,
   (e) $A \cup (B \cap C) = (A \cup B) \cap (A \cup C)$.

4. Prove that $A' \cap B' = (A \cup B)'$ and $C' \cup D' = (C \cap D)'$.

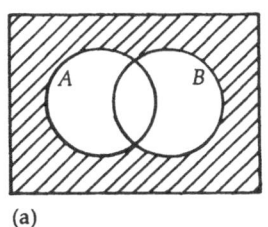 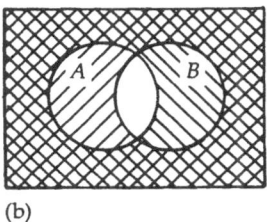

(a)   (b)

FIGURE 1.4 (a) $(A \cup B)'$; (b) $A' \cap B'$

5. Can the events A and $(A \cup B)'$ be simultaneous?
   ANSWER. No, since $(A \cup B)' = A' \cap B'$.

6. Prove that A, $A' \cap B$, and $(A \cup B)'$ form a complete set of events.
   HINT. Use the equality $(A \cup B)' = A' \cap B'$.

7. Two chess players play one game. Let the event A be that the first player wins, and let B be that the second player wins. What event should be added to these events to obtain a complete set?
   ANSWER. The event C of a tie.

8. An installation consists of two boilers and one engine. Let the event A be that the engine is in good condition, let B_k ($k = 1$, 2) be that the kth boiler is in good condition, and let C be that the installation can operate if the engine and at least one of the boilers are in good condition. Express the events C and C' in terms of A and B_k.
   ANSWER. $C = A \cap (B_1 \cup B_2)$, $C' = A \cup (B_1' \cap B_2')$.

9. A device is made of two units of the first type and three units of the second type. Let A_k ($k = 1$, 2) be that the kth unit of the first type is in good condition, let B_j ($j = 1$, 2, 3) be that the jth unit of the second type is in good condition, and let C be that the device can operate if at least one unit of the first type and at least two units of the second type are in good condition. Express the event C in terms of A_k and B_j.
   ANSWER. $C = (A_1 \cup A_2) \cap [(B_1 \cap B_2) \cup (B_1 \cap B_3) \cup (B_2 \cap B_3)]$.

10. (a) What meaning can be assigned to the events $A \cup A$ and $A \cap A$?
    (b) When does the equality $A \cap B = A$ hold?
    (c) When do the following equalities hold true: (i) $A \cup B = A$, (ii) $A \cap B = A'$, (iii) $A \cup B = A \cap B$?

11. The shorthand notation

$$\bigcup_n A_n = A_1 \cup A_2 \cup \cdots$$

is used to denote the union of all the sets of the sequence, and

$$\bigcap_n A_n = A_1 \cap A_2 \cap \cdots$$

to denote the intersection of all the sets of the sequence. De-Morgan's laws state that if $\{A_n\}$, $n \geq 1$, is any sequence of sets, then

$$\left(\bigcup_n A_n\right)' = \left(\bigcap_n A_n'\right) \quad \text{and} \quad \left(\bigcap_n A_n\right)' = \left(\bigcup_n A_n'\right).$$

Establish these laws.

12. A target consists of 10 concentric circles of increasing radius r_k ($k = 1, 2, 3, \cdots, 10$). An event A_k means hitting the interior of a circle of radius r_k ($k = 1, 2, \cdots, 10$). What do the following events mean:

$$B = \bigcup_{k=1}^{6} A_k, \quad C = \bigcap_{k=5}^{10} A_k.$$

ANSWER. $B = A_6$, $C = A_5$.

AXIOMS OF PROBABILITY

In this section we shall define probability using the concept of a *set function*, an additive set function, to be exact. A set function is a function such that the elements of the domain are sets while the elements of the range are real numbers. In other words, a set function represents a correspondence, which assigns real numbers to the subsets of a given set (to the subsets of a sample space if we are concerned with the outcomes of a given experiment). A set function is said to be *additive* if the number it assigns to the union of two subsets that have no elements in common equals the sum of the numbers assigned to the individual subsets.

Using the concept of an additive set function, let us now define the probability of an event. Given a sample space U and an event A in U, we define $P(A)$, the probability of A, to be a value of an additive set function called a probability function. For a set function to be a probability function, it must satisfy the following three conditions:

AXIOM 1.1. $0 \leq P(A) \leq 1$ for each event A in U.

AXIOM 1.2. $P(U) = 1$.

AXIOM 1.3. If A and B are any mutually exclusive events in U, that is, if $A \cap B = \phi$, then

$$P(A \cup B) = P(A) + P(B).$$

The first axiom states that the probability function assigns to every event A in U some real number from 0 to 1, inclusive. The second axiom states that the sample space as a whole is assigned the number 1 and it expresses the idea that the probability of a certain event, an event that must happen, is equal to 1. The third axiom states that the probability function for two disjoint events must be additive, and with the use of mathematical induction it can be extended to include any finite number of disjoint events. In other words, it can be shown that

$$P(A_1 \cup A_2 \cup \cdots \cup A_n) = P(A_1) + P(A_2) + \cdots + P(A_n),$$

where A_1, A_2, $\cdots$, A_n are any mutually exclusive events in U. However, the third axiom must be modified when U is not finite. Axiom 1.3 must be replaced by

AXIOM 1.3′. If A_1, A_2, A_3, $\cdots$ is a finite or infinite sequence of mutually exclusive events in U, then

$$P(A_1 \cup A_2 \cup A_3 \cdots) = P(A_1) + P(A_2) + P(A_3) + \cdots.$$

The sample space, or universal set, U, equipped with its probability measure P is called a *probability space*. A formal way of saying the same thing is that the pair of objects (U,P) is called a probability space. If U is finite or countable, (U,P) is called a *discrete probability space*; otherwise it is called a *continuous probability space*.

In applying probability theory to concrete random experiments, one starts by assigning a "reasonable" probability measure to the sample space. Finite equally likely probability theory only considers experiments with a finite number of possible outcomes with the probability value assigned to each possible outcome being the same. In other words, for reasons of symmetry or because of some other theoretical argument, we conclude that each sample point is equiprobable, and hence we assign to each of the N sample points x_1, $\cdots$, x_N the probability $1/N$. Then $P(A)$, the probability of a set A, is evaluated by counting the number of sample points that make up A and dividing this count by N. This so-called case of *equal likelihood* is important and it provides many of the examples for discrete probability theory.

EXAMPLE. An honest six-faced die is thrown. The word *honest* means that each of the six outcomes 1, 2, 3, 4, 5, and 6 is assigned the same probability $1/6$. If A is the event that the result of the throw is an odd number, then $P(A) = 3/6$, because $A = \{1,3,5\}$.

If a sample space has N possible outcomes (i.e., sample points) it can be shown that it contains 2^N subsets, including the sample space as a whole and the null set ϕ. Thus, if $N = 20$ there are over a million events in U, and the problem of specifying a probability for each event can become a formidable task. Fortunately, the work can be simplified considerably by the use of the following theorem:

THEOREM 1.2. If A is an event in a discrete sample space U, then $P(A)$ equals the sum of the probabilities of the individual sample points comprising A.

PROOF. By definition, the individual sample points are mutually exclusive. Let $x_1, x_2, \cdots, x_N$ be the N sample points making up the event A, so that we can write $A = x_1 \cup x_2 \cup \cdots \cup x_N$. By Axiom 1.3' we have

$$P(A) = P(x_1 \cup x_2 \cup \cdots \cup x_N)$$
$$= P(x_1) + P(x_2) + \cdots + P(x_N),$$

which completes the proof.   Q.E.D.

If we apply this theorem to the equilikely case, we immediately obtain:

THEOREM 1.3. If an experiment has N possible outcomes that are equiprobable and if n of these are labeled "success," then the probability of a "success" is n/N.

This theorem is particularly helpful in problems dealing with games of chance, where it is assumed that if a deck of cards is properly shuffled each card has the same chance of being selected, if a coin is properly flipped each face has the same chance of coming up, and if a die is properly rolled each face is equilikely to come up. We find that the probability of drawing an ace from an ordinary deck of 52 playing cards is $4/52$, the probability of heads with flipping a balanced coin is $1/2$, and the probability of an odd number with rolling a die is $3/6$.

So far so good. However, it is important to stress that the three axioms do not tell us how to assign probabilities to the various outcomes of an experiment; they merely restrict the ways in which it can be done. In practice, probabilities can be assigned either on the basis of estimates obtained from past experience, on the basis of an analysis of conditions underlying the experiment, or on the basis of various assumptions such as the common assumption that various outcomes are equiprobable.

The following examples (1), (2), and (3) are three illustrations of permissible ways of assigning probabilities in an experiment where there are three possible and mutually exclusive outcomes A, B, and C:

1. $P(A) = \frac{1}{3}$, $P(B) = \frac{1}{3}$, $P(C) = \frac{1}{3}$
2. $P(A) = 0.57$, $P(B) = 0.24$, $P(C) = 0.19$
3. $P(A) = \frac{24}{27}$, $P(B) = \frac{2}{27}$, $P(C) = \frac{1}{27}$

However,

4. $P(A) = 0.64$, $P(B) = 0.38$, $P(C) = -0.02$
5. $P(A) = 0.35$, $P(B) = 0.52$, $P(C) = 0.26$

are not permissible because they violate, respectively, Axioms 1.1 and 1.2.

As an example, let us discuss an important class of probability measures, called uniform probability measures. Some of the oldest problems in probability involve the idea of picking a point "at random" from a set S. Our intuitive ideas on this notion show us that if A and B are two subsets having the same "size" then the chance of picking a point from A should be the same as from B. If S has only finitely many points, we can measure the "size" of a set by the number of points in it. Two sets are then of the same "size" if they have the same number of points. Let us consider the experiment of picking a point at random from a set S having a finite number N of points. We take the universal set U to be S, and under the equiprobable assumption we assign to the set A the probability $P(A) = n/N$, where A is a set having exactly n points. This equally likely probability measure is called uniform, because each one-point set carries the same probability.

Let us now look at a set with an infinite number of points. Suppose now that S is the interval $[a,b]$ on the real line where $-\infty < a < b < +\infty$. It seems reasonable in this case to measure the size of a subset A of $[a,b]$ by its length. Two sets are then of the same size if they have the same length. We will denote the length of a set A by $|A|$. To construct a probability measure for the experiment of "choosing a point at random from S," we proceed as follows. We take the universal set U to be S, and under the equiprobable assumption we now define the probability of an event A as $P(A) = |A|/|S|$ whenever A is an interval. This probability measure is said to be uniform.

We may extend this concept to any number of dimensions. Let S be any subset of r-dimensional Euclidean space having finite, nonzero, r-dimensional volume. For a subset A of S denote the volume of A by $|A|$. An equiprobable probability measure P is defined on A such that $P(A) = |A|/|S|$ for any such set A. Such a probability measure is called uniform.

Let us now consider the problem of assigning nonuniform probabilities, which we must do if we can no longer appeal to the equiprobable assumption. Let us consider the time that it takes a radio-active atom to decay to its stable form. In simple language we will say that a radioactive atom is alive, and when it decays it dies. According to physical experience it is impossible to say with certainty when a specified radioactive atom will decay (die). However, if we observe a large number N of radioactive atoms initially, then we can make some accurate predictions about the remaining number of radioactive atoms $N(t)$ that have not decayed (died) by time t. That is, we can, to a good approximation, predict the fraction of radioactive atoms $N(t)/N$ that are still present (alive) at time t, but we cannot say which of the radioactive atoms will have survived. Since all of the radioactive atoms are identical, observing N radioactive atoms simultaneously is equivalent to N repetitions of the same experiment where, in this case, the experiment consists in observing the time that it takes a single radioactive atom to decay (die). Here we make the crucial assumption that each radioactive atom acts independently of the rest. Now to a first approximation experimental evidence shows that the rate $dN(t)/dt$ at which the radioactive substance decays at time t is proportional to the number of radioactive atoms $N(t)$ present at time t. This relation can be expressed by the differential equation

$$\frac{dN(t)}{dt} = -\lambda N(t) \qquad N(0) = N.$$

Here $\lambda > 0$ is a fixed constant of proportionality, and the minus sign occurs on the right side because $N(t)$ is decreasing with time t. The unique solution of this equation is $N(t) = Ne^{-\lambda t}$. Thus the fraction of radioactive atoms that have not decayed (died) by time t is given approximately by $N(t)/N = e^{-\lambda t}$. If $0 \leq t_0 \leq t_1$, the fraction of radioactive atoms that decay in the time interval $[t_0, t_1]$ is $[N(t_0) - N(t_1)]/N = (e^{-\lambda t_0} - e^{-\lambda t_1})$. Consequently, in accordance with the relative frequency interpretation of probability we take $(e^{-\lambda t_0} - e^{-\lambda t_1})$ as the probability that a radioactive atom decays (dies) between times t_0 and t_1. An outcome of the experiment is the time that a radioactive atom takes to decay (die). This can be any positive real number, so we take the universal set U as the entire nonnegative time axis $0 \leq t < \infty$. We then assign to the interval $[t_0, t_1]$ the probability $(e^{-\lambda t_0} - e^{-\lambda t_1})$. In particular if $t_0 = t_1 = t$, then the interval degenerates to the set consisting of the single point $\{t\}$ and the probability assigned to this set is 0. Suppose A and B are two disjoint intervals. Then the proportion of radioactive atoms that decay in the time interval $A \cup B$ is the sum of the proportion that decay in the time interval A and the proportion that decay in the time interval B. Thus we have

$$P(A \cup B) = P(A) + P(B)$$

whenever A and B are two disjoint intervals. This probability measure, namely the probability that a radioactive atom decays (dies) in the time period from t_0 to t given by

$$P(t_0 \leq t \leq t_1) = e^{-\lambda t_0} - e^{-\lambda t_1},$$

is said to be *exponential*. During the course of this book we will have much to say about this important probability distribution. For a radioactive substance for which λ is a large number, the radioactivity goes away quickly, whereas for a substance with a small λ, the radioactivity stays around for a long time, a very long time.

Often in scientific problems where probabilities are difficult to determine mathematically, the relative frequency theory can be used to obtain an empirical approximation. The actual experiment or, if more feasible, a simulation of it is repeated a large number of times in an identical and independent way. The observed proportion of the outcomes in the event of interest is determined and is used as an approximation of the required theoretical probability. In practice, it is useful to try to motivate theoretical facts about probability theory by means of the relative frequency theory. Such motivations can be extremely helpful in applications. In fact the axioms of probability are motivated by the relative frequency interpretation of probability. In summary, the relative frequency interpretation is an empirical theory that considers what happens when a random experiment is repeated a large number of times with each repetition being identical and independent of all other repetitions. Among these independent repetitions the proportion of times that an observed outcome occurs in the set A is computed. According to the relative frequency interpretation, this proportion will approach the limit $P(A)$ as the experiment is repeated indefinitely. Thus, the theoretical probability $P(A)$ is an idealization of the proportion of times an experiment results in outcomes in A over a large number of independent repetitions of the experiment.

EXERCISES

1. For the following random experiments, give a mathematical model, with a description of a listing of the sample points and the assignment of reasonable probabilities to them:
   (a) A box contains five balls numbered 1, 2, 3, 4, 5. A ball is drawn at random.
   (b) A box contains five balls numbered 1, 2, 3, 4, 5. A ball is drawn at random, replaced, and a second drawing is made at random.

(c) A box contains five balls numbered 1, 2, 3, 4, 5. A ball is drawn at random, is not replaced, and a second drawing is made at random.

2. An experiment consists of flipping a coin and then flipping it a second time if a head occurs. If a tail occurs on the first flip, then a die is tossed once.
   (a) List the elements of the sample space U.
   (b) List the elements of U corresponding to event A that a number less than 4 occurred on the die.
   (c) List the elements of U corresponding to event B that two tails occurred.
   ANSWERS. (a) $U = \{HH,HT,T1,T2,T3,T4,T5,T6\}$. (b) $A = \{T1,T2,T3\}$.
   (c) $B = \phi$.

3. Give mathematical models for the following random experiments. In each case, describe a listing of the sample points and reasonable probabilities to assign to them. Unless otherwise stated, the word die will always refer to a conventional six-faced die with faces marked 1, 2, 3, 4, 5, and 6.
   (a) A coin is tossed five times.
   (b) A die is tossed five times.
   (c) A die having faces marked 1, 1, 2, 2, 3, 4 is tossed five times.
   (d) A coin is tossed until a head appears.
   (e) A die is tossed until a 1 appears.
   (f) A die is tossed until either a 1 or a 2 appears.

4. Suppose that an experiment has exactly four possible outcomes: A, B, C, and D. Which of the following assignments of probabilities are permissible:
   (a) $P(A) = 0.36$, $P(B) = 0.18$, $P(C) = 0.21$, $P(D) = 0.25$,
   (b) $P(A) = 0.29$, $P(B) = 0.35$, $P(C) = 0.18$, $P(D) = 0.15$,
   (c) $P(A) = 0.42$, $P(B) = 0.17$, $P(C) = -0.08$, $P(D) = 0.49$,
   (d) $P(A) = {}^{17}/_{80}$, $P(B) = {}^{11}/_{40}$, $P(C) = {}^{1}/_{2}$, $P(D) = {}^{1}/_{80}$.
   ANSWERS. (a) Yes. (b) No. (c) No. (d) Yes.

5. A model for a random spinner can be made by taking uniform probability on the circumference of a circle of radius 1, so that the probability that the pointer of the spinner lands in an arc of length s is $s/2\pi$. Suppose the circle is divided into 37 zones numbered 1, 2, $\cdots$, 37. Compute the probability that the spinner stops in an even zone.

6. Let a point be picked at random in the unit square, i.e., the square bounded by the lines $x = 0$, $x = 1$, $y = 0$, $y = 1$. Compute the

probability that it is in the triangle bounded by $x = 0$, $y = 0$, and $x + y = 1$.

7. Let a point be picked at random in the disk of radius 1. Find the probability that it lies in the angular sector from 0 to $\pi/4$ radians.

8. A cube whose faces are colored is split into 1000 small cubes of equal size. The cubes thus obtained are mixed thoroughly. Find the probability that a cube drawn at random will have two colored faces.

   HINT. The total number of small cubes is $N = 1000$. A cube has 12 edges and there are 8 small cubes with two colored faces on each edge.

   ANSWER. $P = (12)(8)/1000 = 0.096$.

9. By a "random number" here we mean a k-digit number ($k > 1$) such that any of its digits may range from 0 to 9 with equal probability. Find the probability that the last two digits of the cube of a random integer k will be 1.

   ANSWER. Represent k in the form $k = a + 10b + \cdots$, where a, b, $\cdots$ are arbitrary numbers ranging from 0 to 9. Then $k^3 = a^3 + 30a^2b + \cdots$. From this we see that the last two digits of k^3 are affected only by the values of a and b. Therefore the number of possible values is $N = 100$. Since for a favorable outcome the last digit of k^3 must be 1, there is one favorable value $a = 1$. In addition, the last digit of $(k^3 - 1)/10$ must also be 1; i.e., the product $3b$ must end with 1. This occurs only if $b = 7$. Thus the favorable value ($a = 1$, $b = 7$) is unique and, therefore, $P = 0.01$.

10. If a book is picked at random from a shelf containing four novels, three books of poems, and a dictionary, what is the probability that
    (a) the dictionary is selected?
    (b) a book of poems is selected?

    ANSWER. (a) $^1/_8$. (b) $^3/_8$.

2

Basic Probability Theorems

*Nature is pleased with simplicity,
and affects not the pomp of
superfluous causes.*

SIR ISAAC NEWTON (1642–1727)

Theorems of Probability

Combinatorial Analysis

Conditional Probability and
Independence

Bayes' Theorem

THEOREMS OF PROBABILITY

Using the axioms of probability, it is possible to derive many theorems which play an important role in applications.

THEOREM 2.1. If A is any event in the universe U, then $P(A') = 1 - P(A)$.

> **PROOF.** Observe that A and A' are mutually exclusive by definition, and that $A \cup A' = U$ (that is, among them, A and A' contain all of the elements of U). Hence, we have
>
> $$P(A \cup A') = P(A) + P(A')$$
>
> according to Axiom 1.3, and
>
> $$P(A \cup A') = P(U) = 1$$
>
> according to Axiom 1.2; and it follows that
>
> $$P(A) + P(A') = 1,$$
>
> which completes the proof.

As a special case we find that $P(\phi) = 1 - P(U) = 1 - 1 = 0$, since the empty set ϕ is the complement of U.

In the so-called Bernoulli trial, the probability of success is denoted by p and the probability of a failure is denoted by q. According to Theorem 2.1, it follows that $p = 1 - q$.

We have already seen that Axiom 1.3 can be extended to include more than two mutually exclusive events. However, now let us consider the extension of this axiom to find the probability of the union of any two events regardless of whether they are mutually exclusive. This extension leads to the following theorem usually called the *general law of addition*:

THEOREM 2.2. If A and B are any events in U, then

$$P(A \cup B) = P(A) + P(B) - P(A \cap B).$$

> **PROOF.** Observe from the Venn diagram in Figure 2.1 that
>
> $$A \cup B = (A \cap B) \cup (A \cap B') \cup (A' \cap B),$$
>
> and also that
>
> $$A = (A \cap B) \cup (A \cap B'),$$
> $$B = (A \cap B) \cup (A' \cap B).$$
>
> Since $A \cap B$, $A \cap B'$, and $A' \cap B$ evidently are mutually exclusive, the extension of Axiom 1.3 gives

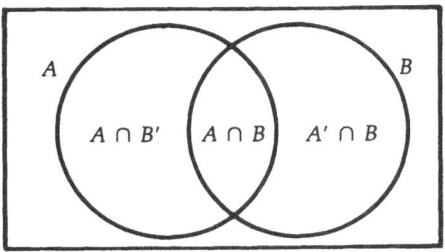

FIGURE 2.1 Partition of $A \cup B$. We see that $P(A) + P(B)$ adds the probability of $A \cap B$ twice when it should only be added once. $P(A) + P(B) - P(A \cap B)$ gives the probability of $A \cup B$.

$$P(A \cup B) = P(A \cap B) + P(A \cap B') + P(A' \cap B).$$

If we add and subtract $P(A \cap B)$, we obtain

$$P(A \cup B) = [P(A \cap B) + P(A \cap B')] + [P(A' \cap B) + P(A \cap B)]$$
$$- P(A \cap B)$$
$$= P(A) + P(B) - P(A \cap B). \qquad \text{Q.E.D.}$$

If A and B are mutually exclusive, this theorem reduces to Axiom 1.3, since in that case $P(A \cap B) = 0$. For this reason, we often refer to Axiom 1.3 as the special law of addition whereas Theorem 2.2 is called the general law of addition. From Theorem 2.2, we have

$$P(A \cup B) \leq P(A) + P(B).$$

EXERCISES

1. For any set A, we have the relation $A \cup A' = U$. For any two sets A and B establish the decomposition of B given by

$$B = U \cap B = (A \cup A') \cap B = (A \cap B) \cup (A' \cap B).$$

Since $A \cap B$ and $A' \cap B$ are disjoint, show that

$$P(B) = P(A \cap B) + P(A' \cap B).$$

As an application set $B = U$; recall that $P(U) = 1$; and thus conclude that $P(A') = 1 - P(A)$. In particular $P(\phi) = 1 - P(U)$, so that $P(\phi) = 0$. As a second application suppose that $A \subset B$. Then $A \cap B = A$ and hence show that $P(B) = P(A) + P(A' \cap B)$ if $A \subset B$. Since $P(A' \cap B) \geq 0$ show that $P(A) \leq P(B)$ if $A \subset B$.

2. Establish the inequality

$$P(A_1 \cup A_2 \cup \cdots \cup A_n) \leqslant \sum_{i=1}^{n} P(A_i).$$

HINT. To prove this, observe that if $n \geqslant 2$, then

$$P(A_1 \cup \cdots \cup A_n) = P((A_1 \cup \cdots \cup A_{n-1}) \cup A_n)$$
$$\leqslant P(A_1 \cup \cdots \cup A_{n-1}) + P(A_n).$$

Hence, if the required inequality holds for $n - 1$ sets, it holds for n sets. Since the required inequality clearly holds for $n = 1$, the result is proved by induction.

3. If $P(A) = \frac{1}{3}$, $P(A \cup B) = \frac{1}{2}$, and $P(A \cap B) = \frac{1}{4}$, find $P(B)$.

4. Suppose events A and B are such that $P(A) = \frac{2}{5}$, $P(B) = \frac{2}{5}$, and $P(A \cup B) = \frac{1}{2}$. Find $P(A \cap B)$.

5. Suppose a point is picked at random in the unit square. Let A be the event that it is in the triangle bounded by the lines $y = 0$, $x = 1$, and $x = y$, and B be the event that it is in the rectangle with vertices $(0,0)$, $(1,0)$, $(1,\frac{1}{2})$, $(0,\frac{1}{2})$. Compute $P(A \cup B)$ and $P(A \cap B)$.

6. Show that $P(\bigcup_n A_n) = 1 - P(\bigcap_n A_n')$.

Now $\bigcup_n A_n$ is the event that at least one of the events A_n occurs, while $\bigcap_n A_n'$ is the event that none of these events occurs. In words, this equation asserts that the probability that at least one of the events A_n will occur is 1 minus the probability that none of the events A_n will occur. The advantage of this equation is that in some instances it is easier to compute $P(\bigcap_n A_n')$ than to compute $P(\bigcup_n A_n)$. [Note that since the events A_n are not necessarily disjoint it is not true that $P(\bigcup_n A_n) = \Sigma_n P(A_n)$.]

COMBINATORIAL ANALYSIS

In finite probability theory the equally likely assumption is used very often. Thus there are many examples that involve finite sample spaces where each of n sample points has the same probability $1/n$. In such cases, evaluating probabilities of events often requires knowledge about permutations and combinations, that is, combinatorial analysis.

Let us now give some rules. The six possible pairs made up from (a_1,a_2,a_3) and (b_1,b_2) are (a_1,b_1), (a_1,b_2), (a_2,b_1), (a_2,b_2), (a_3,b_1), and (a_3,b_2). This is an example of:

RULE 1. Let $(a_1,\cdots,a_n)$ and $(b_1,\cdots,b_m)$ be two finite sets of objects. The number of pairs of the form (a_i,b_j) is nm.

Actually, Rule 1 is a special case of:

RULE 2. Let $(a_{11}, \cdots, a_{1n_1})$, $\cdots$, $(a_{k1}, \cdots, a_{kn_k})$ be k finite sets of objects. The number of k-tuples of the form $(a_{1i_1}, a_{2i_2}, \cdots, a_{ki_k})$ is $n_1 n_2 \cdots n_k$.

The $3 \cdot 2 = 6$ distinct pairs made up from the three symbols a, b, c without repeating the use of a symbol are (a,b), (a,c), (b,a), (b,c), (c,a), and (c,b). This is an example of:

RULE 3. Let a_1, $\cdots$, a_n be n distinct symbols. Let k be no greater than n. The number of distinct k-tuples of these symbols, repetitions not being allowed, is $n(n - 1) \cdots (n - k + 1)$. This number is written for short as $n_{(k)}$ and is read as the "kth factorial power of n." An alternative notation for $n_{(k)}$ is P_k^n, and is called "the number of permutations of n things taken k at a time."

Rule 4 is a special case of Rule 3 for $k = n$. For example, there are $3!$ = 6 distinct arrangements of the symbols a, b, c: namely, (a,b,c), (a,c,b), (b,a,c), (b,c,a), (c,a,b), (c,b,a).

RULE 4. Let a_1, $\cdots$, a_n be n distinct symbols. The number of distinct arrangements of these symbols is $n(n - 1) \cdots 3 \cdot 2 \cdot 1$, which is written as $n!$ and is read as "n factorial." (Note that $0! = 1$ by convention).

Rule 5 defines an important quantity, known as the binomial coefficient.

RULE 5. Let a_1, $\cdots$, a_n be n distinct symbols. Let k be no greater than n. The number of distinct k-tuples of these symbols, repetitions not being allowed and different orderings of the same k symbols not being counted separately, is

$$\frac{n_{(k)}}{k!} = \frac{n(n - 1) \cdots (n - k + 1)}{k(k - 1) \cdots 3 \cdot 2 \cdot 1}.$$

This is written as $\binom{n}{k}$ and is read "binomial coefficient n,k." An alternative notation for the binomial coefficient is C_k^n or $C(n,k)$, and is called "the number of combinations of n things taken k at a time."

Table 2.1 gives values of the binomial coefficient for the indicated values of n and k. Note that the entries for n, k, and $n, k + 1$ add up to the entry for $n + 1, k + 1$. For instance, in the boxed-in entries, we

TABLE 2.1 The Binomial Coefficients $\binom{n}{k}$

n	k					
	0	1	2	3	4	5
1	1	1				
2	1	2	1			
3	1	3	3	1		
4	1	4	6	4	1	
5	1	5	10	10	5	1

see that $1 + 3 = 4$. The entries across for any n sum to 2^n. For instance, for $n = 4$, we have $1 + 4 + 6 + 4 + 1 = 16 = 2^4$. An alternative way of interpreting the binomial coefficient is given by

RULE 6. The number of distinct k-tuples of the form $(a_{i_1}, \cdots, a_{i_k})$, where $i_1 < i_2 < \cdots < i_k$ are k indexes from among 1, 2, $\cdots$, n, arranged in order, is $\binom{n}{k}$.

For example, suppose ID numbers must be of the form (i,j,k), where the symbols are from the set 0, 1, 2, 3, 4, 5, 6, 7, 8, 9. If they must also be arranged in increasing order, $i < j < k$, then there are $\binom{10}{3} = 120$ ID numbers possible. If the restriction of increasing order is removed, there are $10_{(3)} = 720$ ID numbers possible.

Another application of the binomial coefficient is illustrated by the $\binom{4}{2} = 6$ arrangements of the symbols a, a, b, b: namely, (a,a,b,b), (a,b,a,b), (a,b,b,a), (b,a,a,b), (b,a,b,a), (b,b,a,a). The rule is

RULE 7. Suppose there are n symbols, k of which are indistinguishable a's and $(n - k)$ of which are indistinguishable b's. There are $\binom{n}{k}$ distinct arrangements of these symbols. This rule follows from Rule 5 because there are $\binom{n}{k}$ choices of positions for the a's among the n places.

The binomial coefficient can be expressed in the following ways:

$$\binom{n}{k} = \frac{n_{(k)}}{k!} = \frac{n!}{k!(n-k)!} = \frac{n_{(n-k)}}{(n-k)!} = \binom{n}{n-k}.$$

But usually the binomial coefficient is remembered by the single expression

$$\binom{n}{k} = \frac{n!}{k!(n-k)!}.$$

Any reasonable interpretation of $\binom{n}{n}$ is that it should equal 1. Hence, the convention that $0! = 1$.

Finally, Rule 8 defines the multinomial coefficient.

RULE 8. Suppose there are n symbols consisting of n_1 indistinguishable a's, n_2 indistinguishable b's, $\cdots$, n_r indistinguishable z's. The number of distinct arrangements of these $n_1 + n_2 + \cdots + n_r = n$ symbols is $n!/n_1! \, n_2! \cdots n_r!$. This number is sometimes denoted $\binom{n}{n_1, n_2, \cdots, n_r}$ and is read as the "multinomial coefficient n; $n_1, n_2, \cdots, n_r$." Notice that if $r = 2$, the multinomial coefficient reduces to the binomial coefficient. As an example of this rule, we observe that the $\binom{4}{2,1,1} = 12$ arrangements of the symbols a,a,b,c are (a,a,b,c), (a,a,c,b), (a,b,a,c), (a,c,a,b), (a,b,c,a), (a,c,b,a), (b,a,a,c), (c,a,a,b), (b,a,c,a), (c,a,b,a), (b,c,a,a), (c,b,a,a).

EXERCISES

1. If n is a positive integer, show that

$$(a + b)^n = \sum_{k=0}^{n} \binom{n}{k} a^k b^{n-k}.$$

This is called the binomial expansion. Hence, interpret the coefficients in

$$(a + b)^4 = b^4 + 4ab^3 + 6a^2b^2 + 4a^3b + a^4.$$

2. If n is a positive integer, then

$$(a_1 + a_2 + \cdots + a_r)^n = \sum \binom{n}{n_1, n_2, \cdots, n_r} a_1^{n_1} a_2^{n_2} \cdots a_r^{n_r},$$

where the sum is over all r-tuples of nonnegative integers, $n_1, \cdots, n_r$ that sum to n. This is called the "multinomial expansion." Hence, interpret the coefficients in

$$(a + b + c)^3 = a^3 + b^3 + c^3 + 3a^2b + 3ab^2 + 3a^2c$$
$$+ 3ac^2 + 3b^2c + 3bc^2 + 6abc.$$

3. In this exercise, we want to generalize the relationship

$$P(A \cup B) = P(A) + P(B) - P(A \cap B).$$

Let $A_1, \cdots, A_n$ be events and introduce the following notation:

$$S_1 = \sum_i P(A_i),$$

$$S_2 = \sum_{i<j} P(A_i \cap A_j),$$

$$\vdots$$

$$S_k = \sum_{i_1<i_2<\cdots<i_k} P(A_{i_1} \cap A_{i_2} \cap \cdots \cap A_{i_k}), \qquad k = 1, \cdots, n.$$

In other words, S_k is the sum of the probabilities of intersections of k of the events; it is summed out over all $\binom{n}{k}$ choices of k sets out of the available $A_1, \cdots, A_n$. The restriction $i_1 < \cdots < i_k$ means that each k-tuple appears just once. The required generalization is

$$P(A_1 \cup A_2 \cup \cdots \cup A_n) = S_1 - S_2 + \cdots + (-1)^{n+1}S_n.$$

This is called the *inclusion-exclusion formula*. Establish by means of a Venn diagram this formula for $n = 3$, that is, show that

$$\begin{aligned}
P(A \cup B \cup C) &= P(A) + P(B) + P(C) \\
&\quad - P(A \cap B) - P(A \cap C) - P(B \cap C) \\
&\quad + P(A \cap B \cap C) \\
&= S_1 - S_2 + S_3.
\end{aligned}$$

Suggest why the name inclusion-exclusion formula is appropriate.

CONDITIONAL PROBABILITY AND INDEPENDENCE
The formal definition of conditional probability is:

DEFINITION. If A and B are any events in U and $P(B) \neq 0$, the conditional probability of A given B is

$$P(A|B) = \frac{P(A \cap B)}{P(B)}.$$

If $P(B) = 0$ the conditional probability of A given B is undefined.

As a consequence of the definition of conditional probability we have the following theorem, usually called the *general law of multiplication*:

THEOREM 2.3. If A and B are any events in U, then

$$\begin{aligned}
P(A \cap B) &= P(B) P(A|B) & \text{if } P(B) \neq 0, \\
&= P(A) P(B|A) & \text{if } P(A) \neq 0.
\end{aligned}$$

PROOF. The first relation is obtained from the definition by multiplying both sides by $P(B)$. The second relation follows from the first by interchanging the letters A and B and using the fact that $P(A \cap B) = P(B \cap A)$.   Q.E.D.

The definition of conditional probability can be motivated by the relative frequency interpretation of probabilities. Consider an experiment that is repeated a large number of times. Let the number of times the events A, B, and $A \cap B$ occur in n trials of the experiment be denoted by $N_n(A)$, $N_n(B)$, and $N_n(A \cap B)$ respectively. For n large we expect that $N_n(A)/n$, $N_n(B)/n$, and $N_n(A \cap B)/n$ should be close to $P(A)$, $P(B)$, and $P(A \cap B)$ respectively. If now we just record those experiments in which B occurs then we have $N_n(B)$ trials in which the event A occurs $N_n(A \cap B)$ times. Thus the proportion of times that A occurs among these $N_n(B)$ experiments is $N_n(A \cap B)/N_n(B)$. Since

$$\frac{N_n(A \cap B)}{N_n(B)} = \frac{N_n(A \cap B)/n}{N_n(B)/n},$$

we see that as n becomes infinite we obtain the conditional probability $P(A \cap B)/P(B)$.

Suppose a box contains N red balls labeled 1, 2, $\cdots$, N and M black balls labeled 1, 2, $\cdots$, M. Assume that the probability of drawing any particular ball is $1/(M + N)$. If the ball drawn from the box is known to be red, what is the probability that it was the red ball labeled 1? Another way of stating this problem is as follows. Let B be the event that the selected ball was red, and let A be the event that the selected ball was labeled 1. The problem is then to determine the "conditional" probability that the event A occurred, given that the event B occurred. Since there are $M + N$ points each of which carries the probability $1/(M + N)$, we see that $P(B) = N/(M + N)$ and $P(A \cap B) = 1/(M + N)$. Thus

$$P(A|B) = \frac{1}{N}.$$

This should be compared with the "unconditional" probability of A: namely, $P(A) = 2/(M + N)$.

PROBLEM. Compute the probability that a randomly selected item is of first grade if it is known that 4 percent of the entire production is defective, and 75 percent of the nondefective items satisfy the first-grade requirements.

SOLUTION. It is given that $P(A) = 1 - 0.04 = 0.96$, $P(B|A) = 0.75$. The required probability $p = P(A \cap B) = (0.96)(0.75) = 0.72$.

Consider a box having four distinct balls a, b, c, d and an experiment consisting of selecting a ball from the box. We assume that the balls are equally likely to be drawn, so we assign probability $1/4$ to each point, a, b, c, and d in the sample space $\{a,b,c,d\}$. Let A and B be two events. For some choices of A and B, knowledge that A occurs increases the probability that B occurs. For example, if $A = \{a,b\}$ and $B = \{a\}$, then $P(A) = 1/2$, $P(B) = 1/4$, and $P(A \cap B) = 1/4$. Consequently, $P(B|A) = 1/2$, which is greater than $P(B)$. On the other hand, for other choices of A and B, knowledge that A occurs decreases the probability that B will occur. For example, if $A = \{a,b,c\}$, and $B = \{a,b,d\}$, then $P(A) = 3/4$, $P(B) = 3/4$, and $P(A \cap B) = 1/2$. Hence $P(B|A) = 2/3$, which is less than $P(B)$. However, there are cases when knowledge that A occurs does not change the probability that B occurs. As an example of this, let $A = \{a,b\}$ and $B = \{a,c\}$; then $P(A) = 1/2$, $P(B) = 1/2$, $P(A \cap B) = 1/4$, and therefore $P(B|A) = 1/2$. Events such as these, for which the conditional probability is the same as the unconditional probability, are said to be independent.

In general, if A and B are any events in a sample space, we say that A is independent of B if and only if $P(A|B) = P(A)$. Using Theorem 2.3, it can easily be seen that if A is independent of B then B is also independent of A; that is, $P(A|B) = P(A)$ implies $P(B|A) = P(B)$ provided $P(A) \neq 0$, and it is customary to say simply that A and B are *independent*.

In the special case where A and B are independent, Theorem 2.3 leads to the following, which is usually called the *special law of multiplication*:

THEOREM 2.4. If A and B are independent events, then

$$P(A \cap B) = P(A) \cdot P(B).$$

For example, the probability of getting two heads in two successive flips of a balanced coin is $(1/2)(1/2) = 1/4$ and the probability of drawing two kings in succession from a standard deck of 52 playing cards is $(4/52)(4/52) = 1/169$, provided the first card is replaced before the second is drawn. The special law of multiplication can be extended to apply to more than two independent events: if three or more events are mutually independent, the probability that they will all occur is given by the product of their respective probabilities. Thus the probability of four heads in four successive flips is $(1/2)^4 = 1/16$.

In dealing with more than two events, we must be careful; for in order for the special law of multiplication to hold, the events must be *mutually independent*. Let us give an example. We can consider a problem for three sets A, B, and C. Take the universal set a, b, c, d and assign probability $1/4$ to each point. Let $A = \{a,b\}$, $B = \{a,c\}$, and $C = $

$\{a,d\}$. Then it can be shown that the pairs of events A and B, A and C, and B and C are independent. We say that the events A, B, and C are pairwise independent. On the other hand, $P(C) = \frac{1}{2}$ and $P(C|A \cap B) = 1$. Thus a knowledge that the event $A \cap B$ occurs increases the odds that C occurs. In this sense, the events A, B, and C fail to be mutually independent. In general, three events A, B, and C are mutually independent if they are pairwise independent and if

$$P(A \cap B \cap C) = P(A)P(B)P(C).$$

Furthermore it can be shown that if A, B, and C are mutually independent and $P(A \cap B) \neq 0$, then $P(C|A \cap B) = P(C)$.

In general we define n events A_1, A_2, $\cdots$, A_n to be mutually independent where $n > 2$ if

$$P(A_1 \cap \cdots \cap A_n) = P(A_1) \cdots P(A_n),$$

and if any subcollections containing at least two but fewer than n events are mutually independent.

PROBLEM. A break in an electric circuit occurs when at least one out of three elements connected in series is out of order. Compute the probability that a break in the circuit will not occur, given that the elements may be out of order with the respective probabilities 0.3, 0.4, and 0.6. How does the probability change if the first element is never out of order?

SOLUTION. The required probability equals the probability that all three elements are working. Let A_k ($k = 1, 2, 3$) denote the event that the kth element functions. Then $p = P(A_1 \cap A_2 \cap A_3)$. Since the events may be assumed to be independent,

$$p = P(A_1)P(A_2)P(A_3) = (0.7)(0.6)(0.4) = 0.168.$$

If the first element is not out of order, then

$$p = P(A_2 \cap A_3) = (0.6)(0.4) = 0.24.$$

Let us summarize. Two events A and B are said to be independent if $P(A \cap B) = P(A)P(B)$. Three events A, B, and C are said to be mutually independent if

$P(A \cap B \cap C) = P(A)P(B)P(C)$ and
$P(A \cap B) = P(A)P(B), P(A \cap C) = P(A)P(C), P(B \cap C) = P(B)P(C).$

Thus, independence is a notion relative to a given probability measure. In contrast, the notion of disjointness is strictly a set property and does not depend upon any probability measure. In effect, events are independent if they have nothing to do with each other. For example, if we first toss a coin, then throw a die, and finally draw a card from a deck, we believe that any one of these three experiments is in no way influenced by the other two. Accordingly, in our probability model we would require that any three events A, B, C, such that A is alone determined by the coin, B by the die alone, and C by the card drawn alone, would be independent. If we apply the usual equally likely probability model, we can readily verify that the above equations for the mutual independence of A, B, C in fact do hold. For example, if A is a head, B is a 5 or 6, and C is a spade, then $P(A) = {}^{312}/_{624} = {}^{1}/_{2}$, $P(B) = {}^{208}/_{624} = {}^{1}/_{3}$, $P(C) = {}^{156}/_{624} = {}^{1}/_{4}$, $P(A \cap B) = {}^{104}/_{624} = {}^{1}/_{6}$, $P(A \cap C) = {}^{78}/_{624} = {}^{1}/_{8}$, $P(B \cap C) = {}^{52}/_{624} = {}^{1}/_{12}$, $P(A \cap B \cap C) = {}^{26}/_{624} = {}^{1}/_{24}$.

EXERCISES

1. Suppose two identical and perfectly balanced coins are tossed once.
   (a) Find the conditional probability that both coins show a head given that the first shows a head.
   (b) Find the conditional probability that both are heads given that at least one of them is a head.
   ANSWERS. (a) $^{1}/_{2}$. (b) $^{1}/_{3}$.

2. Show that if A, B, and C are three events such that $P(A \cap B \cap C) \neq 0$ and $P(C|A \cap B) = P(C|B)$, then $P(A|B \cap C) = P(A|B)$.

3. If two mutually exclusive events A and B are such that $P(A) \neq 0$ and $P(B) \neq 0$, are these events independent?
   ANSWER. From the incompatibility of the events, it follows that $P(A|B) = 0$ and $P(B|A) = 0$; that is, the events are dependent.

4. A box has 10 balls numbered 1, 2, $\cdots$, 10. A ball is picked at random and then a second ball is picked at random from the remaining nine balls. Find the probability that the numbers on the two selected balls differ by two or more.

5. The probability that the voltage of an electric circuit will exceed the rated value is p_1. For an increase in the voltage, the probability that the device will stop is p_2. Find the probability that the device will stop as a result of an increase in the voltage.
   ANSWER. $p_1 p_2$.

6. If a point selected at random in the unit square (i.e., the square bounded by $x = 0$, $x = 1$, $y = 0$, $y = 1$) is known to be in the triangle

bounded by $x = 0$, $y = 0$, and $x + y = 1$, find the probability that it is also in the triangle bounded by $y = 0$, $x = 1$, and $x = y$.

7. Let S be the unit square $0 \leqslant x \leqslant 1$, $0 \leqslant y \leqslant 1$ in the plane. Consider the probability as being uniform on the square, and let A be the event

$$\{(x,y): 0 \leqslant x \leqslant 1/2, 0 \leqslant y \leqslant 1\}$$

and B be the event

$$\{(x,y): 0 \leqslant x \leqslant 1, 0 \leqslant y \leqslant 1/4\}.$$

Show that A and B are independent events.

HINT. To show this, compute $P(A)$, $P(B)$, and $P(A \cap B)$, and show that $P(A \cap B) = P(A)P(B)$.

8. The probability that the kth unit of a computer is out of order during a time T equals p_k ($k = 1, 2, \cdots, n$). Find the probability that during the given interval of time at least one of n units of this computer will be out of order if all the units run independently.

ANSWER. $1 - (1 - p_1)(1 - p_2) \cdots (1 - p_n)$.

9. Two marksmen whose probabilities of hitting a target are 0.7 and 0.8, respectively, fire one shot each. Find the probability that at least one of them will hit the target.

ANSWER. 0.94.

10. The probability that an item made on the first machine is of first grade is 0.7. The probability that an item made on the second machine is first grade is 0.8. The first machine makes two items and the second machine three items. Find the probability that all items made will be of first grade.

ANSWER. 0.251.

11. A device stops as a result of damage to one tube of a total of N. To locate this tube, one successively replaces each tube with a new one. Find the probability that it will be necessary to check n tubes if the probability is p that a tube will be out of order.

ANSWER. $p(1 - p)^{n-1}$.

12. The probability of the occurrence of an event in each performance of an experiment is 0.2. The experiments are carried out successively until the given event occurs. Find the probability that it will be necessary to perform a fourth experiment.

ANSWER. $(1 - 0.2)^3 = 0.512$.

13. Under what conditions does the following equality hold:

$$P(A) = P(A|B) + P(A|B')?$$

ANSWER. $P(A) = P(A \cap B) + P(A \cap B') = P(B)P(A|B) + P(B')P(A|B')$. The equality is valid only in the particular cases: (a) $A = \phi$, (b) $B = U$, (c) $B = A$, (d) $B = A'$, (e) $B = \phi$.

BAYES' THEOREM

The general law of multiplication is useful in solving many problems in which the ultimate outcome of an experiment depends on the outcomes of various intermediate stages. Suppose we are interested in the performance of missiles received from two different suppliers, B_1 and B_2, in the proportion 3 to 1. In other words, the probability that any one missile received comes from supplier B_1 is $3/4$ and the probability that it comes from supplier B_2 is $1/4$. Suppose, furthermore, that 95 percent of the missiles supplied by B_1 and 80 percent of those supplied by B_2 perform according to specifications. What we would like to know is the probability of the event A that any one missile received will perform according to specifications. We will use the relation $A = (A \cap B_1) \cup (A \cap B_2)$ and that $(A \cap B_1)$ and $(A \cap B_2)$ are mutually exclusive (see Figure 2.2). The special law of addition yields

$$P(A) = P(A \cap B_1) + P(A \cap B_2).$$

If we apply the general law of multiplication to $P(A \cap B_1)$ and $P(A \cap B_2)$, we obtain

$$P(A) = P(B_1)P(A|B_1) + P(B_2)P(A|B_2).$$

We now substitute the given probabilities $P(B_1) = 3/4$, $P(B_2) = 1/4$, $P(A|B_1) = 0.95$, and $P(A|B_2) = 0.80$, to get

$$P(A) = (3/4)(0.95) + (1/4)(0.80) = 0.9125$$

for the desired probability that any one missile received will perform according to specifications.

The above development leads to a general formula. Instead of two suppliers B_1 and B_2 at the intermediate stage, suppose there are n mutually exclusive suppliers B_1, B_2, $\cdots$, B_n. Then the probability of the final outcome A is

$$P(A) = \sum_{i=1}^{n} P(B_i)P(A|B_i).$$

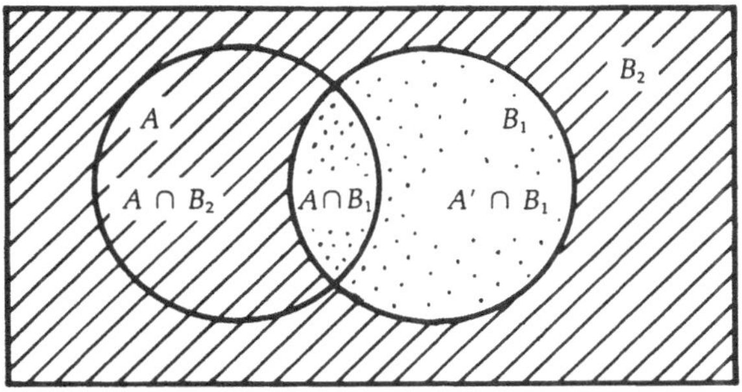

FIGURE 2.2. B_1 is the dotted area, and B_2 is the shaded area.

This formula for the total probability $P(A)$ is known as the *rule of elimination*. The easiest way to visualize this rule is by a tree diagram, as shown in Figure 2.3. The probability of the final outcome A is given by the sum of the products of the probabilities corresponding to each individual branch.

Let us now consider the so-called inverse problem. For example, suppose we have n suppliers of missiles, say $B_1, B_2, \cdots, B_n$. Suppose we want to know the probability that a particular missile came from supplier B_r, when it is known that it performs according to specifications. Referring to Figure 2.3, we shall use the same information as before, but now we must find $P(B_r|A)$ instead of $P(A)$. To solve this problem we write the equation

$$P(B_r|A) = \frac{P(A \cap B_r)}{P(A)};$$

and then substitute into it the general law of multiplication

$$P(A \cap B_r) = P(B_r)P(A|B_r)$$

and the rule of elimination

$$P(A) = \sum_{i=1}^{n} P(B_i)P(A|B_i).$$

The result can be summed up in the following theorem, called *Bayes' Theorem*.

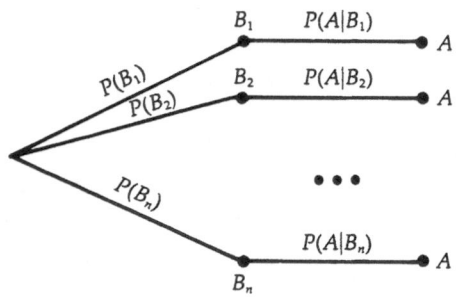

FIGURE 2.3 Rule of elimination.

THEOREM 2.5. If B_1, B_2, $\cdots$, B_n are mutually exclusive events of which one
must occur, that is,

$$\sum_{i=1}^{n} P(B_i) = 1,$$

then

$$P(B_r|A) = \frac{P(B_r)P(A|B_r)}{\displaystyle\sum_{i=1}^{n} P(B_i)P(A|B_i)}$$

for $r = 1, 2, \cdots, n$.

This formula finds the probability that the "effect" A was "caused"
by the event B_r. For example, in our illustration the required probability is that an acceptable missile was made by supplier B_r. The probabilities $P(B_i)$ are called the "prior" or "a priori" probabilities of the
"causes" B_i. In practice it is often difficult to assign numerical values
to the prior probabilities.

As an illustration of the use of Bayes' rule we consider the following
classical problem. Suppose there are three chests each having two
drawers. The first chest has a gold coin in each drawer, the second
chest has a gold coin in one drawer and a silver coin in the other
drawer, and the third chest has a silver coin in each drawer. A chest
is chosen at random and a drawer opened. If the drawer contains a
gold coin, what is the probability that the other drawer also contains
a gold coin?

Often in this problem the erroneous answer $1/2$ is given. However,
this problem is easily and correctly solved using Bayes' rule once the

description is deciphered. We can think of a probability space being constructed in which the events B_1, B_2, and B_3 correspond, respectively, to the first, second, and third chest being selected. These events are disjoint and their union is the whole sample space since exactly one chest is selected. Moreover, it is assumed that the three chests are chosen equally likely so that $P(B_i) = \frac{1}{3}$, $i = 1, 2, 3$. Let A be the event that the coin observed was gold. Then, from the composition of the chests it is clear that

$$P(A|B_1) = 1, \qquad P(A|B_2) = \frac{1}{2}, \qquad \text{and} \qquad P(A|B_3) = 0.$$

The problem asks for the probability that the second drawer has a gold coin given that there was a gold coin in the first. This can only happen if the chest selected was the first, so the problem is equivalent to finding $P(B_1|A)$. We now can apply Bayes' theorem to compute the answer, which is

$$P(B_1|A) = \frac{(\frac{1}{3})(1)}{(\frac{1}{3})(1) + (\frac{1}{3})(\frac{1}{2}) + (\frac{1}{3})(0)} = \frac{2}{3}.$$

As another illustration of Bayes' Theorem, let us consider the following example. There are two lots of items; it is known that all the items of one lot satisfy the technical standards and $\frac{1}{4}$ of the items of the other lot are defective. Suppose that an item from a lot selected at random turns out to be good. Find the probability that a second item of the same lot will be defective if the first item is returned to the lot after it has been checked.

Consider the hypotheses: H_1 that the lot with defective items was selected, and H_2 that the lot with nondefective items was selected. Let A denote the event that the first item is nondefective. By the assumption of the problem, $P(H_1) = P(H_2) = \frac{1}{2}$, $P(A|H_1) = \frac{3}{4}$, $P(A|H_2) = 1$. Thus, using the formula for the total probability, we find that the probability of the event A will be $P(A) = \frac{1}{2}[(\frac{3}{4}) + 1] = \frac{7}{8}$. After the first trial, the probability that the lot will contain defective items is

$$P(H_1|A) = \frac{P(H_1)P(A|H_1)}{P(A)} = \frac{(\frac{1}{2}) \cdot (\frac{3}{4})}{\frac{7}{8}} = \frac{3}{7}.$$

The probability that the lot will contain only good items is given by

$$P(H_2|A) = \frac{4}{7}.$$

Let B be the event that the item selected in the second trial turns out to be defective. The probability of this event can also be found from the formula for the total probability. If p_1 and p_2 are the probabilities of the hypotheses H_1 and H_2 after a trial, then according to the preceding computations $p_1 = {}^3/_7$, $p_2 = {}^4/_7$. Furthermore, $P(B|H_1) = {}^1/_4$, $P(B|H_2) = 0$. Therefore the required probability is $P(B) = ({}^3/_7)({}^1/_4) = {}^3/_{28}$.

EXERCISES

1. Suppose that the population of a certain city is 40 percent male and 60 percent female. Suppose also that 50 percent of the males and 30 percent of the females smoke. Find the probability that a smoker is male.

   ANSWER. ${}^{20}/_{38} \approx 0.53$.

2. A telegraphic communications system transmits the signals dot and dash. Assume that the statistical properties of the obstacles are such that an average of ${}^2/_5$ of the dots and ${}^1/_3$ of the dashes are changed. Suppose that the ratio between the transmitted dots and the transmitted dashes is 5:3. What is the probability that a received signal will be the same as the transmitted signal if (a) the received signal is a dot, (b) the received signal is a dash.

   ANSWERS. (a) ${}^3/_4$. (b) ${}^1/_2$.

3. Consider 10 urns, identical in appearance, of which nine contain two black and two white balls each and one contains five white and one black ball. An urn is picked at random and a ball drawn at random from it is white. What is the probability that the ball is drawn from the urn containing five white balls?

   ANSWER. 5/32.

4. Consider 18 marksmen, of whom five hit a target with the probability 0.8, seven with the probability 0.7, four with the probability 0.6, and two with the probability 0.5. A randomly selected marksman fires a shot without hitting the target. To what group is it most probable that that marksman belongs?

   ANSWER. The second group.

5. In an urn, there are n balls whose colors are white or black with equal probabilities. One draws k balls from the urn, successively, with replacement. What is the probability that the urn contains only white balls if no black balls are drawn?

   ANSWER. $n^k/(1 + 2^k + \cdots + n^k)$.

6. The first born of a set of twins is a boy. What is the probability that the other twin is also a boy if, among twins, the probabilities of two boys or two girls are a and b, respectively, and among twins of dif-

ferent sexes the probabilities of being born first are equal for both sexes?

ANSWER. $2a/(1 + a - b)$.

7. Consider that the probability of the birth of twins of the same sex is twice that of twins of different sexes; that the probabilities of twins of different sexes are equal in any succession; and that the probabilities of a boy and a girl are, respectively, 0.51 and 0.49. Find the probability of a second boy if the first born is a boy.

ANSWER. $^{103}/_{153}$.

8. Two sharpshooters fire successively at a target. Their probabilities of hitting the target on the first shots are 0.4 and 0.5 and the probabilities of hitting the target in the next shots increase by 0.05 for each of them. What is the probability that the first shot was fired by the first sharpshooter if the target is hit by the fifth shot?

ANSWER. $^{5}/_{11}$.

3

Univariate Distributions

*Everything existing in the universe is the fruit of
chance and necessity.*

DEMOCRITOS OF ABDERA (ca 460–370 B.C.)

Random Variables
Probability Mass Functions
Distribution Functions
Probability Density Functions
Random Sampling
Change of Variable

RANDOM VARIABLES

The basic concepts of probability are conventionally defined in terms of qualitative events, and for expository purposes this is desirable. However, a great many applications of probability theory concern quantitative variables rather than qualitative events.

Most of the statistical applications of probability theory concern the numerical values of a given variable rather than various classes of objects in general. For this reason, we introduce the term *random variable* or *chance variable*. Roughly speaking, a random variable is a variable that varies over certain numerical values with given probabilities. In order to develop some insight, it is convenient to think of an experiment such as a game of chance. All sorts of things may go on during the experiment but in the end specific (generally unpredictable) outcomes occur. These possible outcomes are called the sample points or elements of the experiment. By definition, the sample points are (1) *elemental* (i.e., part of a sample point is not admitted), (2) *exhaustive* (i.e., one sample point does occur), and (3) *exclusive* (i.e., only one sample point does occur.) The set of all sample points is the *universal set* (or *universe*) of the experiment. Alternative terms for the universal set are either the *sample space* or the *population*. Each subset of this universe is called an *event*. As we know, each event has a probability of occurrence, which is called the probability of the event. Thus far we have not suggested anything new. An experiment is represented by a universe of possible outcomes (or sample points), these sample points make up various events, and a probability is attached to each event. Now we are ready to define *random variable*. Let there be a "book" of payoff instructions which states how much should be given to the player when a certain outcome occurs. This book designates a payoff for each of the possible outcomes of the experiment. This payoff book is the intuitive version of the random variable. To each possible outcome (i.e., sample point), which we denote by s, the payoff book specifies a payoff we will call $X(s)$. That is, the payoff X is a function of the sample point s. In fact, a random variable X is defined as a function that maps the universe of sample points into real numbers. As an example, let the experiment be the tossing of a coin. The two sample-point cases (or possible results) are H and T, which make up the universe. Let the payoff be 1 if the coin shows H and -1 if the coin shows T. Thus the random variable is

$$X(H) = 1, \qquad X(T) = -1.$$

As another example, let the experiment be the choosing of a card at random from a well-shuffled standard deck. Let the payoff book state that the payoff is 0 if the card is neither a club nor the ace of spades.

If a club is chosen, the payoff is 1, and if the ace of spades is chosen, the payoff is -13. Here the universe is made up of the 52 cards. The random variable is

$$X(\text{club}) = 1, \qquad X(\text{ace of spades}) = -13, \qquad X(\text{other card}) = 0.$$

As we have seen, a random variable is defined as a function that assigns a numerical value to each possible result in the universe of an experiment. These assigned numerical values (i.e., the values of the random variable) thus represent the experimental outcomes. Each performance of the experiment generates one such numerical value of the random variable. The value of a random variable, then, is given by the numerical value associated with the sample point that occurred. For example, one random variable X associated with a person's health would be the numerical value of the person's temperature. Another random variable Y might be how the person feels on a scale from one to ten. A random variable is a function that transforms some physical outcome (or sample point) into a number. For example, if a person feels fairly good, we might have

$$Y(\text{fairly good}) = 8.$$

This represents an example of a random variable transforming a non-number (feeling fairly good) into a number (8). In the case of a person's temperature, the sample point is itself a number (say 98.6°), so the random variable is simply the function

$$X(98.6°) = 98.6°$$

or, in symbols,

$$X(x) = x.$$

Because this situation is often the case, we generally do not bother to differentiate between the symbols X and x, and merely write x for the random variable itself.

As familiar examples of random variables drawn from every-day life, we cite: (1) the number of points scored by a basketball team from game to game, (2) the daily number of absentees from an office, (3) the weekly grocery bill of a family, (4) the volume of gasoline used, and (5) the amount of time spent by an individual driver in daily commuting.

In connection with random variables, we are usually interested in numerical relationships, such as $X \geqslant a$ or $a < X < b$. That is, where random variables are involved, it is usually necessary to consider the probabilities of numerical relationships. By way of illustration, consider a random variable X and let a and b (where $a < b$) stand for two of its possible values. Most of the interesting problems relating to random variables require a suitable notation for the probabilities of relatinships such as $X > a$, or perhaps $a \leqslant X \leqslant b$. Extending our notation to cover cases of this sort, we define the symbol $P(X > a)$ to mean "the probability that X exceeds a." Similarly, we define the symbol $P(A \leqslant X \leqslant b)$ to mean "the probability that X lies in the interval (a,b) including a and b." Likewise, the symbols

$$P(X \geqslant a) \qquad \text{and} \qquad P(a < X < b)$$

are read, respectively, as "the probability that X equals or exceeds a," and "the probability that X lies between a and b."

A random variable is completely defined if the set of all its admissible values is specified together with the probability associated (in an appropriate sense) with each value.

PROBABILITY MASS FUNCTIONS

A random variable is said to be discrete if in the nature of its definition there is a finite separation between all of its possible values. An example of a discrete random variable is furnished by the annual incomes of all wage earners in the United States for a given year. The members (or sample points) of the universe are all the wage earners, and the payoff function (or random variable) assigns a numerical income to each member. Such a random variable is inherently discrete, for its possible values are integral multiples of one cent. A more common situation, however, is exemplified by the heights, weights, or ages of the same wage earners. Here the random variables themselves are defined on a continuous scale; but discrete random variables can be constructed from them by dividing their ranges into a finite number of intervals and substituting the midpoint of each interval for all values included within the interval. For practical purposes, a continuous random variable is often replaced by a discrete one constructed in this manner. However, it is obvious that the substitution of a discrete random variable would serve merely as a crutch, and that our definition of probability ought to be made broad enough to admit of continuous random variables. This is easier said than done, but we shall try to bridge the gap in the next section.

Consider a discrete random variable X defined for an experiment in which the universal set of possible distinct outcomes $x_1, x_2, \cdots, x_N$ is finite. Suppose that the outcomes are themselves numerical values.

For example, in tossing a die, the possible outcomes {1,2,3,4,5,6} are numerical. Let us now define a payoff function (i.e., random variable) that assigns the same numerical value as the possible outcome. Thus, in our example of a die, the random variable would be $X(1) = 1$, $X(2) = 2$, $X(3) = 3$, $X(4) = 4$, $X(5) = 5$, $X(6) = 6$. or $X(i) = i$ for $i = 1, 2, \cdots, 6$. Generally, we would have

$$X(x_i) = x_i \qquad \text{for } i = 1, 2, \cdots, N.$$

Let p_i be the probability that x_i occurs. In our example of a die, $p_i = 1/6$ for $i = 1, 2, \cdots, 6$. It therefore follows that the probability that the random variable X is equal to x_i is simply

$$P(X = x_i) = p_i \qquad \text{for } i = 1, 2, \cdots, N.$$

The function $P(X = x_i)$, which is usually written simply as $f(x_i)$ is known as the *probability mass function* for the discrete random variable. In the case of the die, the probability mass function is

$$f(i) = P(X = i) = 1/6 \qquad \text{for } i = 1, 2, \cdots, 6.$$

It is perfectly possible to define a discrete random variable that has an infinite set of distinct values, and in that case the probability mass function would have to be suitably defined as a function of the index i such that

$$\sum_{i=1}^{\infty} f(x_i) = 1.$$

The probability distribution of a random variable represents the mathematical specification of that random variable. Hence it is a statement of the possible values that random variable can assume, together with the probability associated with each value. For a discrete random variable, the probability distribution is given as the probability mass function $f(x_i)$ for the discrete values x_i that the random variable can assume. As we will see, for a continuous random variable, the probability measure can be represented by the cumulative distribution function $F(x)$ or, in many cases by the probability density $f(x)$. Eleven important *examples of probability mass functions* are now given here, and we shall see in the following chapters how they are derived and used in applications:

1. An example of a *uniform,* or *rectangular, distribution* is given by the probability mass function

$$f(x) = \frac{1}{N} \qquad \text{for } x = 1, 2, \cdots, N.$$

EXAMPLE. The number that comes up with one toss of a die has a uniform distribution with $N = 6$.

2. Suppose a sample of n items is to be drawn without replacement from a population of N items, of which N_1 are defective. The probability of n_1 defective items in the sample is

$$h(n_1;n,N_1,N) = \frac{\binom{N_1}{n_1}\binom{N-N_1}{n-n_1}}{\binom{N}{n}} \qquad \text{for } n_1 = 0, 1, \cdots, n.$$

This probability mass function is that of a *hypergeometric distribution*. The parameters of this family of distributions are the sample size n, the number N_1 of defectives in the population, and the population size N. The random variable is the number n_1 of defectives in the sample.

EXAMPLE. In a milk case with 5 bottles of milk, 4 are fresh and 1 is sour. Let us randomly select 2 bottles without replacement, and let n_1 be the number of sour ones we obtain. Then n_1 is a hypergeometric random variable with $n = 2$, $N_1 = 1$, $N = 5$.

3. In one trial of an experiment, there are only two possible outcomes, arbitrarily called "failure" or "success." The random variable x is equal to 0 for a failure and 1 for a success. The probability of a success is denoted by the letter p and hence the probability of a failure is $q = 1 - p$. The trial is called a *Bernoulli trial*. The probability of x success in 1 trial is

$$b(x;p) = p^x q^{1-x} \qquad \text{for } x = 0, 1.$$

The probability mass function defined by this equation is that of a *Bernoulli distribution* with parameter p. Since p is a probability, it must satisfy $0 \leq p \leq 1$.

EXAMPLE. Let success be the occurrence of the face "1" in a single toss of a fair die. Then the toss is a Bernoulli trial with $p = \frac{1}{6}$ and $q = \frac{5}{6}$.

4. There is a sequence of Bernoulli trials, with p and q constant from trial to trial. The n trials are mutually independent. Such a sequence is called a *Bernoulli process*. The probability of x successes in n trials is given by

$$b(x;n,p) = \binom{n}{x} p^x q^{n-x} \qquad \text{for } x = 0, 1, 2, \cdots, n.$$

This is the probability mass function of the *binomial distribution*. The random variable is x, and the parameters are n and p, where n is a positive integer. When $n = 1$, the distribution reduces to a Bernoulli distribution.

EXAMPLE. Let success be the occurrence of either of the faces "2" or "4" in a single toss of a fair die. Then a sequence of 5 tosses is a Bernoulli process with $n = 6$, $p = \frac{1}{3}$, $q = \frac{2}{3}$. The probability of x successes in the six tosses is binomial with $n = 6$, $p = \frac{1}{3}$, $q = \frac{2}{3}$.

5. In a Bernoulli process, the probability that the first success occurs on trial number k is

$$f(k;p) = pq^{k-1} \qquad \text{for } k = 1, 2, 3, \cdots.$$

This probability mass function is that of a *Pascal-geometric* distribution. The random variable is k and the parameter is p.

EXAMPLE. Suppose that one copy in ten of the *Tulsa Tribune* daily newspaper bears a special prize-winning number. Let k be the number of papers you must buy to get one prize. Then k is a Pascal-geometric random variable with $p = 0.1$ and $q = 0.9$.

6. In a Bernoulli process, the probability that the rth success occurs on trial number k is

$$f(k;r,p) = \binom{k-1}{r-1} p^r q^{k-r} \qquad \text{for } k = r, r + 1, r + 2, \cdots$$

This is the probability mass function of the *Pascal distribution*. The random variable is k, and the parameters are r and p, where r is a positive integer. If $r = 1$, the distribution reduces to the Pascal-geometric distribution.

EXAMPLE. In a contest, the probability of a winning number is 0.05 for each trial. A grand prize is given for three winning numbers. Let k be the number of trials required to obtain a grand prize. Then k has a Pascal distribution with $r = 3$, $p = 0.05$, $q = 0.95$.

7. In a Bernoulli process, the probability that s failures occur before the first success is

$$f(s;p) = pq^s \qquad \text{for } s = 0, 1, 2, \cdots.$$

This is the probability mass function of a *geometric distribution* with random variable s and parameter p.

EXAMPLE. In drilling for oil, the probability that a new wildcat well is a
dry hole is 0.95. Let s be the number of dry holes drilled before
the first successful oil well. Then s has a geometric distribution
with $p = 0.05$ and $q = 0.95$.

8. In a Bernoulli process, the probability that s failures occur before the
rth success is

$$b^-(s;r,p) = \binom{s+r-1}{r-1} p^r q^s \qquad \text{for } s = 0, 1, 2, \cdots .$$

This is the probability mass function of a *negative binomial distribution*.
The random variable is s, and the parameters are r and p, where r is
a positive integer. When $r = 1$, the negative binomial distribution reduces to the geometric distribution.

EXAMPLE. Assume that every time you drive your car the probability is
0.001 that you will get a ticket for speeding. Also assume that
you will lose your license once you have received two tickets.
Let s be the number of times you drive your car without getting
a ticket and before you lose your license. Then s has a negative
binomial distribution with $r = 2$, $p = 0.001$, $q = 0.999$.

9. A generalization of the binomial distribution arises when each trial
can have more than two possible outcomes. Let each trial permit k
mutually exclusive and exhaustive outcomes with respective probabilities $p_1, p_2, \cdots, p_k$, where $p_1 + p_2 + \cdots + p_k = 1$. In n independent
trials, the probability $f(n_1,n_2,\cdots,n_k)$ of getting n_1 outcomes of the first
kind, n_2 outcomes of the second kind, $\cdots$, and n_k outcomes of the kth
kind, is

$$f(n_1,n_2,\cdots,n_k) = \frac{n}{n_1!n_2! \cdots n_k!} p_1^{n_1} p_2^{n_2} \cdots p_k^{n_k}$$

for $n_i = 0, 1, \cdots, n$, subject to the restriction that $n_1 + n_2 + \cdots + n_k$
$= n$. This is the probability mass function of the *multinomial distribution*. When $k = 2$, it reduces to the binomial distribution with $p = p_1$,
$q = p_2$, $x = n_1$, $n - x = n_2$.

EXAMPLE. A college freshman is taking five courses in a semester. She
assumes that she will get an A with probability 0.1, a B with
probability 0.7, and a C with probability 0.2 in each of the
courses. She defines n_1, n_2, n_3 to be the number of A's, B's,
C's, respectively. Then n_1, n_2, n_3 is a multinomial random variable with parameters $n = 5$, $p_1 = 0.1$, $p_2 = 0.7$, $p_3 = 0.2$.

10. The *Poisson distribution* has the probability mass function

$$p(x;\mu) = \frac{\mu^x e^{-\mu}}{x!} \qquad \text{for } x = 0, 1, 2, \cdots.$$

The random variable is x and the parameter is μ.

EXAMPLE. (The raisin bun problem): Suppose N raisin buns of equal size are baked from a batch of dough into which n raisins have been carefully mixed. The number of raisins will vary from bun to bun, and the average number of raisins per bun is $\mu = n/N$. It is natural to assume that the raisins move around freely and virtually independently during the mixing, and hence whether or not a given raisin ends up in a given bun does not depend upon what happens to the other raisins. The raisins are approximately uniformly distributed throughout the dough after careful mixing, so that every raisin has the same probability $p = 1/N$ of ending up in a given bun. Thus, we can interpret the problem in terms of a series of n independent Bernoulli trials, so that the probability of exactly x raisins in a given bun is binomial with parameters n, $1/N$. However, we suppose that both the number of raisins n and the number of buns N are quite large, so that in particular $p = 1/N$ is small. Then the probability of exactly x raisins in a given bun is given approximately by the Poisson probability mass function with parameter $\mu = n/N$. Hence the probability of at least one raisin in the bun is $1 - p(0;n/N) = 1 - \exp(-n/N)$.

11. A *Poisson process* with parameter λ is described by the following model. The probability of a "success" during a very small interval of time Δt is given by $\lambda \Delta t$. The probability of two or more successes during Δt equals zero. The occurrence or nonoccurrence of a success during the interval from t to $t + \Delta t$ does not depend on what happened prior to time t. From this model it can be concluded that the probability of x successes during a time interval of length t is

$$p(x;\lambda,t) = \frac{(\lambda t)^x e^{-\lambda t}}{x!} \qquad \text{for } x = 0, 1, 2, \cdots.$$

We see that this distribution is a Poisson distribution with parameter $\mu = \lambda t$.

EXERCISES

NOTE. In the answers to these exercises we write the combination symbol $\binom{n}{x}$ as $C(n,x)$.

1. A box contains ten light bulbs, two of which are burned out.
    (a) Show that the distribution of the number of good bulbs in a

random sample of eight is hypergeometric.

(b) What is the probability of drawing no defective bulb?

(c) Two defective bulbs?

ANSWERS. (a) $f(x) = C(8,x)C(2,8-x)/C(10,8)$ for $x = 6,7,8$. (b) $^1/_{45} \approx$ 0.022, (c) $^{28}/_{45} \approx 0.622$.

2. Assuming that the sexes are equally likely, what is the distribution of the number of girls in a family of six children? Plot the probability mass function. What is the probability that a family of six children will have at least two girls?

ANSWER. Binomial with $n = 6$, $p = ^1/_2$; hence $f(x) = C(6,x)(^1/_2)^6$ for $x = 0, 1, \cdots, 6$; $P(x \geqslant 2) = ^{57}/_{64} \approx 0.89$.

3. For an unbiased coin, the probability of heads is $^1/_2$. Letting x denote the number of heads in five independent tosses, write an equation for the probability mass function and indicate the values of x for which the equation is valid. Show by actual addition that the separate probabilities add up to unity.

ANSWER. Binomial with $n = 5$, $p = ^1/_2$; hence $f(x) = C(5,x)/32$ for $x = 0, 1, \cdots, 5$.

4. In the game of odd-man wins, three people toss coins. The game continues until someone has an outcome different from the other two. The person with a different outcome wins. Let k be the number of games needed before a decision is reached. Find the probability mass function of k.

ANSWER. Pascal-geometric with $p = ^3/_4$.

5. A box contains four red and two black balls. Two balls are drawn. Let x be the number of red balls obtained. Find the probability mass function of x.

ANSWER. $f(0) = ^1/_{15}$, $f(1) = ^8/_{15}$, $f(2) = ^6/_{15}$.

6. If A, B, and C match pennies and odd man wins, what is the distribution of the number of times that A wins in four matches? Assume that there must be an odd man to constitute a match.

(a) Plot the probability mass function.

(b) Compute the probability: (i) that A wins exactly twice, (ii) that A loses at least twice, (iii) that A either wins all matches or loses all matches.

ANSWERS. (a) $f(x) = C(4,x)(^1/_3)^x(^2/_3)^{4-x}$ for $x = 0, 1, \cdots, 4$. (b) (i) 0.296; (ii) 0.889; (iii) 0.210.

7. Human blood types have been classified into four exhaustive, mutually exclusive categories O, A, B, AB. In a certain very large population, the respective probabilities of these four types are 0.46, 0.40,

0.11, 0.03. Given that the population is so large that sampling depletions may be ignored, find the probability that a random sample of five individuals will contain
(a) two cases of type O and one case of each of the others;
(b) three cases of type O and two of type A;
(c) no case of type AB.
ANSWERS. (a) 0.0168. (b) 0.1557. (c) 0.8587.

8. Using the information of Exercise 7, compute the entire distribution for random samples of two.
(a) How many distinct allocations are there?
(b) What is the total probability of the two most likely allocations?
(c) What is the total probability of the four most likely allocations?
(d) What is the total probability of the two least likely allocations?
ANSWERS. (a) 10. (b) 0.5796. (c) 0.8408. (d) 0.0075.

9. A five-card hand contains three aces. If another player is allowed to withdraw any two cards sight unseen, what is the distribution
(a) of the number of aces withdrawn?
(b) of the number of aces left? Compute each distribution in full.
ANSWERS. (a) $f(x) = C(3,x)C(2,2-x)/C(5,2)$ for $x = 0, 1, 2$. (b) $f(x) = C(3,x)C(2,3-x)/C(5,3)$ for $x = 1, 2, 3$.

10. Given the probability mass function $f(x) = e^{-1}/x!$ for $x = 0, 1, 2, \cdots$:
(a) Calculate $P(X = 2)$ and $P(X < 2)$.
(b) Graph the probability mass function $f(x)$.

ANSWERS. (a) 0.814, 0.736. (b) $f(0) = 0.368$, $f(1) = 0.368$, $f(2) = 0.184$, $f(3) = 0.061, \cdots$.

11. A coin is tossed until a head appears.
(a) What is the probability that a head will first appear on the fourth toss?
(b) What is the probability that x tosses will be required to produce a head?
(c) Graph the probability mass function of x.
ANSWERS. (a) $^1/_{16}$. (b) $f(x) = (^1/_2)^x$ for $x = 1, 2, \cdots$. (c) $f(1) = ^1/_2$, $f(2) = ^1/_4$, $f(3) = ^1/_8$, $f(4) = ^1/_{16}, \cdots$.

12. The probability of "making the green light" at a certain intersection is 0.2.
(a) In a series of independent trials, what is the distribution of the number of trials required for the first success?
(b) What is the probability of obtaining a success in less than four trials?

ANSWERS. (a) $f(x) = (0.2)(0.8^{x-1})$ for $x = 1, 2, 3, \cdots$. (b) 0.488.

13. Under the usual assumptions as to randomness, what is the distribution of the number of diamonds in a 13-card hand?

ANSWER. $f(x) = C(13,x)C(39,13-x)/C(52,13)$ for $x = 0, 1, 2, \cdots, 13$.

DISTRIBUTION FUNCTIONS

From the mathematical standpoint, a random variable is completely defined by stating its distribution; that is, the specification of all of its admissible values together with the probability associated, in an appropriate sense, with each value. Ordinarily, the specification of admissible values is perfectly straightforward and can be done by simple enumeration. In case a random variable x is capable of assuming only a finite number of distinct values, say $x_1, x_2, \cdots, x_N$, there is no problem with its probability function. For we may regard the occurrence of each possible value x_i as constituting an event E_i, and thus apply the theory already developed for qualitative events. As we have seen, we thus arrive at the concept of the probability mass function

$$f(x_i) = P(X = x_i).$$

Continuous random variables, however, are more subtle, because their admissible values are not only infinitely many but infinitely dense. However, a simple device proves powerful enough to cope with the most general situation.

First let us observe that a distinction should be made between the random variable itself and the real variable which includes all of its possible values; thus X is used for the former and x for the latter. However, usually the distinction is only implied, and one writes x both for the random variable and the real variable. Once the reader is used to the difference between these two concepts, it is actually easier in most situations to use the same symbol x for both. The situation is similar to the statement: "I saw John Wayne on television." Certainly, we know the difference between the electronic image John Wayne (x) and the person John Wayne (X); but there is no need to make a verbal distinction because the meaning is clear from the context.

Choosing an arbitrary number x, let us define two mutually exclusive and complementary classes of values: (I) $X \leqslant x$, and (II) $X > x$. With Class I we associate the event E_x in the sense that any value of the random variable belonging to Class I corresponds to the occurrence of E_x; similarly, we associate the complementary event E_x' with Class II. Thus $P(E_x)$ and $P(E_x')$ are rigorously definable, but since E_x and E_x' are complementary, it is sufficient to know $P(E_x)$. This con-

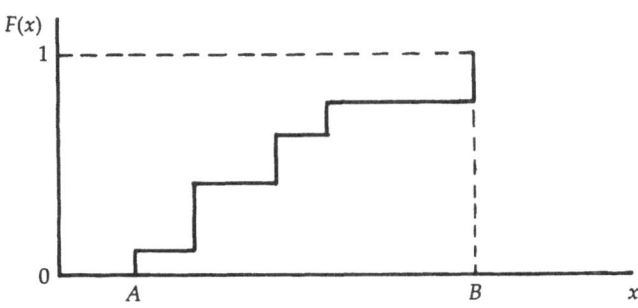

FIGURE 3.1 Typical distribution function of a discrete random variable.

struction can be carried out for any value of x whatever, and the result will obviously be a function of x. Hence we would replace $P(E_x)$ by the equivalent functional notation

$$F(x) \equiv P(E_x) \equiv P(X \leqslant x).$$

We are going to allow x to take on all possible values. That is, $F(x)$ denotes the probability that the random variable in question assumes any value less than or equal to x. A table, graph, or mathematical formula giving $F(x)$ for all possible values of x uniquely defines the distribution of the random variable X, and for this reason $F(x)$ is called the *distribution function* of X. A more precise term is *cumulative distribution function*, abbreviated *cdf*. As we proceed, we shall elaborate on this concept, but for the present we shall merely point out a few general features of distribution functions. If the least possible value that a certain random variable X can assume is denoted by A and the greatest possible value by B, then it is evident that the graph of $F(x)$ will coincide with the x-axis for all values of x less than A and will coincide with the horizontal line $F = 1$ for all values of x greater than or equal to B. We recall that random variables capable of assuming only a finite or denumerable number of distinct values are called discrete. Discrete random variables necessarily yield distribution functions of the discontinuous type (step function) shown in Figure 3.1. The reason is that inadmissible values of x add nothing to the probability; consequently, the graph of $F(x)$ is horizontal between admissible values but takes a finite step at each point on the x-axis that corresponds to an admissible value. Discontinuities may occur also in the graphs of the distribution functions of continuous random variables, but most of the ones commonly encountered are continuous, as indicated in Figure 3.2.

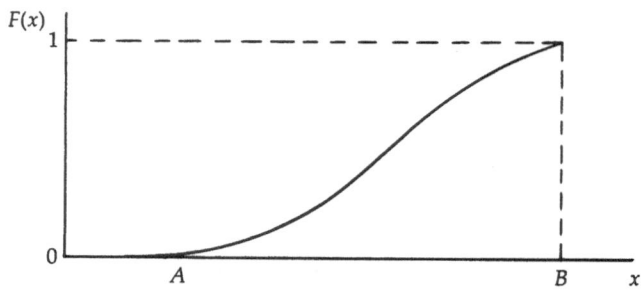

FIGURE 3.2 Typical distribution function of a continuous random variable.

EXAMPLE 3–1. As a simple example of a discrete random variable having only a small number of admissible values, consider the toss of an unbiased die where the occurrence of an ace or a six may be considered a success. If the die is tossed three times (or if three different dice are tossed once), it is possible to obtain either 0, 1, 2, or 3 successes. The probabilities of these four events can be easily calculated from classical probability theory. The probability of success on a single trial is $2/6 = 1/3$, and the probability of failure is $2/3$. The numbers of distinct sequences of successes and failures that yield 0, 1, 2, 3 successes in three tosses are 1, 3, 3, 1 respectively. Therefore, if X denotes the number of successes, the probability mass function is binomial and is given by

$$P(X = 0) = (2/3)^3 = 8/27,$$

$$P(X = 1) = (3)(2/3)^2(1/3) = 12/27,$$

$$P(X = 2) = (3)(2/3)(1/3)^2 = 6/27,$$

$$P(X = 3) = (1/3)^3 = 1/27.$$

Hence the distribution of X may be expressed in terms of the binomial probability mass function, as in Table 3.1, in which $f(x)$ denotes the probability of $X = x$, or alternatively in terms of the binomial cumulative distribution function $F(x)$, as in Table 3.2. These two representations of the distribution of X are exhibited graphically with different scales in Figures 3.3 and 3.4. We see that the cumulative distribution function represents an accumulation of the probability mass as the value of x increases, that is,

TABLE 3.1 Binomial Probability Mass Function

Possible value x:	0	1	2	3
Probability mass function $f(x)$:	$8/27$	$12/27$	$6/27$	$1/27$

TABLE 3.2 Binomial Cumulative Distribution Function

Possible value x:	0	1	2	3
Cumulative distribution function $F(x)$:	$8/27$	$20/27$	$26/27$	$27/27$

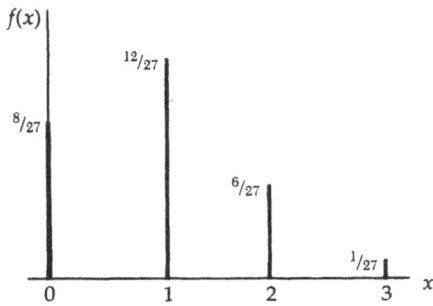

FIGURE 3.3 The probability mass function $f(x)$ of a discrete random variable X where $f(x) = P(X = x)$.

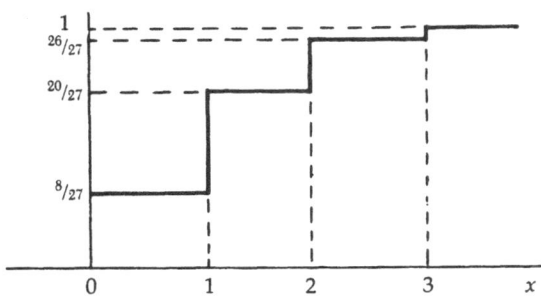

FIGURE 3.4 The cumulative distribution function $F(x)$ of a discrete random variable X where $F(x) = P(X \leqslant x)$.

$$F(x) = \sum_{k=0}^{x} f(k).$$

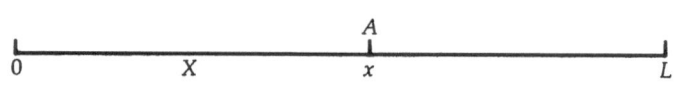

FIGURE 3.5 Permissible region.

EXAMPLE 3–2. Let us give an example of a continuous random variable. The permissible region, shown in Figure 3.5, consists of the line segment 0L. If points are picked at random on 0L, their distances from 0 will generate a continuous random variable X, the admissible values of which will range from 0 to L inclusive. Assuming all points equally likely, let us find the distribution of this random variable. Let A denote an arbitrary point on 0L and x its distance from 0. The probability that any point X picked at random will fall somewhere on the segment 0A is

$$P(X \text{ on } 0A) = \frac{\overline{0A}}{\overline{0L}} = \frac{x}{L}.$$

But having X fall on 0A is precisely the condition that the random variable X be less than or equal to x. Hence by definition of the cumulative distribution function, we have

$$F(x) = P(X \leqslant x) = \frac{x}{L},$$

an equation that holds everywhere in the admissible range of X; namely, $0 \leqslant x \leqslant L$. Adopting the convention that the cumulative distribution function is identically 0 for all values of x below the minimum admissible value of the random variable and identically 1 for all values of x above the maximum admissible value of the random variable, we may express the distribution of X in one statement by combining the range of definition with the equation of F(x). Thus the distribution of X is

$$F(x) = \frac{x}{L} \qquad \text{for } 0 \leqslant x \leqslant L.$$

The graph of the distribution function is shown in Figure 3.6. Although this distribution function is linear over the allowable range of x, geometric problems readily give rise to curved distribution functions, as illustrated by the next example.

EXAMPLE 3–3. The permissible region, indicated by the shading in Figure 3.7, is bounded below by the x-axis and above by the parabola $y = 2x - x^2$. If all points within the region are equally likely, let us find the distribution of the distance X of a random point

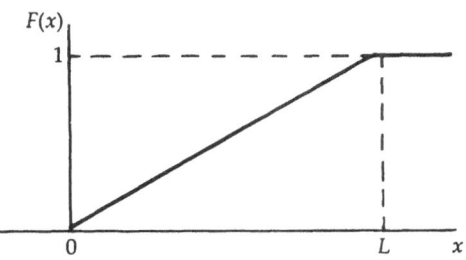

FIGURE 3.6 Distribution function.

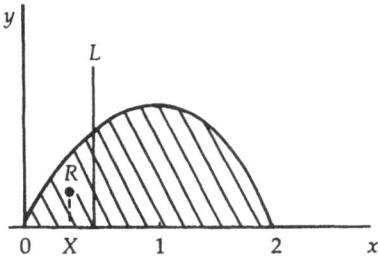

FIGURE 3.7 Permissible region given by the shaded area.

to the y-axis. At any distance $x \leqslant 2$ to the right of the y-axis, pass the straight line L parallel to the y-axis. The distance of a random point R from the y-axis is equal to its abscissa X, and the probability that $X \leqslant x$ is the same as the probability that R lies on or to the left of L. Hence,

$$F(x) = P(X \leqslant x) = \frac{\text{area to left of } L}{\text{total area}} = \frac{\displaystyle\int_0^x y \, dx}{\displaystyle\int_0^2 y \, dx} = \frac{3x^2 - x^3}{4}.$$

Thus the distribution of the random variable X is

$$F(x) = \frac{3x^2 - x^3}{4} \qquad \text{for } 0 \leqslant x \leqslant 2.$$

The graph of $F(x)$ is shown in Figure 3.8.

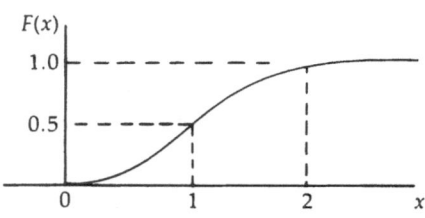

FIGURE 3.8 Distribution function.

Let us now discuss the computation of probabilities, given that we know the distribution function. The probability that X lie in any particular range $a < X \leqslant b$ may be computed from the distribution function of X in the following manner. Since the event $X \leqslant b$ can be decomposed into two mutually exclusive events, $X \leqslant a$ and $a < X \leqslant b$, it follows from the law of total probability that

$$P(X \leqslant b) = P(X \leqslant a) + P(a < X \leqslant b), \qquad (3.1)$$

whence

$$P(a < X \leqslant b) = P(X \leqslant b) - P(X \leqslant a) = F(b) - F(a). \qquad (3.2)$$

For example, from Table 3.2, the probability that X exceeds 0 but does not exceed 2 is

$$P(0 < X \leqslant 2) = F(2) - F(0) = {}^{26}/_{27} - {}^{8}/_{27} = {}^{18}/_{27} = f(1) + f(2).$$

In passing, we note that Equation (3.2) implies that $F(b) \geqslant F(a)$ whenever $b > a$, for the left side of this equation can never be negative, since it is the probability of an admissible event.

Suppose now that $F(x)$ is continuous, and let $a = b - \epsilon$, where ϵ is an arbitrarily small, positive number. From (3.2) we then have

$$P(b - \epsilon < X \leqslant b) = F(b) - F(b - \epsilon). \qquad (3.3)$$

In the limit as $\epsilon \to 0$, the left side of Equation (3.3) reduces to $P(x = b)$, and on the right side, the hypothesized continuity requires that $F(b - \epsilon) \to F(b)$. Hence, for a *continuous* distribution function,

$$P(X = b) = F(b) - F(b) = 0 \qquad \text{for all } b. \qquad (3.4)$$

Consequently, if $F(x)$ is continuous, any interval may be so defined as to include or exclude one or both end points without affecting the probability that a random point will belong to the interval; that is,

$P(a < X \leq b) = P(a \leq X \leq b) = P(a \leq X < b)$

$$= P(a < X < b) = F(b) - F(a).$$

Thus, in Example 3–3 the probability that a random value X lies in the open interval $1 < X < 2$ and the probability that it lies in the closed interval $1 \leq X \leq 2$ are both given by the same quantity, namely

$$F(2) - F(1) = 1 - \frac{1}{2} = \frac{1}{2}.$$

The zero probability given by Equation (3.4) for $X = b$ does not mean that the value of $X = b$ is impossible, but merely that it fades into immeasurable insignificance as compared to the total, inexhaustible number of distinct possibilities.

EXERCISES

1. A point is picked at random between -1 and 2. If the number selected is negative, it is rounded off to 0; if it is greater than 1, it is rounded off to 1; if it is between 0 and 1, it is not changed.
   (a) Show that the cumulative distribution function (cdf) of the final number selected is

   $$F(x) = \begin{cases} 0 & \text{for } x < 0, \\ \dfrac{1 + x}{3} & \text{for } 0 \leq x < 1, \\ 1 & \text{for } 1 \leq x. \end{cases}$$

   (b) Represent F as a mixture of a discrete cdf and a continuous cdf.

2. Points are uniformly distributed along the periphery of a semicircle of radius R.
   (a) Find the cumulative distribution function of the shortest distance X from one end of the diameter to random points on the periphery.
   (b) Find $P(X \leq R)$.
   ANSWERS. (a) $F(x) = (2/\pi) \sin^{-1} (x/2R)$ for $0 \leq x \leq 2R$. (b) $P = \frac{1}{3}$.

3. Points are uniformly distributed inside a circle of radius R.
   (a) Find the cumulative distribution function of the distance X of random interior points from the center.
   (b) For what distance x are the chances even that a random interior

point will be nearer to the center than that distance or farther away?

ANSWERS. (a) $F(x) = x^2/R^2$ for $0 \leqslant x \leqslant R$. (b) $x = R/\sqrt{2}$.

4. Points are uniformly distributed on the periphery of the circle $x^2 + y^2 = R^2$.
   (a) Find the cumulative distribution function $F(x)$.
   (b) Sketch the cumulative distribution function.
   (c) Evaluate $P(R/2 \leqslant X \leqslant R)$.
   (d) Evaluate $P(X \geqslant 3R/2)$.
   ANSWERS. (a) $F(x) = (1/\pi) \cos^{-1} [(R - x)/R]$ for $0 \leqslant x \leqslant 2R$. (c) $1/6$.
   (d) $1/3$.

PROBABILITY DENSITY FUNCTIONS

The concept of density has proved fruitful and convenient in physics. It also aids in the study of random variables. Consider the physical notion of density for a long, thin wire. For a homogeneous wire, the linear mass density is constant and is simply the mass per unit length. For a nonhomogeneous wire, it varies as a function of the distance x measured from the origin, taken at one end of the wire. Hence if $M(x)$ represents the mass of the segment of wire from the origin to any point x, the linear mass density $\rho(x)$ is defined as the limit of the ratio of the mass increment $\Delta M(x)$ to the length increment Δx as $\Delta x \to 0$. Thus $\rho(x)$ is the derivative of the mass function $M(x)$, and in turn

$$M(x) = \int_0^x \rho(x')dx'.$$

The function $M(x)$ gives at point x the cumulative mass of the wire up to and including the point x. As an analog to $M(x)$, the distribution function $F(x)$ is a function that gives the cumulative probability up to and including the point x.

Provided that the distribution function is differentiable (hence continuous), we may define probability density in a similar fashion. The probability that a value X of the random variable will lie between x and $x + \Delta x$ may be interpreted as the probability increment at the point x. This increment is represented by ΔP and is given by

$$\Delta P = P(x < X \leqslant x + \Delta x) = F(x + \Delta x) - F(x) = \Delta F(x).$$

At the same time, the average "concentration" of probability in the neighborhood of x will be given by the relative increment

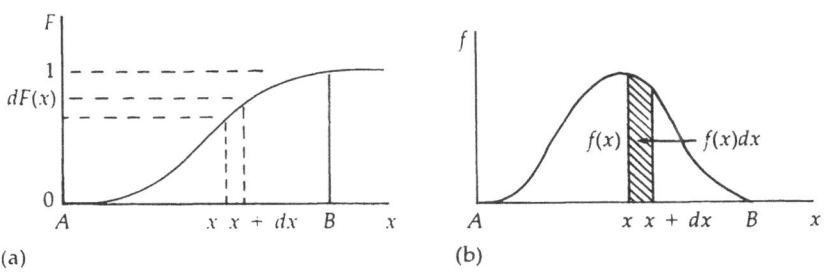

Fig. 3.9 (a) Typical distribution function and (b) corresponding density function.

$$\frac{\Delta P}{\Delta x} = \frac{\Delta F(x)}{\Delta x},$$

and under the assumption that $F(x)$ is differentiable, this ratio will approach a limit as $\Delta x \to 0$. This limit, which is the derivative of the distribution function at the point x, is called the *probability density function* (pdf) at x and will be denoted by $f(x)$. That is,

$$f(x) = \frac{dF(x)}{dx}.$$

Often, the term *density function* is used instead of the full term probability density function. By the same token, the infinitesimal element of probability $dF(x)$ is given by

$$dF(x) = f(x)dx$$

and represents the probability that the random variable X will assume a value lying somewhere in the range x to $x + dx$. The geometric connection between $F(x)$ and $f(x)$ is shown schematically in Figure 3.9. From its definition, the distribution function is cumulative; i.e., $F(x + \Delta x) \geqslant F(x)$ for all $\Delta x \geqslant 0$. It follows at once that the density function $f(x)$, being the derivative of the distribution function, is inherently non-negative. If A and B represent, respectively, the minimum and maximum admissible values of any random variable X, then $F(A) \equiv P(X \leqslant A) = P(X = A)$ and $F(B) \equiv P(X \leqslant B) = 1$. Now if the random variable possesses a density function, its distribution function is certainly continuous, because continuity is a necessary condition for the existence of the derivative. For any random variable with a continuous distribution function, the probability that the random variable will take on a value equal to any preassigned number (such as A) is vanishingly small. Thus we have $P(X = A) = 0$, whence $F(A) = 0$. Consequently, we arrive at the following general conclusions with regard to density functions:

1. $f(x) \geqslant 0$ for all x.

2. $\displaystyle\int_A^{x'} f(x)dx = F(x') - F(A) = F(x')$.

3. $\displaystyle\int_A^B f(x)dx = F(B) = 1$.

4. $P(a \leqslant X \leqslant b) = \displaystyle\int_a^b f(x)dx = P(a < X < b)$.

In words: The probability density function $f(x)$ of a random variable X is a non-negative function having unit area when integrated over the admissible range of x. When integrated from the minimum admissible value A up to any point x', it yields the value of the distribution function at x'; and, when integrated between any two limits (a,b), it yields the probability that the random variable X lies in the interval (a,b).

Let us now give some examples of density functions. Examples 3–2 and 3–3 previously given can serve to yield illustrations of density functions. The distribution function found in Example 3–2 was $F(x) = x/L$ in the range $0 \leqslant x \leqslant L$. Therefore, in the same range,

$$f(x) = \frac{dF(x)}{dx} = \frac{1}{L}$$

and $f(x) = 0$ for $x < 0$ and $x > L$. Again, the distribution function found in Example 3–3 was $F(x) = (3x^2 - x^3)/4$ in the range $0 \leqslant x \leqslant 2$; therefore, in the same range,

$$f(x) = \frac{dF(x)}{dx} = \frac{6x - 3x^2}{4} = \frac{3}{4}x(2 - x)$$

and $f(x) \equiv 0$ for $x < 0$ and $x > 2$. These two density functions are exhibited in Figure 3.10.

Both of these density functions admit of simple geometric interpretations. In Example 3–2, for instance, the probability that a random

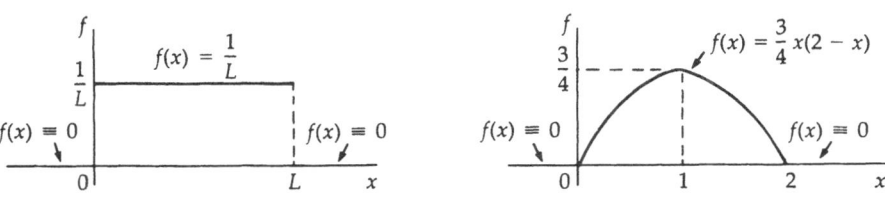

FIGURE 3.10 Density functions corresponding to the random variables defined in Examples 3–2 and 3–3.

point X falls in any segment of length S is equal to S/L. If the segment is of length dx, we have

$$f(x)\, dx = P(x \leqslant X \leqslant x + dx) = dx/L,$$

whence $f(x) = 1/L$. In Example 3–3 (see Figure 3.11), let M and M' denote two straight lines parallel to the y-axis and cutting the x-axis at x and $x + dx$ respectively. The probability that a random point R will fall between M and M' is equal to the ratio of the area included between these lines to the total area. We recall that X denotes the abscissa of the point R. We have

$$f(x)\, dx \equiv P(x \leqslant X \leqslant x + dx)$$
$$= \frac{\text{area between } M \text{ and } M'}{\text{total area}}$$
$$= \frac{y\, dx}{\displaystyle\int_0^2 y\, dx}$$
$$= \frac{(2x - x^2)\, dx}{{}^4/_3} = \frac{3}{4} x(2 - x)\, dx,$$

whence $f(x) = {}^3/_4\, x(2 - x)$ for $0 \leqslant x \leqslant 2$. Note that the density function in this interval is simply proportional to the parabola $y = x(2 - x)$, which defined the probability in the first place.

When the density function exists, its equation together with the specification of the admissible range can be used to define the distribution of the random variable. The range $A \leqslant x \leqslant B$ includes both limits. However, the admissible range can exclude one or both of the limits, for example, $A < x \leqslant B$, $A < x < B$, etc. Thus the distributions

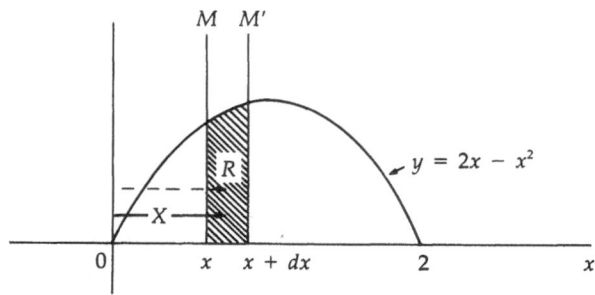

FIGURE 3.11 Probability as the ratio of the shaded area to the total area under the curve.

of the random variables considered in Examples 3–2 and 3–3 may be expressed in terms of density functions as follows:

Example 3–2: $f(x) = 1/L$ for $0 \leqslant x \leqslant L$.

Example 3–3: $f(x) = \frac{3}{4} x(2 - x)$ for $0 \leqslant x \leqslant 2$.

Outside of the expressed range of definition, it shall be taken for granted that the density function is identically zero.

In defining a distribution either by the distribution function or the density function, the specification of the admissible range of the random variable must always be included. Otherwise, the definition of the distribution is incomplete, and substitution of inadmissible values into the formulas could lead to absurdities such as negative probabilities or probabilities greater than one.

From the elementary standpoint, the density function is more convenient to work with and more appealing to the intuition than the distribution function; but, whereas the latter always exists for any definable distribution, the density function exists only when the distribution function is differentiable. This condition, however, is not very restrictive in practice, inasmuch as most applications of continuous random variables are confined to those that do have differentiable distribution functions. Even the few exceptions to this rule are usually amenable to rather obvious modifications of the ordinary analysis. A case in point is the poundage of brass rod shipped daily by a dealer in nonferrous metals. Since the minimum value, zero, has a finite probability of occurrence, the distribution function takes a finite jump at $x = 0$ as we cross from the left to the right of the origin; but thereafter it behaves in a perfectly regular manner. Therefore, a density function could be defined in the range $x > 0$. A similar situation is presented by the amount of rainfall over a particular city in a 24-hour period. Here there is a finite probability of a zero amount and also of a detectable but nonmeasurable amount called a "trace"; but, in the measurable range beyond, a density function is definable.

Later on, in the solution of some fairly simple problems, we shall encounter a few instances of another type of irregularity, such that the density function must be defined by two or more equations, each of which holds for a specified interval. Strictly speaking, the distribution function might not be differentiable at boundary points between adjacent intervals, and so the density function would be undefined there. However, because of the continuity of the distribution function, it is not necessary to distinguish between the integral of the density function over the rigorously correct open interval which excludes the boundary points and over the more convenient closed interval which includes them. Therefore, it is permissible to define the density function at a boundary point arbitrarily by the equation that holds for the interior of either interval. This convention applies also when the distribution function becomes nondifferentiable at the extremes of the admissible region. Let us consider two examples.

EXAMPLE 3-4. (See Figure 3.12.) Let the density function be given by

$$f(x) = \begin{cases} x & \text{for } 0 \leq x \leq 1 \\ 2 - x & \text{for } 1 \leq x \leq 2. \end{cases}$$

Here both equations yield the same value, $f(1) = 1$, at the boundary point $x = 1$; hence the distribution function is differentiable at $x = 1$, and the density function is rigorously defined as the derivative of the distribution function. To put it another way, we note that for any value $0 < a \leq 1$, the distribution function is given by

$$F(a) = \int_0^a x \, dx = \frac{a^2}{2},$$

whereas for any value $1 \leq b \leq 2$, it is given by

$$F(b) = \int_0^1 x \, dx + \int_1^b (2 - x) \, dx = 2b - \frac{b^2}{2} - 1.$$

Both formulas yield the same value at their common point; that is, $F(1) = 1/2$, and both have the same derivative, $f = 1$.

EXAMPLE 3-5. (See Figure 3.13.) Let the density function be given by

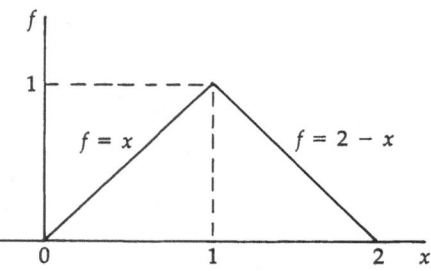

FIGURE 3.12 Triangular density function.

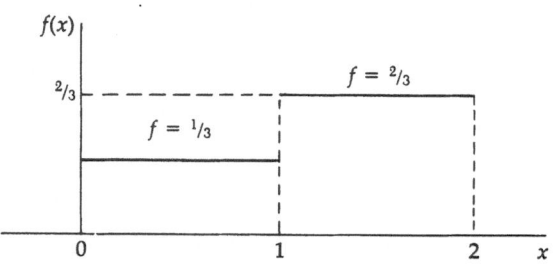

FIGURE 3.13 Block density function.

$$f(x) = \begin{cases} \frac{1}{3} & \text{for } 0 \leqslant x \leqslant 1, \\ \frac{2}{3} & \text{for } 1 < x \leqslant 2. \end{cases}$$

Here we exclude $x = 1$ from the definition of $f(x)$ in the second interval in order to avoid a contradiction of the definition in the first interval. However, in the integrals by which probabilities are computed, it is permissible to include $x = 1$ in both formulas. For if the lower limit of the second one were taken as $1 + \varepsilon$, the ultimate value of the integral as $\varepsilon \to 0$ would equal that obtained by taking the lower limit as 1 to begin with. For any value $0 \leqslant a \leqslant 1$, the distribution function is given by

$$F(a) = \int_0^a \frac{1}{3}\, dx = \frac{a}{3},$$

while for any value $1 \leqslant b \leqslant 2$ it is given by

$$F(b) = \int_0^1 \frac{1}{3} dx + \int_1^b \frac{2}{3} dx = \frac{2b}{3} - \frac{1}{3},$$

both of which yield $F(1) = 1/3$. However, the two functions yield different derivatives at $x = 1$, and so the density function is not defined at this point except by the stated arbitrary convention. In this example, the strict definition of the density function as a derivative also breaks down at the extremes of the admissible region $0 \leqslant x \leqslant 2$. At the lower extreme, we have

$$F(x) = 0 \quad \text{if } x \leqslant 0, \qquad \text{but} \qquad F(x) = \frac{x}{3} \quad \text{if } 0 \leqslant x \leqslant 1;$$

and although $F(0) = 0$ in either case, the two functions have different derivatives at $x = 0$. At the upper extreme,

$$F(x) = \frac{2}{3}x - \frac{1}{3} \quad \text{if } 1 \leqslant x \leqslant 2, \qquad \text{but} \qquad F(x) = 1 \quad \text{if } x \geqslant 2;$$

and as before, the functions agree at their common point, yielding $F(2) = 1$, but their derivatives are different. Consequently, the definition of the density function at $x = 0$ and at $x = 2$ must rest upon convention. Finally, we note the same situation in the density function associated with Example 3–2. Here the equation

$$f(x) = \frac{1}{L}$$

represents the derivative of $F(x)$ only in the range $0 < x < L$, but no inconsistency is caused by setting

$$f(0) = \frac{1}{L} = f(L),$$

thereby extending the definition of the density function to the end points of the range. So much for this type of irregularity; hereafter we shall take it in stride.

Let us now review some of the main points. Continuous random variables require the apparatus of differential and integral calculus. Let X be a continuous random variable and x is one of its permissible values. We can speak consistently of the probability that X lies in the neighborhood of x, but our definition breaks down when we try to evaluate the probability of the exact equality $X = x$. The correct procedure is to define the probability in terms of an infinitesimal neighborhood.

We take the total probability of the X-universe to be unity. Let $F(x)$ be the probability of the subset $X \leqslant x$. Then $F(x + dx)$ is the probability of the subset $X \leqslant x + dx$, where $dx > 0$. It now follows that the probability of the subset $x < X \leqslant x + dx$ is given by the difference

$$F(x + dx) - F(x).$$

This difference is inherently non-negative, because the subset $X \leqslant x$ is wholly included in the subset $X \leqslant x + dx$. In accordance with ordinary notation we may write

$$F(x + dx) - F(x) = dF(x).$$

Therefore the probability that X lies in an infinitesimal neighborhood of x may be represented as

$$P(x < X \leqslant x + dx) = dF(x).$$

It is to be noted that the symbol $dF(x)$ is well defined even if $F(x)$ is discontinuous; in fact, the definition applies to both continuous and discrete random variables. This is the starting point of the Stieltjes integral, by which discrete and continuous random variables are unified under one all-inclusive theory.

There are, accordingly, two legitimate ways of assigning a probability measure to a given value of a continuous random variable. We may use $F(x)$, which is the measure of the subset including all values of X less than or equal to x, or we may use $dF(x)$, which is the measure of the subset including all values of X within the infinitesimal interval $(x, x+dx)$. The function $F(x)$ is called the distribution function (or, more explicitly, the cumulative distribution function or cdf) of the random variable X. Although the distribution function is of fundamental importance, it is usually more convenient in elementary work to employ a function based upon $dF(x)$. Since there is a probability increment $dF(x)$ associated with each value of the random variable, it is natural to look upon the probability increment $dF(x)$ as a function of x and to define a related $f(x)$ function which gives the same result when multiplied by dx. This latter function, when it exists, is called the *probability density*. Accordingly, $f(x)$ is defined by the equation

$$dF(x) = f(x)\, dx.$$

As a geometric interpretation, imagine a curve having the values of x as the abscissa and so chosen that the element of area erected above an infinitesimal segment anywhere on the abscissa yields the probability that the random variable X assumes some value within that infinitesimal interval. The ordinate of such a curve is the probability density. We know the distribution function $F(x)$ and its differential $dF(x)$ exist whenever probability can be defined for any random variable, continuous or discrete. The existence of the probability density, on the other hand, presupposes that $F(x)$ be differentiable; in other words, it presupposes that a limiting ratio

$$\lim_{\Delta x \to 0} \frac{F(x + \Delta x) - F(x)}{\Delta x}$$

exists; for by definition $f(x) = dF(x)/dx$. For this reason, the distribution function is more general than the probability density.

Let us now give some *examples of probability density functions*:

1. *Triangular*: A simple case of this type of distribution is specified by the probability density function

$$f(x) = 2x \qquad \text{for } 0 \leqslant x \leqslant 1.$$

2. *Rectangular or uniform*: The probability density function is

$$f(x) = \frac{1}{B - A} \qquad \text{for } A \leqslant x \leqslant B.$$

With $A = 0$ and $B = 1$, we have $f(x) = 1$ for $0 \leqslant x \leqslant 1$, and this $f(x)$ is called the *standard uniform density*.

3. *Normal*: Also called *Gaussian* (with mean μ and standard deviation σ), the probability density function is

$$f(x) = \frac{1}{\sigma\sqrt{2\pi}} \exp\left[-\frac{(x - \mu)^2}{2\sigma^2}\right] \qquad \text{for } -\infty < x < \infty.$$

The normal distribution is often abbreviated $N(\mu,\sigma^2)$. If $\mu = 0$ and $\sigma^2 = 1$, then we have the *standard normal distribution* $N(0,1)$.

4. *Beta*: With $r > 0$ and $s > 0$, the density function is

$$f(x) = \frac{\Gamma(r + s)}{\Gamma(r)\Gamma(s)} x^{r-1}(1 - x)^{s-1} \qquad \text{for } 0 \leqslant x \leqslant 1.$$

5. *Cauchy*: With $a > 0$ and arbitrary μ, the probability density function is

$$f(x) = \frac{a/\pi}{a^2 + (x - \mu)^2} \qquad \text{for } -\infty < x < \infty.$$

6. *Laplace*, or *two-sided exponential*: With $\lambda > 0$ and arbitrary μ, the probability density function is

$$f(x) = \frac{\lambda}{2} e^{-\lambda|x-\mu|} \qquad \text{for } -\infty < x < \infty.$$

7. *Exponential*: With $\lambda > 0$, the probability density function is

$$f(x) = \lambda e^{-\lambda x} \qquad \text{for } 0 \leqslant x < \infty.$$

8. *Gamma*: With $r > 0$ and $\lambda > 0$, the probability density is

$$f(x) = \frac{\lambda(\lambda x)^{r-1}e^{-\lambda x}}{\Gamma(r)} \qquad \text{for } 0 \leqslant x < \infty.$$

9. *Erlang*: If r is a positive integer, then the Gamma distribution is called the Erlang distribution.

10. *Chi-square* (with n degrees of freedom): If $r = n/2$ and $\lambda = 1/2$, then the Gamma distribution is called the chi-square distribution with n degrees of freedom. Its density function is

$$f(x) = \frac{1}{2^{n/2}\Gamma(n/2)} x^{(n/2)-1}e^{-x/2} \qquad \text{for } 0 \leqslant x < \infty.$$

11. *F-distribution* (with m and n degrees of freedom): With $m \geqslant 1$ and $n \geqslant 1$, the probability density function is

$$f(x) = \frac{m^{m/2}\, n^{n/2}\, \Gamma[(m + n)/2]}{\Gamma(m/2)\Gamma(n/2)} \frac{x^{(m-2)/2}}{(mx + n)^{(m+n)/2}} \qquad \text{for } 0 \leqslant x < \infty.$$

12. *t-distribution* (with n degrees of freedom): With $n \geqslant 1$, the probability density function is

$$f(x) = \frac{1}{\sqrt{n\pi}} \frac{\Gamma[(n + 1)/2]}{\Gamma(n/2)} \left(1 + \frac{x^2}{n}\right)^{-(n+1)/2} \qquad \text{for } -\infty < x < \infty.$$

Much will be said about these distributions as the course of the book develops. The function $\Gamma(r)$ which appears in these definitions is the gamma function, which is treated in detail in Chapter 6.

EXERCISES

1. Given $f(x) = ce^{-2x}$, $x > 0$,
   (a) find the value of c that will make $f(x)$ a density;
   (b) calculate $P(2 < X < 4)$;

(c) calculate $P(\frac{1}{2} < X < 1)$.

ANSWERS. (a) 2. (b) 0.018. (c) 0.233.

2. Given $f(x) = cx(1 - x)^2$, $0 \le x \le 1$, and 0 elsewhere,
(a) find the value of c that will make $f(x)$ a density;
(b) calculate $P(X > \frac{1}{2})$.

ANSWERS. (a) 12. (b) $\frac{5}{16}$.

3. Given the uniform density $f(x) = c$, $1 < x < 3$,
(a) determine the value of c;
(b) calculate $P(X < 2)$;
(c) calculate $P(X > 1.5)$.

ANSWERS. (a) $\frac{1}{2}$. (b) $\frac{1}{2}$. (c) $\frac{3}{4}$.

4. If $f(x) = \frac{1}{2} e^{-x/2}$ for $x > 0$, find the number x_0 such that $P(X > x_0) = \frac{1}{2}$.

ANSWER. $x_0 = 2 \ln 2 = 1.39$.

5. If $f(x)$ is the density of the random variable X, find and graph the distribution function $F(x)$ in each case.
(a) $f(x) = 1/3$, $x = 0$, 1, 2.
(b) $f(x) = 3(1 - x)^2$, $0 < x < 1$.
(c) $f(x) = 1/x^2$, $x > 1$.

ANSWERS. (a) $F(x) = 0$ for $x < 0$, $\frac{1}{3}$ for $0 \le x < 1$, $\frac{2}{3}$ for $1 \le x < 2$, 1 for $x \ge 2$. (b) $F(x) = 0$ for $x \le 0$, $1 - (1 - x)^3$ for $0 < x \le 1$, 1 for $x > 1$. (c) $F(x) = 0$ for $x < 1$, $1 - 1/x$ for $x \ge 1$.

6. Suppose the probability that an atom of a radioactive material will disintegrate in time t is given by $1 - e^{-at}$, where a is a constant depending on the material.
(a) Find the density function of X, the length of life of such an atom.
(b) If $\frac{1}{2}$ of the radioactive material will disintegrate in 1000 units of time, calculate the probability that the life of an atom of this material will exceed 2000 units of time.

ANSWERS. (a) $f(x) = ae^{-ax}$ for $x \ge 0$. (b) $\frac{1}{4}$.

7. Suppose the life in hours, X, of a type of radio tube has the density $f(x) = c/x^2$, $x > 100$.
(a) Evaluate c.
(b) Find $F(x)$.
(c) Calculate $P(X > 500)$.

ANSWERS. (a) $c = 100$. (b) $F(x) = 0$ for $x < 100$, $1 - 100/x$ for $x \ge 100$. (c) $P(x > 500) = \frac{1}{5}$.

8. Find the distribution function $F(x)$ and graph it if the density of X is

(a) $f(x) = \frac{1}{2}$, $0 \leqslant x \leqslant 2$;
(b) $f(x) = x$ for $0 \leqslant x \leqslant 1$ and $f(x) = -x + 2$ for $1 < x \leqslant 2$;
(c) $f(x) = [\pi(1 + x^2)]^{-1}$.

ANSWERS. (a) $F(x) = 0$ for $x \leqslant 0$, $x/2$ for $0 < x \leqslant 2$, 1 for $x > 2$. (b) $F(x) = x^2/2$ for $0 \leqslant x \leqslant 1$, $-1 + 2x - x^2/2$ for $1 < x \leqslant 2$, 1 for $x > 2$. (c) $F(x) = \frac{1}{2} + (1/\pi) \tan^{-1} x$.

9. Scaled to convenient units, the daily electric power consumption in a certain locality is a gamma variable with $r = 2.5$ and $\lambda = 0.3$. How much power should a plant be able to supply in order that stand-by equipment will not be drawn upon more than 1 percent of the time?

ANSWER. 25.1.

10. The random variable x is distributed as follows:

$$f(x) = \frac{6x}{(1 + x)^4} \qquad \text{for } 0 \leqslant x < \infty.$$

(a) Show that the total area is unity and find the distribution function $F(x)$.
(b) Find $P(x \leqslant \frac{1}{4})$.
(c) Find $P(x \geqslant 4)$.
(d) Find $P(1 \leqslant x \leqslant 2)$.

ANSWERS. (a) $F(x) = 1 - 3/(1 + x)^2 + 2/(1 + x)^3$; $F(\infty) = 1$. (b) 0.104. (c) 0.104. (d) 0.241.

RANDOM SAMPLING

Basic to the applications of probability theory and statistics is the subject of random sampling. Random sampling requires that we make explicit reference to the distribution $F(x)$ of a random variable X. More will be said about random sampling in Chapter 5; here we want to introduce the topic.

The concept of random sampling is the operational link between the abstract theory of probability and its practical application to observable phenomena. From the sampling viewpoint, an observation X_i represents a particular value of a random variable X, which has a distribution $F(x)$ we call the "theoretical distribution function." Sampling is said to be *independent* when one observation, X_i, does not affect another, X_j, in a probability sense. Typically, observations are made on individual physical objects belonging to a definite set or collection of objects, technically termed a population, or universe. It often happens that the act of taking the observation involves the withdrawal of the object observed. If the withdrawal of an object disturbs the theoretical distribution $F(x)$ (as in sampling without replacement from a deck of cards), then in order to keep the probabilities constant, the observed item must be put back before another drawing is made. With an infinite population, we may disregard sampling depletions, but in sampling from a finite population we have to take them into account. If replacement in the initial finite population is ruled out by the conditions of the problem (as in dealing card hands), then we may imagine a synthetic population of all possible distinct samples and proceed from that point to theoretical replacement of the sample as a whole. Thus the effect of depletion can be overcome through an appropriate model.

One basic postulate of random sampling is that the empirical distribution function of a sequence of independent observations will tend toward the theoretical distribution function $F(x)$ as the number of observations increases without limit. We postulate the existence of an observational process having the properties of (1) independence among observations and (2) convergence of the empirical distribution to the theoretical distribution. We name the process *random sampling*. Of course, there are observational processes that yield limiting theoretical distributions in systematic nonrandom ways. An example of such a nonrandom process is one in which we deal one card at a time from the top of a deck and return it to the bottom, thus picking 52 different cards in a fixed order, and the cycle is repeated indefinitely. In contrast, a random process has no preassigned order of individual values, nor is the order definable in any way except by complete enumeration of the probabilities involved. Random sampling is a process of obtaining independent observations in such a way that (1) any observation has the same probability $P(x < X \leq x + dx)$ as any other observation of falling in the interval $x < X \leq x + dx$; and (2) the limiting relative frequency of those observations that do fall in the interval $x < X \leq x + dx$ in an infinite sequence of observations is equal to the probability $F(x + dx) - F(x)$ as given by the theoretical distribution. The logical role of the theoretical distribution $F(x)$ is to provide a foundation for the applications of probability theory. Since all observations are governed independently by the same theoretical distribution, all can be identified with (in the sense of being particular values of) independent, identically distributed random variables. In summary, a set of observations X_1, X_2, $\cdots$, X_n constitutes a *random sample* of size n from a population characterized by a distribution function $F(x)$ if

1. each of the X_1, X_2, $\cdots$, X_n is a value of a random variable X with distribution $F(x)$, and
2. these n random variables are independent.

EXERCISES

1. From five cards numbered 1, 2, 3, 4, 5, a card is drawn at random. Suppose that the number drawn is k. Then, from among the cards 1 through k, a second card is drawn at random. Suppose that the number drawn on the second trial is r. Then, from among the cards 1 through r, a third card is drawn at random.
   (a) List all 35 sample points (why 35?), and determine their probabilities. Make sure that these probabilities add to 1.
   (b) Evaluate the conditional probability of drawing a 1 on the first trial, given the drawing of a 5 on the third trial.
   (c) Evaluate the conditional probability of drawing a 5 on the first trial, given the drawing of a 1 on the third trial.
   ANSWERS. (a) For example, the probabilities of (5, 5, 5), (4, 4, 1), and (1, 1, 1) are $1/125$, $1/80$, and $1/5$. (b) 0. (c) 0.287.

2. Which features of random sampling are satisfied and which are not satisfied if you wish to estimate the distribution of students' grade point averages and do so by taking a sample of 100 students from the registration files by consulting a table of random numbers but always ignoring any grade point average less than 0.8. Assume the student enrollment is (a) large, (b) small.
   ANSWERS. (a) Independence is satisfied, but $F(x_1) = F(x_2) = \cdots = F(x_{100})$ is not equal to theoretical distribution of students' grade point averages. (b) $F(x_1) = F(x_2) = \cdots = F(x_{100})$ is not satisfied because of dependence.

3. The probability density function of length of life (period of satisfactory service) of a certain manufactured product is given by $f(x) = 2xe^{-x^2}$, $(0 \leqslant x < \infty)$.
   (a) Find the value of a such that $P(x \geqslant a) = 0.99$.
   (b) If 1000 articles are sold, what is the probability that the life of at least one will be less than this value a?
   ANSWERS. (a) $a = 0.100$. (b) $p = 0.999956$.

4. A series of forecasts were made by method A and the errors were recorded. For a subset comprising the worst errors, independent forecasts (not utilizing any information that would not have been available to the user of A) were made by method B. The differences between the absolute errors would be judged highly significant in favor of method B, provided that the requisite conditions were met. What is the catch?
   ANSWER. Stacked deck; sample not random.

5. Discuss several ways of assuring the selection of a sample that is at least approximately random,

(a) when dealing with finite populations;
(b) when dealing with infinite populations.

ANSWERS. (a) We can serially number the elements of the population and then select a sample with the aid of a table of random digits. For example, if a population has $N = 500$ elements and we wish to select a random sample of size $n = 10$, we can use three arbitrarily selected columns of the tables to obtain 10 different three-digit numbers less than or equal to 500, which will then serve as the serial numbers of those elements to be included in the sample. If the population size is large, the use of random numbers can become very laborious and at times practically impossible. For instance, if we want to test a sample of five items from among the many thousands of items, we can hardly be expected to number all the items, make a selection with the use of random numbers, and then pull out the ones that were chosen. In a situation like this, we really have very little choice but to make the selection relatively haphazard, hoping that this will not seriously violate the assumption of randomness which is basic to most statistical theory. Suggest other ways. (b) For an infinite population, the situation is somewhat different since we cannot physically number the elements of the population. Efforts can be made to approach conditions of randomness by the use of artificial devices. For example, in selecting a sample from a production line we may be able to approximate conditions of randomness by choosing one unit each half hour; when tossing a coin, we can try to flip it in such a way that neither side is intentionally favored, and so forth. The proper use of artificial or mechanical devices for selecting random samples is always preferable to human judgment, as it is extremely difficult to avoid unconscious biases when making almost any kind of selection. Suggest additional ways.

CHANGE OF VARIABLE

In many problems we are interested in a function of a random variable and in the distribution of that function. For example, we may have the distribution of temperatures at which a certain chemical reaction occurs, and we might want to translate these temperatures from Farenheit F to Celsius C. The equation is $C = (5/9)(F - 32)$. This example represents a linear transformation, but in many scientific applications nonlinear transforms are increasingly required. In this section we will study the probability distributions of functions of random variables. If X represents the random variable, then typical functions of X would be X^2, $(X - 3)^2$, $X^{1/2}$, and log X. Our problem then is to find the distribution of the transformed variable, say $Y = y(X)$ where y represents the function, from the known distribution of X. Another problem would be to find a transformed variable $y(X)$ whose distribution is more convenient to handle than the distribution of X.

Let us first consider a discrete variable X which takes on the values 0, 1, 2, $\cdots$, n with probability mass function $f(0)$, $f(1)$, $\cdots$, $f(n)$. The distribution of the new random variable $Y = y(X)$ is determined directly from these probabilities. In fact, the possible values of Y are determined by substituting the successive values of X into the function $y(X)$. Thus the possible values of the random variable Y are $y_0 = y(0)$, $y_1 = y(1)$, $\cdots$, $y_n = y(n)$. In the case when there is a one-to-one relationship between the values of X and the values of Y, then the probability mass function of Y is simply

$$g(y_0) = f(0), \quad g(y_1) = f(1), \quad \cdots, g(y_n) = f(n).$$

However, if there is a many-to-one relationship (i.e., two or more values of X give the same value of Y), then we must add all the probabilities that give rise to each value of Y. In this case the probability mass function is

$$g(y_i) = \sum_j f(x_j),$$

where the summation is over all values of x_j such that $y(x_j) = y_i$.

EXAMPLE. Suppose X has the Poisson probability mass function

$$f(x) = \frac{\mu^x e^{-\mu}}{x!} \qquad \text{for } x = 0, 1, 2, \cdots .$$

Define a new random variable by $Y = 4X$. We wish to find the probability mass function of Y. The function $y = 4x$ is a one-to-one transformation, since $y = 4x$ is a single-valued function of x, and the inverse function $x = y/4$ is a single-valued function of y. Hence the event $Y = 4X$ can occur when and only when the event $X = Y/4$ occurs. That is, the two events are equivalent and have the same probability. Hence,

$$g(y) = P(Y = 4x) = P(X = y/4) = f(y/4),$$

so the required probability mass function is

$$g(y) = \frac{\mu^{y/4} e^{-\mu}}{(y/4)!} \qquad \text{for } y = 0, 4, 8, \cdots .$$

EXAMPLE. Let X take the values 0, 1, 2, 3, 4, 5 with probabilities p_0, p_1,

p_2, p_3, p_4, p_5. The probability mass function of the new random variable $Y = (X - 2)^2$ is

$$g(0) = p_2, \quad g(1) = p_1 + p_3, \quad g(4) = p_0 + p_4, \quad g(9) = p_5,$$

where 0, 1, 4, 9 are the possible values of Y.

Let us now consider the ideal case for transforming a continuous random variable X into another continuous random variable Y. Let the transformation $y = y(x)$ be single-valued with a continuous positive derivative $dy/dx > 0$. In such a case there is a single-valued inverse transformation $x = x(y)$ whose derivative $dx/dy = 1/(dy/dx)$ is also positive. Thus the transformation is one-to-one, and the events $x < X \leqslant (x + dx)$ and $y < Y \leqslant (y + dy)$ have the same probability, namely

$$f(x)\, dx = g(y)\, dy,$$

where $f(x)$ and $g(y)$ are the probability density functions of X and Y respectively. Thus the required probability density function is

$$g(y) = f(x) \frac{dx}{dy},$$

where x as a function of y is given by the inverse function $x = x(y)$.

EXAMPLE. Suppose X has the density function

$$f(x) = 2x^3 e^{-x^2} \qquad \text{for } 0 \leqslant x < \infty.$$

Find the density function of the random variable $Y = X^2$. The function $y = x^2$ is single valued with a positive derivative $dy/dx = 2x$ over the range $0 < x < \infty$. The inverse function is $x = y^{1/2}$. Thus $dx/dy = \frac{1}{2} y^{-1/2}$. The required density function is

$$g(y) = f(x) \frac{dx}{dy} = f(y^{1/2})(\tfrac{1}{2})\, y^{-1/2} = 2y^{3/2} e^{-y}\, (\tfrac{1}{2})\, y^{-1/2},$$

which is

$$g(y) = y e^{-y} \qquad \text{for } 0 \leqslant y < \infty.$$

Let us now consider the case where again the transformation $y = y(x)$ is single valued, but now with a continuous negative derivative $dy/dx < 0$. In such a case there is a single-valued inverse transformation $x = x(y)$ whose derivative dx/dy is also negative. Thus the transformation is one-to-one, but now we must write

$$g(y) = f(x)\left(-\frac{dx}{dy}\right),$$

so the right hand side will necessarily be positive. The point is that, in this case, y is large when x is small, and vice versa.

EXAMPLE. Suppose X has the uniform density $f(x) = 1$ in the interval $0 < x < 1$ and $f(x) = 0$ elsewhere. Let us find the distribution of the random variable defined by $Y = -2 \ln X$. Here the transformation is $y = y(x) = -2 \ln x$, so the inverse is $x = x(y) = e^{-y/2}$. The transformation is one-to-one, mapping the admissible region $0 < x < 1$ into the admissible region $0 < y < \infty$. The derivative is

$$\frac{dx}{dy} = -\frac{1}{2} e^{-y/2},$$

which is negative. Accordingly the required density function is

$$g(y) = f(e^{-y/2})(^1/_2)\, e^{-y/2} = (^1/_2)\, e^{-y/2} \qquad \text{for } 0 < y < \infty,$$

which we recognize to be a chi-square distribution with two degrees of freedom.

The case of a one-to-one transformation, regardless of whether the derivative is positive over the admissible region or negative over the region can be incorporated in the equation

$$g(y) = f(x)\left|\frac{dx}{dy}\right|.$$

In other words, we must use the absolute value of the derivative. This procedure insures that the density function $g(y)$ is positive, which it must be.

In those cases when the cumulative distribution function $F(x)$ is mathematically tractable, then it may be used in the change-of-variable technique, as illustrated by the following two examples.

EXAMPLE. Let X have a uniform distribution for $0 \leqslant x \leqslant 1$. Thus its cdf is $F(x) = x$ for the admissible range, $F(x) = 0$ for $x < 0$, and $F(x) = 1$ for $x \geqslant 1$. Consider the new random variable $Y = \sqrt{X}$. To find the cdf of Y we note that

$$P(Y \leqslant y) = P(\sqrt{X} \leqslant y)$$
$$= P(X \leqslant y^2) = y^2 \qquad \text{for } 0 \leqslant y \leqslant 1.$$

Thus we conclude that $G(y) = y^2$ for $0 \leqslant y \leqslant 1$, with $G(y) = 0$ or 1, otherwise, is the cdf of Y. The pdf of Y is simply $g(y) = G'(y) = 2y$ for $0 \leqslant y \leqslant 1$, and $g(y) = 0$ otherwise.

EXAMPLE. Let X be normal with mean zero and variance one, i.e., $N(0,1)$ and $Y = X^2$. We wish to determine the density function of Y. Now for $y > 0$,

$$G(y) = P(Y \leqslant y) = P(X^2 \leqslant y)$$
$$= P(-\sqrt{y} \leqslant X \leqslant \sqrt{y})$$

$$= \frac{1}{\sqrt{2\pi}} \int_{-\sqrt{y}}^{\sqrt{y}} \exp\left(\frac{-x^2}{2}\right) dx = \frac{2}{\sqrt{2\pi}} \int_{0}^{\sqrt{y}} \exp\left(\frac{-x^2}{2}\right) dx$$

is the cdf of Y. Differentiating the integral (Appendix B) gives

$$g(y) = G'(y) = \frac{2}{\sqrt{2\pi}} \exp\left(\frac{-y}{2}\right) \frac{d\sqrt{y}}{dy}$$

$$= \frac{1}{\sqrt{2\pi y}} \exp\left(\frac{-y}{2}\right) \qquad \text{for } y > 0,$$

with $g(y) = 0$ for $y < 0$, is the density of Y.

In the previous two examples we have seen how the cdf enables one to find the pdf of a new random variable. Let X be a random variable with a pdf given by $f(x)$ for $-\infty < x < \infty$. Now let $Y = y(X)$ represent a new random variable, and let us assume that the transformation $y(x)$ is single valued with a continuous positive derivative $y'(x) > 0$. Then to each y there exists a unique x, given by the inverse transformation $x = x(y)$. Thus

$$G(y) = P(Y \leqslant y) = P[y(X) \leqslant y] = P[X \leqslant x(y)] = \int_{-\infty}^{x(y)} f(x) \, dx$$

is the cdf of Y, so the density function is

$$g(y) = G'(y) = f[x(y)](dx/dy).$$

This result may be obtained from the following graphical point of view. From Figure 3.14 we note that the probability that Y will lie in the region $(y, y + dy)$ is precisely the probability that X will lie in the region $(x, x + dx)$, which is $f(x) \, dx$. Hence

$$g(y)dy = f(x) \, dx.$$

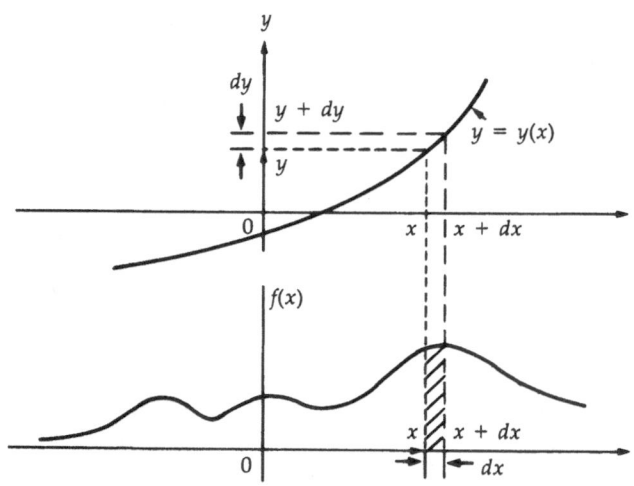

FIGURE 3.14 Case of single-valued transformation. The probability that Y lies in interval $(y,y+dy)$ is equal to the probability that X lies in the interval $(x,x+dx)$; that is, $g(y)\,dy$ is equal to the shaded area, namely $f(x)\,dx$.

The inverse transformation of $y = y(x)$ is denoted by $x = x(y)$. Thus we have $dx = x'(y)\,dy$. Hence $g(y) = f[x(y)]x'(y)$, which is the result given above. We note also that $dy = y'(x)\,dx$, so that

$$g(y) = \frac{f(x)}{y'(x)}\bigg|_{x=x(y)}$$

is an alternative expression for $g(y)$.

If the inverse transformation of $y = y(x)$ is not single valued—i.e., if for a given value of y there exist k values of x, say $x_1, x_2, \cdots, x_k$, such that $y = y(x_j)$, $j = 1, 2, \cdots, k$, then we must modify the change-of-variable equation to the form

$$g(y) = \sum_{j=1}^{k} f(x_j) \left|\frac{dx}{dy}\right|_{x=x_j}.$$

The absolute-value bars enclosing the derivatives insure that all the contributions are positive. Each x_j, for $j = 1, 2, \cdots, k$, yields a contribution to $g(y)$. In this equation, in each $f(x_j)$ we replace x_j by $x(y)$, so $f(x_j)$ is in terms of y. This process is illustrated in Figure 3.15, where we have

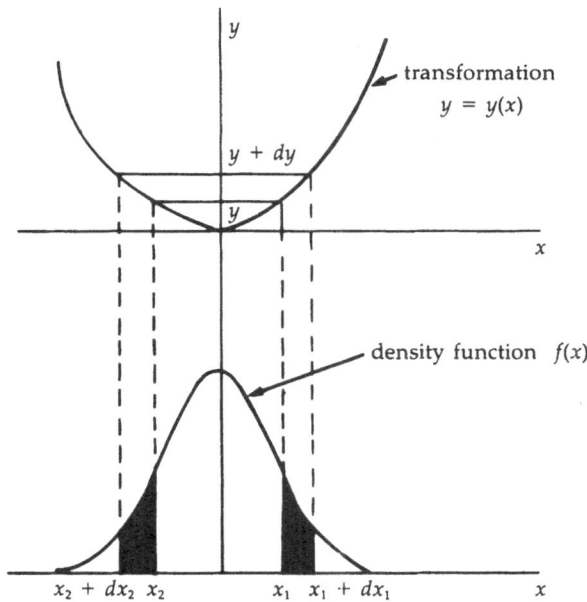

FIGURE 3.15 Case of a double-valued transformation. The probability that Y lies in the interval $(y, y+dy)$ is equal to the sum of the probabilities that X lies in the intervals (x_1, x_1+dx_1) and (x_2, x_2+dx_2); that is, $g(y)\, dy$ is equal to the sum of the shaded areas. The right shaded area is $f(x_1)\, dx_1$ whereas the left shaded area is $f(x_2)\, |dx_2|$.

$$g(y)\, dy = f(x_1)\, |dx_1| + f(x_2)\, |dx_2|,$$

so that

$$g(y) = f(x_1) \left| \frac{dx_1}{dy} \right| + f(x_2) \left| \frac{dx_2}{dy} \right|$$

$$= \sum_{j=1}^{2} f(x_j) \left| \frac{dx_j}{dy} \right|.$$

Note that for x_1 the derivative is positive, and that for x_2 the derivative is negative. In other words, for positive dy, the differential dx_1 is positive and the differential dx_2 is negative, so we have

$$g(y)\, dy = f(x_1)\, dx_1 + f(x_2)\, |dx_2|.$$

In summary, when the transformation $y = y(x)$ has a multivalued inverse $x = x(y)$, the $g(y)\, dy$ is obtained as the sum of transformed elements $f(x_j)\, |dx_j|$ for all values of $x(y)$ corresponding to any one y. Take for illustration $y = x^2$ which often arises in statistical theory. If x can take on negative as well as positive values, i.e., $y = x^2 = (-x)^2$, then the inverse transformation is bivalued, i.e., $x = \pm y^{1/2}$. Because the inverse function is symmetrical about zero, we have $dx = \pm \tfrac{1}{2}\, y^{-1/2}\, dy$, and thus $|dx| = \tfrac{1}{2}\, y^{-1/2}\, dy$. We therefore have the density function

$$g(y)\, dy = f(-y^{1/2})(\tfrac{1}{2})\, y^{-1/2}\, dy + f(y^{1/2})(\tfrac{1}{2})\, y^{-1/2}\, dy \qquad \text{for } y > 0.$$

In those cases when the density function $f(x)$ is symmetrical about zero, then both terms on the right of the above equation are the same, and thus by combining them we eliminate the factor $\tfrac{1}{2}$:

$$g(y) = f(y^{1/2})y^{-1/2} \qquad \text{for } y > 0.$$

For example, let X have the uniform density $f(x) = \tfrac{1}{2}$ over $-1 \leqslant x \leqslant 1$. Then

$$g(y) = \tfrac{1}{2}\, y^{-1/2} \qquad \text{for } 0 \leqslant y \leqslant 1.$$

The following example is a case when the density function $f(x)$ is not symmetrical over the admissible range.

EXAMPLE. Let X have the density $f(x) = \tfrac{2}{9}(x + 1)$ for the admissible region $-1 < x < 2$. Find the density of $Y = X^2$. The change of variable function is thus $y = x^2$, and its inverse is $x = \pm y^{1/2}$. We have

$$y = x^2 \quad \text{for } -1 < x < 0; \qquad \text{inverse } x = -y^{1/2} \quad \text{for } 0 < y < 1,$$
$$y = x^2 \quad \text{for } 0 < x < 2; \qquad \text{inverse } x = y^{1/2} \quad \text{for } 0 < y < 4.$$

The absolute value of the derivative is

$$\left| \frac{dx}{dy} \right| = \tfrac{1}{2}\, y^{-1/2}.$$

For y in the interval $0 < y < 1$, we get two contributions, namely

$$g(y) = f(-y^{1/2})(\tfrac{1}{2})\, y^{-1/2} + f(y^{1/2})(\tfrac{1}{2})\, y^{-1/2}$$

whereas for y in the interval $1 < y < 4$ we have

$$g(y) = f(y^{1/2})(\tfrac{1}{2})\, y^{-1/2}.$$

Thus

$$g(y) = \begin{cases} 2/(9y^{1/2}) & \text{for } 0 < y < 1, \\ (y^{1/2} + 1)/(9y^{1/2}) & \text{for } 1 < y < 4. \end{cases}$$

EXERCISES

1. Let X have the uniform density $f(x) = {}^1\!/_4$ for $-2 < x < 2$. Find the density function of $Y = X^2$.

2. Let X be a random variable with density

$$f(x) = {}^1\!/_2\, e^{-|x|} \qquad \text{for } -\infty < x < \infty.$$

Show that the cdf is given by

$$F(x) = \begin{cases} {}^1\!/_2\, e^x & \text{for } x < 0, \\ 1 - \dfrac{e^{-x}}{2} & \text{for } x \geq 0. \end{cases}$$

Show that the density for the random variable $Y = X^2$ is

$$f(y) = {}^1\!/_2\, y^{-1/2}\, e^{-\sqrt{y}} \qquad \text{for } 0 \leq y < \infty.$$

3. Let X be a random variable with density given by $f(x)$ for $-\infty < x < \infty$. Let $Y = \int_{-\infty}^{X} f(x)\, dx$ define a new random variable Y. What is the range of Y? Show that Y has the uniform density function $g(y) = 1$ for $0 < y < 1$. This is a fundamental result which states that the random variable Y defined by the cdf of a random variable with any density $f(x)$ always has uniform density on the interval $(0,1)$.

4. Density functions of the square and of the absolute value. Let X be a random variable with density $f(x)$. Show that the density of X^2 is

$$g(y) = \frac{f(\sqrt{y}) + f(-\sqrt{y})}{2\sqrt{y}} \qquad \text{for } y \geq 0,$$

and show that the density of $|X|$ is

$$g(y) = f(y) + f(-y) \qquad \text{for } y \geq 0.$$

5. Let $f(x)$ be any probability density function and define

$$Y = \left[\int_{-\infty}^{X} f(x)\, dx \right]^2.$$

Find the density of Y.

ANSWER. $g(y) = {}^1\!/_2\, y^{-1/2}$, $0 < y < 1$.

6. Given two random variables X and Y with respective densities:

$$f(x) = 1 - x/2 \quad \text{for } 0 \leqslant x \leqslant 2;$$
$$f(y) = e^{-y} \quad \text{for } 0 \leqslant y < \infty$$

Find the transformation $y = u(x)$ that relates X and Y.

HINT. Find the functions $z_1 = z_1(x)$ and $z_2 = z_2(y)$ that respectively transform X and Y into random variables Z_1 and Z_2, each uniformly distributed between 0 and 1. Then substitute $z_1 = z_2$ into the inverse of $z_2 = z_2(y)$, so $y = z_2^{-1}[z_1(x)] = y(x)$.

7. Let X have the density $f(x) = x^2/9$ for the admissible region $0 < x < 3$. Find the density of $Y = X^3$.

ANSWER. $1/27$ for $0 < y < 27$.

8. If the density of X is $f(x) = 2xe^{-x^2}$ for the admissible region $0 < x < \infty$, find the density of $Y = X^2$.

9. Given that X is $N(0,1)$,
   (a) find the density of $Y = X^2$
   (b) Find the density of $y = |X|$.
   ANSWERS. (a) $g(y) = (1/\sqrt{2\pi})y^{-1/2} e^{-y/2} \quad$ for $0 \leqslant y < \infty$.
   (b) $g(y) = (2/\sqrt{2\pi})e^{-y^2/2} \quad$ for $0 \leqslant y < \infty$.

10. Let X be a random variable with density $f(x)$. Find the cdf of $Y = X^2$.

ANSWER. $G(y) = 0$ for $y < 0$, $G(y) = \displaystyle\int_{-\sqrt{y}}^{\sqrt{y}} f(x)\, dx$ for $y \geqslant 0$.

11. Suppose that X has uniform density

$$f(x) = \begin{cases} 1 & \text{for } 0 \leqslant x \leqslant 1, \\ 0 & \text{otherwise.} \end{cases}$$

   (a) Show that $1/X$ has the density function $g(y) = 1/y^2$ for $y > 1$ and 0 otherwise.
   (b) Show that X^2 has the density function $g(y) = 1/(2y^{1/2})$ for $0 \leqslant y \leqslant 1$.
   (c) Show that X^3 has the density function $g(y) = 1/(3y^{2/3})$ for $0 \leqslant y \leqslant 1$.
   (d) Show that X^4 has the density function $g(y) = 1/(4y^{3/4})$ for $0 \leqslant y \leqslant 1$.
   (e) Show that $X^{1/2}$ has the density function $g(y) = 2y$ for $0 \leqslant y \leqslant 1$.
   (f) Show that $X/(1 + X)$ has the density function $g(y) = 1/(1 - y)^2$ for $0 \leqslant y \leqslant 1/2$.
   (g) Show that $-\ln X$ has the density function $g(y) = e^{-y}$ for $y \geqslant 0$.

12. Given that

$$f(x) = \begin{cases} x & \text{for } 0 \leqslant x \leqslant 1, \\ 2-x & \text{for } 1 \leqslant x \leqslant 2, \end{cases}$$

find $P(\frac{1}{2} \leqslant X \leqslant \frac{3}{2})$.

ANSWER. 0.75

13. Suppose that X has the density function

$$f(x) = \begin{cases} e^{-x} & \text{for } x \geqslant 0, \\ 0 & \text{for } x < 0. \end{cases}$$

(a) If c is a positive constant, show that X/c has the density function

$$g(y) = \begin{cases} ce^{-cy} & \text{for } y \geqslant 0, \\ 0 & \text{for } y < 0. \end{cases}$$

(b) Show that $X/(1 + X)$ has the density function

$$g(y) = \begin{cases} \dfrac{\exp\left[-y/(1-y)\right]}{(1-y)^2} & \text{for } 0 \leqslant y < 1, \\ 0 & \text{otherwise.} \end{cases}$$

(c) Show that $X + c$ has the density function

$$g(y) = \begin{cases} e^{-(y-c)} & \text{for } y \geqslant c, \\ 0 & \text{for } y < c. \end{cases}$$

(d) Show that $-X$ has the density function

$$g(y) = \begin{cases} e^y & \text{for } y \leqslant 0, \\ 0 & \text{for } y > 0. \end{cases}$$

14. Given that X is uniform over the range $-\frac{1}{2} \leqslant x \leqslant \frac{3}{2}$, find the density function of X^2.

ANSWER.

$$g(y) = \begin{cases} (\frac{1}{2})y^{-1/2} & \text{for } 0 \leqslant y \leqslant \frac{1}{4}, \\ (\frac{1}{4})y^{-1/2} & \text{for } \frac{1}{4} < y \leqslant \frac{9}{4}. \end{cases}$$

Note that $g(y)$ has a discontinuity at $y = \frac{1}{4}$.

4

Applications of Mathematical Expectation

No human investigation can be called real science if it cannot be demonstrated mathematically.

LEONARDO DA VINCI (1452–1519)

Mathematical Expectation

Chebyshev Inequality

Applications to Operations Research

MATHEMATICAL EXPECTATION

The idea of an average is especially pertinent to the subject of random variables and readily lends itself to broad development. By the ordinary rule, the arithmetic average of a set of N numbers $x_1, x_2, \cdots, x_N$ is obtained by computing their sum and then dividing by N; that is, $\bar{x} = (x_1 + x_2 + \cdots + x_N)/N$. Now since it is not necessary that these numbers all be different, let us suppose, in general, that there are n distinct values, $x_1, x_2, \cdots, x_n$ respectively occurring $N_1, N_2, \cdots, N_n$ times, where $N_1 + N_2 + \cdots + N_n = N$. Then the sum of the N numbers could be found by adding up the products $N_1x_1, N_2x_2, \cdots, N_nx_n$ and the arithmetic average would be obtained by dividing the result by N. Therefore,

$$\bar{x} = \frac{\sum_{i=1}^{n} N_i x_i}{N} = \sum_{i=1}^{n} (N_i/N)x_i = \sum_{i=1}^{n} R_i x_i,$$

where $R_i = N_i/N$. We recognize R_i as the relative frequency of x_i. We realize that the average of a random variable can be defined by an equation similar to that above but with relative frequencies replaced by probabilities. Hence we arrive at the following definitions:

1. Given a discrete random variable X with an exhaustive set of admissible values x_i ($i = 1, 2, \cdots, n$) and corresponding probability mass function $f(x_i)$, then the average of X is

$$\mu = \sum_{i=1}^{n} x_i f(x_i).$$

2. Given a continuous random variable X with probability density function $f(x)$ defined in the range $A \leqslant x \leqslant B$, then the average of X is

$$\mu = \int_{A}^{B} xf(x)\, dx.$$

Because $f(x)$ is zero outside the interval $A \leqslant x \leqslant B$, we generally use the limits $-\infty$ to ∞ on the integral sign whether A and B are infinite or finite.

In technical terminology, this average is denoted by the Greek letter μ and is called the *arithmetic mean*, or briefly the *mean*, of the random variable X. The definition of the average of a random variable was stated formally for the first time in connection with games of chance and was then interpreted as the expected amount of winnings in a long series of trials. Hence the average μ came to be called the *expected value* of a random variable X and is customarily denoted by the symbol $E(X)$. While this notation is well adapted to mathematical generalization, the symbol μ for the "mean" $E(X)$ is more convenient when the context is clear.

Let us look at the definition

$$E(X) = \int_{-\infty}^{\infty} xf(x)\,dx.$$

The symbol $E(\ )$ represents a mathematical operator which acts on the random variable X. This definition says that this operation is carried out by weighting the real variable x by the probability density $f(x)$, and then integrating from $-\infty$ to ∞. The real variable x that appears under the integral sign is a dummy variable of integration, and so does not appear in the final result. Specifically, $E(X)$ is not a function of the real variable x as, for example, is the density function $f(x)$. Instead, $E(X)$ is a constant, namely the mean value μ. However, if we are dealing with several random variables, and we want explicitly to indicate that this mean is the mean of X we would represent it as either μ_x or $E(X)$. As is well known, mathematicians like simplicity, and they do not like to use a great number of symbols when fewer will do. In this spirit, after the concepts of a random variable X and its admissible values x are well understood, there is really no reason to use both symbols, and so the lower case x is usually used for both concepts. Thus we would write

$$E(x) = \int_{-\infty}^{\infty} xf(x)\,dx,$$

where the x appearing after the expectation operator is the random variable x, and the x appearing under the integral sign is a dummy variable of integration. Any other dummy variable would serve equally as well, so we could just as well write

$$E(x) = \int_{-\infty}^{\infty} tf(t)\,dt.$$

But generally we would not want to do so, because we like to see just the symbol x when we are dealing with the random variable x. This convention keeps things neat. Under this convention we would write

$$E(y) = \int_{-\infty}^{\infty} yg(y)\,dy$$

for the mean of the random variable y. Everything matches, and this economy of symbols is extremely important in order to see the forest instead of the trees when working on involved problems. In mathematics, the fewer the symbols the better, and often the carrying of both symbols X and x is more compulsive than efficient when it comes to solving problems in a hurry, as on examinations.

There is a need for a broadened definition of the mean, for in addition to the mean of a random variable itself, we shall be interested in determining the mean of a function of the random variable. Appealing to first principles, suppose that we wish to compute the average of a function $\alpha(X)$, where as before the distinct values of X are $x_1, x_2, \cdots, x_n$. Suppose as before that we have a set of N numbers in which x_1 occurs N_1 times, x_2 occurs N_2 times, and so on, where

$$N = N_1 + N_2 + \cdots + N_n.$$

For each distinct value x_i of X we could compute the corresponding value $\alpha(x_i)$ of the function. Even though these n functional values need not be distinct, the sum of all N values of $\alpha(X)$ would be given by

$$N_1\alpha(x_1) + N_2\alpha(x_2) + \cdots + N_n\alpha(x_n).$$

The average is then

$$\alpha(x_1)\frac{N_1}{N} + \alpha(x_2)\frac{N_2}{N} + \cdots + \alpha(x_n)\frac{N_n}{N}.$$

If we let $R_i = N_i/N$ denote the relative frequency, then the average can be written as

$$\sum_{i=1}^{N} \alpha(x_i)R_i.$$

As we know, the relative frequency corresponds to the probability. Therefore, by direct extension, we define the mean or expected value of a (continuous) function $\alpha(X)$ of a random variable X by the following equations:

1. Given a discrete random variable X with admissible values x_i for $i = 1, 2, \cdots, n$ and probability mass function $f(x_i)$, then

$$E[\alpha(X)] = \sum_{i=1}^{n} \alpha(x_i)f(x_i).$$

2. Given a continuous random variable X with range $A \leqslant x \leqslant B$ and probability density function $f(x)$, then

$$E[\alpha(X)] = \int_{A}^{B} \alpha(x)f(x)\,dx.$$

In reference to our previous remarks about simplicity, we would often write $E[\alpha(x)]$ instead of $E[\alpha(X)]$.

As an alternative method of determining $E[\alpha(X)]$, we can first compute the distribution of $\alpha(X)$, where $\alpha(X)$ is considered as a random variable in its own right. Then we can compute the expected value of $\alpha(X)$ as the average value of $\alpha(X)$ taken with respect to its own distribution. Thus we have:

1. Given a discrete random variable $Y = \alpha(X)$ with admissible values $y_i = \alpha(x_j)$ for $i = 1, 2, \cdots, n'$ and probability mass function $g(y_i)$, then

$$E[\alpha(X)] = \sum_{i=1}^{n'} y_i g(y_i).$$

2. Given a continuous random variable $Y = \alpha(X)$ with range $A' \leqslant y \leqslant B'$ and probability density function $g(y)$, then

$$E[\alpha(X)] = \int_{A'}^{B'} yg(y)\,dy.$$

PROBLEM. Find the expected value, hence the average, of the discrete random variable X, with the geometric probability mass function

$$f(x) = (1 - q)q^x \qquad 0 < q < 1$$
$$x = 0, 1, 2, \cdots.$$

SOLUTION.

$$E(X) = \sum_{0}^{\infty} x(1 - q)q^x = (1 - q)\sum_{0}^{\infty} xq^x.$$

To evaluate this summation, put

$$S_n = \sum_{0}^{n} xq^x.$$

If we multiply by q and subtract, we obtain

$$S_n - qS_n = \sum_0^n xq^x - \sum_0^n xq^{x+1}$$

$$= (0q^0 + 1q^1 + 2q^2 + \cdots + nq^n)$$
$$- (0q^1 - 1q^2 - \cdots - (n-1)q^n - nq^{n+1})$$
$$= 0q^0 + (1-0)q^1 + (2-1)q^2 + \cdots + [n - (n-1)]q^n - nq^{n+1},$$

which is

$$(1-q)S_n = q + q^2 + \cdots + q^n - nq^{n+1}.$$

As $n \to \infty$, we have

$$q + q^2 + \cdots + q^n \to q(1 + q + q^2 + \cdots) = \frac{q}{1-q}$$

and

$$nq^{n+1} \to 0.$$

We finally obtain

$$E(X) = \lim_{n \to \infty} (1-q)S_n = \frac{q}{1-q}.$$

PROBLEM. Show that the mean of the Poisson distribution $p(x;\mu)$ is equal to its parameter μ.

SOLUTION.

$$E(X) = \sum_{x=0}^{\infty} (xe^{-\mu}\mu^x/x!) = \mu \sum_{x=1}^{\infty} [e^{-\mu}\mu^{x-1}/(x-1)!].$$

If we let $k = x - 1$, this expression becomes

$$E(X) = \mu \sum_{k=0}^{\infty} (e^{-\mu}\mu^k/k!) = \mu \sum_{k=0}^{\infty} p(k;\mu) = \mu.$$

PROBLEM. Show that the mean of the binomial distribution $b(x;n,p)$ is equal to the product of its parameters, np.

SOLUTION.

$$E(X) = \sum_{x=0}^{n} \frac{xn!}{x!(n-x)!} p^x q^{n-x} = np \sum_{x=1}^{n} \frac{(n-1)!}{(x-1)!(n-x)!} p^{x-1}q^{n-x}.$$

If we let $k = x - 1$, this becomes

$$E(X) = np \sum_{k=0}^{n-1} b(k;n-1,p) = np.$$

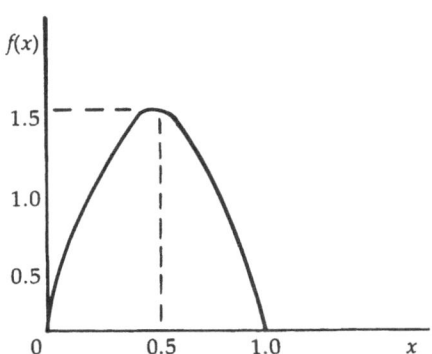

FIGURE 4.1 Parabolic density function.

PROBLEM. Find the mean value of the continuous random variable X, where

$$f(x) = 6x(1 - x) \qquad \text{for } 0 \leqslant x \leqslant 1.$$

This density function is shown in Figure 4.1.

SOLUTION.

$$E(X) = \int_0^1 xf(x)\, dx = \int_0^1 6x^2(1 - x)\, dx = \frac{1}{2}.$$

PROBLEM. Find the mean value of the function $Y = X^2$ given that X is a standard normal variable (i.e., normal with zero mean and unit standard deviation).

SOLUTION. By setting $\mu = 0$ and $\sigma = 1$ into the general expression for the normal probability density function, we obtain the standard normal density function

$$f(x) = \frac{1}{\sqrt{2\pi}} e^{-x^2/2} \qquad \text{for } -\infty < x < \infty.$$

Thus the required mean value is

$$E(Y) = E(X^2) = \int_{-\infty}^{\infty} x^2 \frac{1}{\sqrt{2\pi}} e^{-x^2/2}\, dx.$$

This integral can be evaluated by means of the gamma function $\Gamma(n)$, which is defined and explained in Chapter 6. The result is

$$E(Y) = \frac{2\sqrt{2}}{\sqrt{2\pi}} \Gamma(^3/_2) = 1.$$

Alternatively, we may evaluate the expected value of Y from its own density function. From Exercise 9 on page 82, the density of Y is

$$g(y) = \frac{1}{\sqrt{2\pi}} y^{-1/2} e^{-y/2} \qquad \text{for } 0 \le y < \infty.$$

Thus

$$E(Y) = \int_0^\infty y g(y)\, dy = \frac{1}{\sqrt{2\pi}} \int_0^\infty y^{1/2} e^{-y/2}\, dy.$$

This integral may be evaluated by use of the gamma function (see Chapter 6). The result is the same, namely

$$E(Y) = \frac{2\sqrt{2}}{\sqrt{2\pi}} \Gamma(^3/_2) = 1.$$

An important property of the mean is that the mean cannot fall outside the limits for which the random variable is defined. That is, if $A \le X \le B$, then $A \le E(X) \le B$. We see that this statement follows from

$$\int_A^B A f(x)\, dx \le \int_A^B x f(x)\, dx \le \int_A^B B f(x)\, dx.$$

Another important property is the linear property, which can be stated as follows. If $\alpha(X)$ is a linear function of any number of components, then the expected value of $\alpha(X)$ is the same linear function of the expected values of those components. In mathematical symbols, the *linear property* is as follows: If

$$\alpha(X) = a_0 + a_1 \alpha_1(X) + \cdots + a_n \alpha_n(X),$$

then

$$E[\alpha(X)] = a_0 + a_1 E[\alpha_1(X)] + \cdots + a_n E[\alpha_n(X)].$$

To prove this fact, we write

$$E[\alpha(X)] = \int_A^B [a_0 + a_1 \alpha_1(X) + \cdots + a_n \alpha_n(X)] f(x)\, dx$$

and integrate term by term.

As a simple example, let us determine the mean of the function $\alpha(X)$ $= X - \mu$, where $\mu = E(X)$. For a discrete random variable we have,

$$E[\alpha(X)] = \Sigma(x_i - \mu)f(x_i) = \Sigma x_i f(x_i) - \mu\Sigma f(x_i) = \mu - \mu = 0,$$

and the same is true of a continuous random variable. On the other hand, the mean of the function $\beta(X) = (X - \mu)^2$ is not equal to zero unless X is a constant. This particular quantity $E[(X - \mu)^2]$, which is the mean square deviation from the mean, is called the *variance* of X and is denoted by the special symbol σ^2; its positive square root, denoted by σ, is called the *standard deviation*. Other commonly used notations for the variance of the random variable X are var (X), $\sigma^2(X)$, and σ_x^2; the corresponding notations for the standard deviation are $\sqrt{\text{var}(X)}$, $\sigma(X)$, and σ_x. The significance of the variance and standard deviation will become apparent when we begin to put them to use, as we soon shall. Just now, however, further explanation would entail too great a digression. By expanding the quadratic $(X - \mu)^2$ and then summing or integrating term by term replacing $E(X)$ by its equal μ, we obtain the following identity, which often simplifies the evaluation of σ^2:

$$\sigma^2 = E[(X - \mu)^2] = E(X^2) - \mu^2.$$

PROBLEM. Find the mean, variance, and standard deviation of the discrete random variable with probability mass function $f(0) = \frac{8}{27}$, $f(1) = \frac{12}{27}$, $f(2) = \frac{6}{27}$, $f(3) = \frac{1}{27}$.

SOLUTION. $\mu = 0(\frac{8}{27}) + 1(\frac{12}{27}) + 2(\frac{6}{27}) + 3(\frac{1}{27}) = 1$, $\sigma^2 = (0 - 1)^2(\frac{8}{27}) + (1 - 1)^2(\frac{12}{27}) + (2 - 1)^2(\frac{6}{27}) + (3 - 1)^2(\frac{1}{27}) = \frac{2}{3}$,

$$\sigma = \sqrt{\frac{2}{3}} = 0.816.$$

PROBLEM. Find the mean, variance, and standard deviation of the uniform probability density function $f(x) = 1/L$ for $0 \leqslant x \leqslant L$ and $f(x) = 0$ otherwise.

$$\mu = \int_0^L x(1/L)\, dx = \frac{L}{2},$$

$$\sigma^2 = \int_0^L \left(x - \frac{L}{2}\right)^2 \frac{1}{L}\, dx = \frac{L^2}{12},$$

$$\sigma = \frac{L}{\sqrt{12}} = 0.289L.$$

PROBLEM. Find the mean, variance, and standard deviation of the probability density function $f(x) = 3x(2 - x)/4$ for $0 \leqslant x \leqslant 2$ and $f(x) = 0$ otherwise.

SOLUTION.

$$\mu = \int_0^2 x\left[\frac{3}{4}x(2 - x)\right] dx = 1,$$

$$\sigma^2 = E(x^2) - \mu^2 = \int_0^2 \left(\frac{3}{2}x^3 - \frac{3}{4}x^4\right) dx - 1 = 0.2,$$

$$\sigma = \sqrt{0.2} = 0.447.$$

While the mathematical properties of a distribution are exhaustively determined by the distribution function or its alternatives (either probability density function or probability mass function) together with the range of admissible values, broad classifications as to the geometric form and analytic character of a distribution aid in its visualization. The principal attributes chosen for descriptive purposes are modality, symmetry, norm, and concentration.

Modality: In scientific or industrial applications of probability, the predominant type of distribution is that of a continuous random variable having a density function with a single rounded peak, to either side of which the density falls off steadily toward zero. Such distributions are called unimodal, and the abscissa of the peak is called the *mode.* If a density function has two rounded peaks, the distribution is called bimodal; if three, trimodal; and if more than three, multimodal. Sometimes a density function is constant over its admissible range and is therefore called rectangular; such a distribution evidently has no peak at all. Other density functions have no rounded peaks but, instead have relative maxima in the forms of cusps. Curves with two cusps of approximately equal height are said to be U-shaped, and curves with one or two of distinctly unequal height are called J-shaped, although the "J" is often backwards. Typical geometric shapes are depicted in Figure 4.2.

For a probability mass function, the *mode* is defined as that value of x having the highest probability of occurrence; and for a probability density function, the mode is defined as that value of x having the greatest density. For example, for J-shaped distributions, the abscissa of the higher cusp is called the mode.

Symmetry: If a probability density function or a probability mass function $f(x)$ has an axis of symmetry $x = \xi$ such that $f(\xi - \epsilon) = f(\xi + \epsilon)$, the distribution is said to be symmetrical about the value $x = \xi$. Nonsymmetrical distributions are said to be skewed. If a skewed dis-

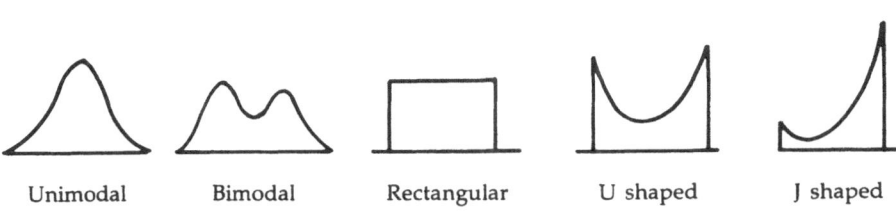

| Unimodal | Bimodal | Rectangular | U shaped | J shaped |

FIGURE 4.2 Typical geometric shapes.

tribution is unimodal, the direction of greater development is visually obvious, and the curve is said to be skewed in the direction of the more pronounced tail. Although the admissible range may well extend equally far to either side of the mode, an unequal division of area constitutes skewness.

Norm: The norm or central tendency of a random variable is usually conceived as the common run or typical value of the variable or sometimes as a medium value. Depending upon the intended use, any one of several measures may be chosen to represent the norm, but the three most widely employed are the *mode,* the *mean,* and the *median.* The first two have already been defined; the *median* is the center of area of the distribution, that is, the value of x such that $F(x) = \frac{1}{2}$. All three of these measures have various drawbacks, but for general serviceability and analytic convenience, the mean is the best.

The *mode* has an intuitive appeal as a norm provided that the distribution is unimodal; otherwise it has little appeal. The mean often lies fairly close to the mode of a unimodal curve and has significance as a medium value for curves other than unimodal. However, the mean has the theoretical limitation of being undefined for certain (rather exceptional) distributions for which the range is infinite and the curve falls off at too slow a rate. For instance, the mean does not exist for the J-shaped curve given by

$$f(x) = \frac{1}{(1 + x)^2} \qquad \text{for } 0 \leq x < \infty,$$

because the integral for the mean diverges. In contrast, the *median* exists for any continuous distribution function and in the case just considered it is given by $x = 1$, since

$$F(x) = 1 - \frac{1}{1 + x}$$

and $F(1) = \frac{1}{2}$. Nevertheless, the mean always does exist whenever the random variable has a finite range and usually exists even when the range is infinite. A more practical objection to the mean is founded upon its sensitivity to extreme values under certain conditions, whereas the median is only slightly affected by extremes. Admittedly, when the curve is highly skewed, the median is preferable to the mean as a representative measure; but moderate skewness does not cause severe distortion, and as a rule the mean is a satisfactory measure. Whereas the equation $F(x) = \frac{1}{2}$ whereby the median is determined always has a solution if $F(x)$ is continuous, the definition sometimes breaks down in the case of a discontinuous distribution function. Consider, for instance, a discrete random variable, the admissible values of which are $x_1 < x_2 < \cdots < x_n$; clearly there might or might not exist an index value k such that $F(x_k) = \frac{1}{2}$. If not, the median is redefined as the lowest value of x for which $F(x)$ exceeds $\frac{1}{2}$. On the other hand, suppose there does exist a value of k such that $F(x_k) = \frac{1}{2}$. Then, although this value itself would satisfy the original definition of the median, it is also true that any number between x_k and x_{k+1} would likewise satisfy this definition; therefore, as a matter of convention the median is then taken as $(x_k + x_{k+1})/2$.

In comparing the relative merits of these three measures, two further points should be kept in mind. First, the analytic convenience of the chosen measure is important, because many practical problems concern composite random variables obtained from the addition of simpler ones; and second, it is ordinarily necessary to estimate the norm and other properties of the distribution from samples. On the first score, the mean is distinctly superior to the other two; and on the second, the mean as a rule can be estimated with the highest precision. Therefore, barring the exceptions already noted, we may say in conclusion that the mean is usually the best representation of the norm.

PROBLEM. Determine the mode, mean, and median of the following skewed, unimodal distribution with infinite range.

$$f(x) = \lambda^2 x e^{-\lambda x} \qquad \text{for } \lambda > 0, 0 \leqslant x < \infty.$$

This is a gamma distribution with parameter $r = 2$.

SOLUTION.

MODE.

$$\frac{df}{dx} = \lambda^2 e^{-\lambda x}(1 - \lambda x); \qquad \frac{df}{dx} = 0 \qquad \text{when } x = \frac{1}{\lambda}.$$

Hence the mode $= 1/\lambda$.

MEAN.

$$\mu = \int_0^\infty xf(x)\,dx = \int_0^\infty \lambda^2 x^2 e^{-\lambda x}\,dx.$$

If we let $z = \lambda x$, so $dz = \lambda\,dx$, the integral becomes

$$\int_0^\infty \lambda^2 x^2 e^{-\lambda x}\,dx = \frac{1}{\lambda}\int_0^\infty z^2 e^{-z}\,dz = \frac{1}{\lambda}\Gamma(3),$$

where $\Gamma(3)$ is the gamma function for argument 3. The gamma function is explained in Chapter 6. Since $\Gamma(3) = 2 \cdot 1 = 2$, the mean is $\mu = 2/\lambda$.

MEDIAN.

$$\int_0^x f(\xi)\,d\xi = 1 - e^{-\lambda x}(1 + \lambda x).$$

If $F(x) = \frac{1}{2}$ then $e^{-\lambda x}(1 + \lambda x) = \frac{1}{2}$ and the median is the root of this equation. Setting $u = \lambda x$ we find that $e^{-u}(1 + u) = \frac{1}{2}$ when $u = 1.679$. Hence the median $= 1.679/\lambda$. This problem bears out the universal fact (Figure 4.3) that if a unimodal distribution is skewed toward the right, the numerical values of the mode, median, and mean are in that order, while if the distribution is skewed to the left, the numerical order is reversed. From the distribution function, we find that the mode in this example is so located that 26.4 percent of the area lies to its left, while 59.4 percent of the area lies to the left of the mean.

Concentration: Once the norm is chosen, there remains the question of concentration, that is, the degree to which the norm represents the distribution as a whole. Intuitively we feel that a measure of concentration should indicate how much area there is in the neighborhood of the norm; but upon reflection, it becomes obvious that unless a definite range about the norm is specified, more or less arbitrarily, what we are really considering is something akin to the distribution function itself. Because an arbitrarily chosen range is of dubious merit, let us propose a measure that is itself an average derived from the entire distribution. As the intuitive concept of concentration applies equally to both sides of the norm, the function to be averaged should depend upon the magnitude of the difference between the norm and an individual value. Moreover, since it should put a penalty upon large deviations in order to distinguish the immediate from the remote, the function should increase with the magnitude of the deviation. We thus

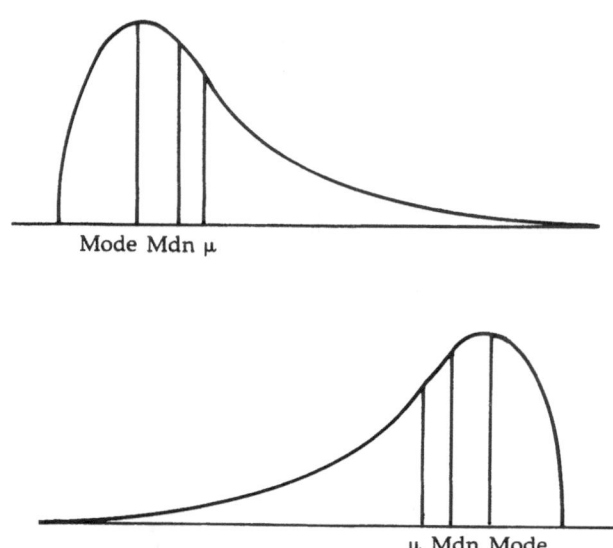

Mode Mdn μ

μ Mdn Mode

FIGURE 4.3 Relative positions of the mean, median, and mode in unimodal skewed distributions.

arrive at a measure of divergence from the norm rather than a direct measure of concentration, but it will serve the same purpose. Initially we might consider using simply the average absolute deviation from the norm, and in fact this measure has gained a limited acceptance; but its exceedingly poor combinative properties render it useless for most purposes, practical and theoretical alike. The simplest measure that satisfies both intuitive and analytical requirements is the mean square deviation from the norm; and when the mean μ is chosen as the norm, this measure becomes the variance σ^2 as defined previously in this section. For general use, therefore, we shall adopt the mean as the norm and the variance or its square root, the standard deviation, as indicative of concentration.

PROBLEM. Find the variance and standard deviation of the gamma distribution (with $r = 2$):

$$f(x) = \lambda^2 x e^{-\lambda x} \qquad \text{for } \lambda > 0 \quad \text{and} \quad 0 \leqslant x < \infty.$$

SOLUTION. From the previous problem, we know that $\mu = 2/\lambda$. We have

$$\sigma^2 = E(x^2) - \mu^2 = \int_0^\infty \lambda^2 x^3 e^{-\lambda x} \, dx - \frac{4}{\lambda^2}.$$

The integral can be evaluated by means of the gamma function. See Chapter 6. If we let $z = \lambda x$, then the integral is

$$\int_0^\infty \lambda^2 x^3 e^{-\lambda x} \, dx = \frac{1}{\lambda^2} \int_0^\infty z^3 e^{-z} \, dz = \frac{1}{\lambda^2} \Gamma(4).$$

Since $\Gamma(4) = 3 \cdot 2 \cdot 1 = 6$, we have

$$\sigma^2 = \frac{6}{\lambda^2} - \frac{4}{\lambda^2} = \frac{2}{\lambda^2} \qquad \sigma = \frac{\sqrt{2}}{\lambda}.$$

The gamma distribution with $r = 2$, namely,

$$f(x) = \lambda^2 x e^{-\lambda x} \qquad \text{for } \lambda > 0 \quad \text{and} \quad 0 \leqslant x < \infty,$$

furnishes a specific example of the existence of families of distributions, the members of which are similar in mathematical structure but differ with respect to norm, concentration, or other characteristics as functions of certain arbitrary quantities such as λ in the gamma distribution. Arbitrary quantities (as distinguished from fixed quantities like 2, π, e, etc.) that occur in the mathematical specification of a distribution are called *parameters*. By an extension of the term, descriptive quantities, such as the mode, mean, median, variance, etc., which are derived from the true distribution, are also called parameters in contradistinction to their respective empirical estimates, called *statistics*, which are derived from samples. Although a large part of current statistical theory is devoted to the problem of estimating parameters by suitable statistics, we shall touch only very lightly upon the subject of statistical estimation in this book.

An important set of parameters are the moments of a distribution. The kth moment about the origin of a random variable X is defined as

$$\mu_k' = E(X^k).$$

Note that one may speak of μ_k' as being the kth moment of X or as the kth moment of the distribution of X. As we know, the first moment μ_1', which is $E(X)$, occurs so often that it is given the special symbol μ. Since moments about the mean are used extensively, they also need to be defined. In terms of the special symbol μ, the kth moment about the mean is defined as

$$\mu_k = E[(X - \mu)^k].$$

The second moment, μ_2, about the mean, as we know, is the variance σ^2.

EXERCISES

1. An insurance company writes a policy for an amount of money A which must be paid if an event E occurs. If the probability of E is p, what should the company charge the customer in order that the expected profit to the company be 10 percent of A?
   ANSWER. $A(p + 0.1)$.

2. Given $f(x) = 6x(1 - x)$, for $0 \leqslant x \leqslant 1$, find $E(y)$ where $y = 1/x$ and check by using the distribution of y.
   ANSWER. 3

3. Find $E(x)$ for the Pascal-geometric probability mass function
   $$f(x) = pq^{x-1} \qquad \text{for } x = 1, 2, \cdots; p > 0, q > 0, p + q = 1.$$
   ANSWER. $1/p$.

4. Given $f(x) = e^{-x}$, for $0 \leqslant x < \infty$, find $E(y)$ where $y = e^{x/2}$ and check by applying the distribution of y.
   ANSWER. 2

5. Given $y = x^3$ and $v = (x - \mu)^4$, find $E(y)$ and $E(v)$ where x is normal $N(\mu, \sigma^2)$.
   ANSWER. $E(y) = 3\sigma^2\mu + \mu^3$; $E(v) = 3\sigma^4$.

6. A useful and simple type of random variable is the indicator random variable. Let A be an event and A' be the complementary event. We define the corresponding *indicator random variable I* as $I = 1$ when A occurs and $I = 0$ when A' occurs. Find the mean and variance of an indicator random variable. In particular, show that the variance of a constant c is var $(c) = 0$.
   ANSWER. $E(I) = P(A)$, var $(I) = P(A)[1 - P(A)]$.

7. Show that if var $(X) = 0$, then $P[X = E(X)] = 1$.

8. Find the variance of the total score in two independent throws of a balanced die.

9. Show that $E[(X - c)^2]$ is a minimum in c when $c = \mu = E(X)$.
   HINT. Write $E[(X - c)^2] = E\{[(X - \mu) + (\mu - c)]^2\}$ and expand the latter.

10. A random variable is said to be standardized if its expectation is zero and its standard deviation (and hence variance) is one. Show that if X is a random variable with expectation μ and standard deviation $\sigma > 0$, then the random variable $(X - \mu)/\sigma$ is standardized.

11. One card is drawn at random from a standard 52-card deck. If the card drawn is the seven of diamonds, we win \$420; if it is any other card, we win nothing. On the basis of expectation, should we be willing to sell our interest in this experiment for (a) \$5, (b) \$10?

    ANSWER. (a) No. (b) Yes (as the expectation is $^{105}/_{13} = \$8.08$.)

12. Show that the variance of the Poisson distribution $p(x;\mu)$ is equal to its parameter μ.

    HINT.
    $$E[X(X - 1)] = \sum_{x=0}^{\infty} [x(x - 1)e^{-\mu}\mu^x/x!]$$
    $$= \mu^2 \sum_{x=2}^{\infty} [e^{-\mu}\mu^{x-2}/(x - 2)!]$$
    $$= \mu^2 \sum_{k=0}^{\infty} p(k;\mu) = \mu^2,$$

    where $k = x - 2$. The variance is then
    $$\sigma^2 = E[X(X - 1)] + \mu - \mu^2 = \mu.$$

13. Show that the variance of the binomial distribution $b(x;n,p)$ is equal to npq.

    HINT. Use the same approach as in Exercise 12.

14. For four independent tosses of a balanced coin, find the expectation of the number of heads that appear. Also find the expectation of the number of heads minus the number of tails.

    ANSWER. 2, 0.

15. For two independent throws of a balanced die, find the expectations of the following random variables:
    (a) the score on the first throw.
    (b) the total score.
    (c) the larger of the two scores.
    (d) the smaller of the two scores.
    (e) the first score minus the second.
    (f) the larger score minus the smaller.
    (g) twice the larger score minus the second.
    (h) the total number of sixes that appear.
    (i) the total number of fives and sixes that appear.

    ANSWERS. (a) $^7/_2$. (b) 7. (c) $^{161}/_{36}$. (d) $^{91}/_{36}$. (e) 0. (f) $^{70}/_{36}$. (g) $^{196}/_{36}$. (h) $^1/_3$. (i) $^2/_3$.

16. Suppose an experiment produces a success with probability $1/4$. What is the expected number of successes in four independent trials of the experiment?

17. For any random variable for which the mean μ is finite, prove the identities
(a) $E(x - \mu) = 0$,
(b) $E(x - a)^2 = E(x - \mu)^2 + (\mu - a)^2$, where $a = $ constant.
HINT. Use the substitution $(x - a) \equiv (x - \mu) + (\mu - a)$.

18. For any sample of N observations $x_1, x_2, \cdots, x_N$, put $\bar{x} = (1/N) \Sigma_i x_i$ and prove as analogs to Exercise 17 the analogous properties of $\bar{x}$:
(a) $\Sigma_i (x_i - \bar{x}) = 0$,
(b) $\Sigma_i (x_i - a)^2 = \Sigma_i (x_i - \bar{x})^2 + N(\bar{x} - a)^2$.
As a special case of (b) notice that

$$\Sigma_i (x_i - \mu)^2 = \Sigma_i (x_i - \bar{x})^2 + N(\bar{x} - \mu)^2.$$

19. If x has the density function $f(x)$, show that the quantity $E(|x - c|)$ is minimized by choosing c as the median.

20. A density function is said to be symmetrical about zero when $f(-x) \equiv f(x)$. If a is any positive value of x, derive the following results, using geometric intuition freely.
(a) $F(0) = 0.5$.
(b) $P(X > a) \equiv P(X \geqslant a) = 0.5 - \int_0^a f(x)\, dx$.
(c) $F(-a) + F(a) = 1$.
(d) $P(-a < X < a) \equiv P(-a \leqslant X \leqslant a) = 2F(a) - 1$.
(e) $P(|X| > a) \equiv P(|X| \geqslant a) = 2F(-a) = 2[1 - F(a)]$.

21. The median of a random variable X that has a continuous distribution function $F(x)$ is defined as that value of c such that $F(c) = 1/2$. If x has a density function $f(x)$ and the lower and upper limits of x are A and B respectively, show that the median satisfies the following equation:

$$\int_A^c f(x)\, dx = \int_c^B f(x)\, dx.$$

22. A distribution is said to be unimodal if its density function $f(x)$ has only one turning point, and this is a maximum value of the ordinate; the abscissa of the maximum is called the mode. In (a) through (e) show that the distributions are unimodal and find the mode in each case. Also determine k in each case so that the total area will be unity.

(a) $f(x) = kx^4 e^{-x}$ for $0 \leqslant x < \infty$.
(b) $f(x) = kxe^{-ax}$ for $a > 0$; $0 \leqslant x < \infty$.
(c) $f(x) = kxe^{-x^2}$ for $0 \leqslant x < \infty$.
(d) $f(x) = kx(a - x)^2$ for $0 \leqslant x \leqslant a$.
(e) $f(x) = k(ax + x^2)e^{-bx}$ for $a > 0$, $b > 0$; $0 \leqslant x < \infty$.

ANSWERS. (a) $k = \frac{1}{24}$; mode $= 4$. (b) $k = a^2$; mode $= 1/a$. (c) $k = 2$; mode $= 1/\sqrt{2}$. (d) $k = 12/a^4$; mode $= a/3$. (e) $k = b^3/(ab + 2)$; mode $= (2 - ab + \sqrt{4 + a^2 b^2})/2b$.

23. The random variable X has the Cauchy density function:

$$f(x) = \frac{k}{a^2 + (x - b)^2} \qquad \text{for } -\infty < x < \infty.$$

(a) Determine k so that the total area will be unity.
(b) Sketch the curve.
(c) Locate the median and mode by inspection, checking the former by integration and the latter by differentiation.

ANSWERS. (a) $F(x) = 1 - [3/(1 + x)^2] + [2/(1 + x)^3]$ for $0 \leqslant x < \infty$. (b) 0.104. (c) 0.104. (d) 0.241.

24. The random variable X has the density function

$$f(x) = \frac{6x}{(1 + x)^4} \qquad \text{for } 0 \leqslant x < \infty.$$

(a) Show that the total area is unity and find the distribution function $F(x)$.
(b) Find $P(X \leqslant \frac{1}{4})$
(c) $P(X \geqslant 4)$
(d) $P(1 \leqslant X \leqslant 2)$.

ANSWERS. (a) $F(x) = 1 - [3/(1 + x)^2] + [2/(1 + x)^3]$ for $0 \leqslant x < \infty$. (b) 0.104. (c) 0.104. (d) 0.241.

25. A random observation X is made. Its density function is

$$f(x) = \lambda^2 x e^{-\lambda x} \qquad \text{for } \lambda > 0, 0 \leqslant x < \infty.$$

(a) What is the probability that X will be less than the mode?
(b) More than twice the mode?

ANSWERS. (a) 0.264. (b) 0.406.

CHEBYSHEV INEQUALITY

The connection between the standard deviation and the concept of concentration is brought out by the fact that, in terms of standard deviation, an upper bound can be set on the probability of extreme deviations from the mean. A very general relation of this sort, applicable to any distribution for which μ and σ are defined is known as the *Chebyshev inequality*. A stronger inequality, but one restricted to unimodal distributions, is known as the *Gauss inequality*.

Any random variable, continuous or discrete, for which the mean is zero and the standard deviation is equal to unity is termed a *standard random variable* or, synonomously, a *standard score*. We shall derive the Chebyshev inequality first of all for the special case of a continuous standard random variable Z with density function $f(z)$ defined in the range $A \leqslant z \leqslant B$. By hypothesis, therefore,

$$\mu = \int_A^B zf(z) \, dz = 0; \quad \text{and} \quad \sigma^2 = \int_A^B z^2f(z) \, dz = 1.$$

Letting t denote an arbitrary positive quantity, the Chebyshev inequality for the standard random variable Z is

$$P(|Z| \geqslant t) < 1/t^2.$$

In words, this inequality states that the probability that the absolute value of Z is greater than or equal to t is less than $1/t^2$. Let us now prove the proposition that this inequality is true. Dividing the interval (A,B) into three ranges $(A,-t)$, $(-t,t)$, (t,B) as shown in Figure 4.4, denote the respective areas by p_i $(i = 1, 2, 3)$ and the integrals of $z^2f(z)$ dz by V_i; that is,

$$p_1 = \int_A^{-t} f(z) \, dz, \qquad p_2 = \int_{-t}^{t} f(z) \, dz, \qquad p_3 = \int_t^B f(z) \, dz,$$

$$V_1 = \int_A^{-t} z^2 f(z) \, dz, \qquad V_2 = \int_{-t}^{t} z^2 f(z) \, dz, \qquad V_3 = \int_t^B z^2 f(z) \, dz.$$

Since the magnitude of z is greater than or equal to t in Ranges 1 and 3, for which the corresponding probability is equal to $p_1 + p_3$, we see that $P(|Z| \geqslant t) = p_1 + p_3$. Thus we wish to show that $p_1 + p_3 < 1/t^2$. From the fact that $z^2 \geqslant t^2$ in these two ranges, it follows that

$$V_1 > \int_A^{-t} t^2f(z) \, dz = t^2p_1, \qquad V_3 > \int_t^B t^2f(z) \, dz = t^2p_3.$$

Since $\sigma^2 = 1$, we have

$$1 = V_1 + V_2 + V_3 \geqslant V_1 + V_3 > t^2(p_1 + p_3),$$

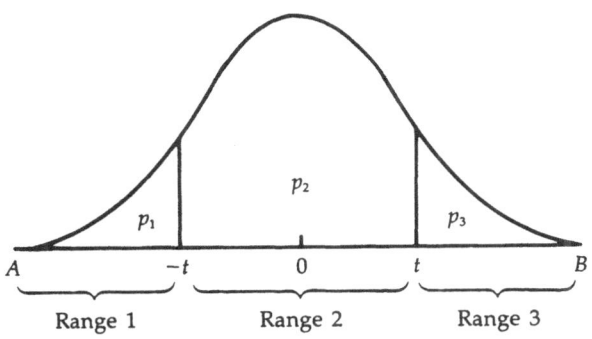

FIGURE 4.4 The three ranges with respective areas p_1, p_2, p_3.

which is

$$t^2(p_1 + p_3) < 1$$

or

$$P(|Z| > t) = p_1 + p_3 < 1/t^2.$$

Thus the Chebyshev inequality is proved.

Extending the proposition to cover a discrete standard random variable Z, we must allow for the fact that conceivably a discrete distribution might consist merely of two equally likely values, and that for an appropriate selection of t the entire distribution could be concentrated at the two points $z = \pm t$. In that case it would follow that

$$P(|Z| \geq t) = P(|Z| = t) = f(-t) + f(t)$$

and

$$\sigma^2 = 1 = t^2 f(-t) + t^2 f(t) = t^2 P(|Z| \geq t),$$

whence

$$P(|Z| \geq t) = \frac{1}{t^2}.$$

Therefore, in order to be perfectly general, the Chebyshev inequality assumes the following form for an unrestricted standard random variable:

$$P(|Z| \geq t) \leq 1/t^2.$$

The *Gauss inequality* for unimodal distributions makes use of the distance between the mean and the mode, and for a standard random variable, it can be stated as follows:

$$P(|Z| \geq t) \leq \frac{4(1 + \lambda^2)}{9(t - \lambda)^2},$$

where λ denotes the magnitude of the difference between the mean and mode, and t must be taken greater than λ. In case the mean and mode coincide, the Gauss result reduces to a special form known as the *Camp-Meidell inequality*, which is

$$P(|Z| \geq t) \leq \frac{1}{2.25t^2}.$$

Observing that any random variable X can be reduced to a corresponding standard random variable Z by the linear transformation $Z = (X - \mu)/\sigma$, provided μ and σ exist, we can restate the foregoing inequalities in general terms. The *Chebyshev inequality*, in a form that holds for all distributions with finite mean and standard deviation, is

$$P\left(\frac{|X - \mu|}{\sigma} \geq t\right) \leq \frac{1}{t^2}.$$

And for a unimodal distribution with coincident mode and mean, the *Camp-Meidell inequality* is

$$P\left(\frac{|X - \mu|}{\sigma} \geq t\right) \leq \frac{1}{2.25t^2}.$$

The probability P is represented graphically as the shaded portion of Figure 4.5. From the quantity $(|X - \mu|)/\sigma$ in these inequalities, it is seen that the probability of exceeding a stated magnitude of deviation from the mean tends to decrease as σ decreases; or, what amounts to the same thing, the probability of keeping within preassigned limits tends to increase with decreasing σ. Accordingly, in a general (but not unique) way, concentration is inversely related to the standard deviation. Whereas the Chebyshev inequality guarantees that no more than one ninth of any distribution with finite μ and σ lies beyond 3σ from its mean, we learn from the Camp-Meidell inequality that this region includes less than 5 percent of any distribution to which that inequality applies. It should be borne in mind, however, that because of their generality, these inequalities cannot specify the true probabilities very

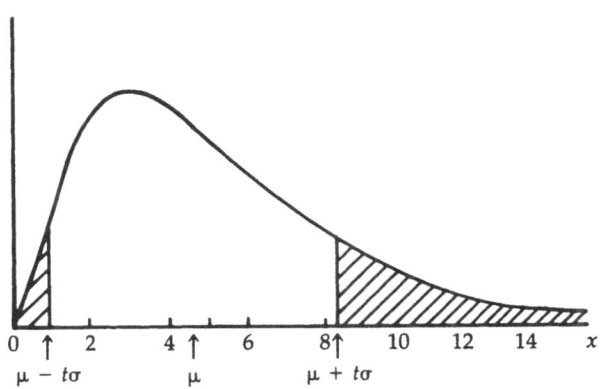

FIGURE 4.5 The sum of the shaded areas gives the probability that the random variable x will deviate from the mean μ by an amount greater than t_σ.

precisely; and as they yield only upper bounds, the actual probabilities are often much lower. In normal circumstances, such overestimates of the tail areas (outlying portions of the curve) would be on the conservative side. Even when the mean and mode do not coincide, the Camp-Meidell inequality frequently leads to an overestimate of tail areas of unimodal distributions. In the problem on page 96, for instance, where the mean is exactly twice the mode, the true probability is 0.039 that a random point will deviate from μ by as much as 2σ and 0.047 that it will deviate by as much as 3σ. The corresponding bounds, as given by the Camp-Meidell inequality, are 0.111 and 0.049 respectively.

In summary, the most important feature of Chebyshev's theorem is that it applies to any probability distribution for which μ and σ exist. So far as applications are concerned, however, this generality is also its greatest weakness; it only provides an upper limit (often a very poor one) to the probability of getting a value that deviates from the mean by more than t standard deviations. Thus, we can state in general that the probability of getting a value which deviates from the mean by more than two standard deviations is less than 0.25. In comparison, the corresponding exact probability for the binomial distribution with the parameters $n = 16$ and $p = 1/2$ is 0.021. Similarly, we can state in general that the probability of getting a value which deviates from the mean by more than three standard deviations is less than 0.112. In comparison the corresponding exact probability for the Poisson distribution with the parameter $\lambda = 9$ is 0.0024.

EXERCISES

1. Show that the Chebyshev inequality may be written as

$$P(|X - \mu| \geq \alpha) \leq \sigma^2/\alpha^2$$

   for any positive number α.

2. Suppose that the discrete random variable X takes the values 0, 1, 2, $\cdots$, n, and that $E(X) = 1$, var $(X) = 1$. Show that $P(X \geq 3) \leq \frac{1}{4}$. In the same way show that for $k = 4, 5, \cdots, n$, $P(X \geq k) \leq 1/(k - 1)^2$.

3. Show that if X is a discrete random variable with expectation μ and if α is a positive number, then $P(|X - \mu| \geq \alpha) \leq E[(X - \mu)^4]/\alpha^4$. Graph the following three functions of $\alpha > 0$ when X has a uniform probability mass function on 1, 2, 3, 4, 5: $P(|X - \mu| \geq \sigma)$, var$(X)/\alpha^2$ and $E[(X - \mu)^4]/\alpha^4$. Comment on the meaning of these curves relative to inequalities on $P(|X - \mu| \geq \alpha)$.

4. Show that at least one of the numbers $P(X - \mu \geq \alpha)$ and $P(X - \mu \leq -\alpha)$ is no larger than var $(X)/2\alpha^2$.

5. If X is a binomial random variable with parameters $n = 4$ and $p = \frac{1}{2}$, compare $P(|X - \mu| \geq \alpha)$ with var $(X)/\alpha^2$ for $\alpha = \frac{1}{2}$, 1 and $\frac{3}{2}$. Repeat when X is instead a binomial random variable with parameters $n = 4$ and $p = \frac{1}{8}$.

6. Show that $P(|X - \mu| \geq \alpha) = $ var $(X)/\alpha^2$ when X is a discrete random variable with probability mass function given by $f(\alpha) = \frac{1}{2} = f(-\alpha)$. Can you see why this should be so in the proof of the Chebyshev inequality?

APPLICATIONS TO OPERATIONS RESEARCH

The term "mathematical expectation" dates back to the early work in the theory of probability, which received impetus from games of chance. The common experience of fortuitous variations in luck prompted gamblers to pose the question as to how much, in the long run, one might expect to win per game under stated rules of play, and mathematicians answered the question by defining "expected value."

From the probability standpoint, a game of chance can be regarded as a process of making a random draw x from a possible set of n distinct scores or n distinct payoffs $x_1, x_2, \cdots, x_n$. If the probability of any particular score (payoff) x_i is $f(x_i)$, where $\Sigma_i f(x_i) = 1$, then the *expected value* $E(x)$ of the scores (payoffs) is defined as the sum of products obtained by weighting each score (payoff) by its own chance of occurrence:

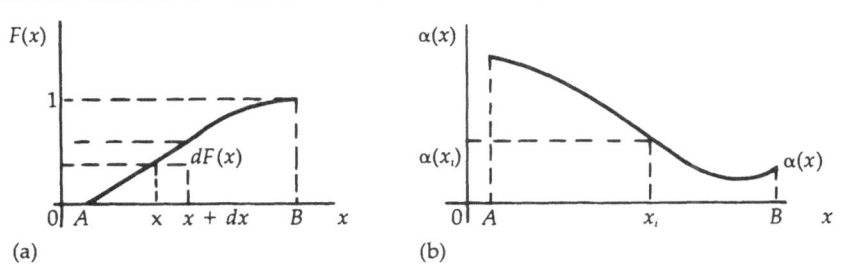

FIGURE 4.6 (a) Cumulative distribution function. (b) Continuous function.

$$E(x) \equiv \sum_{1}^{n} x_i f(x_i).$$

PROBLEM. Let us suppose we are tossing a die and that if the numbers 1, 2, or 3 appear, we will receive 2 cents. If 4 or 5 turn up, we will receive 1 cent and if 6 occurs, we will lose 2 cents. Thus, in ascending order, the payoffs are -2, 1, 2, and their respective probabilities are $\frac{1}{6}$, $\frac{1}{3}$, $\frac{1}{2}$. What is the expected value of the game?

SOLUTION. The expected value of the game to us is obtained by multiplying each payoff by its chance of occurrence and then adding these products over all possibilities. Hence

$$E(x) = (-2)(\tfrac{1}{6}) + (1)(\tfrac{1}{3}) + (2)(\tfrac{1}{2}) = 1 \text{ cent.}$$

If we played this game over and over, we should expect to average one cent on each individual toss.

As we have seen, the concept of expected value can be generalized in the following manner. Let us assume we have a cumulative distribution function $F(x)$ and also any other function of x, say $\alpha(x)$, which we shall assume is continuous over the region of definition $A \leq x \leq B$ of $F(x)$. (See Figure 4.6.) For a particular value of x, $\alpha(x)$ takes on a certain value and also $dF(x)$ has a specific value. In this case $dF(x)$ represents the probability increment associated with a particular value of x, while $\alpha(x)$ represents the value of the function if that value of x does occur. The product of $\alpha(x)$ and $dF(x)$ integrated for all values of x in the region of definition in most cases represents a convergent integral, which has important physical interpretation. The integral, defined as the *expected value* of $\alpha(x)$, is

$$E[\alpha(x)] = \int_A^B \alpha(x) \, dF(x).$$

In the case of a discrete random variable, this definition becomes the form

$$E[\alpha(x)] = \sum_{x_i \geq A}^{x_i \leq B} \alpha(x_i) f(x_i) = \sum_{i=1}^{n} \alpha(x_i) f(x_i);$$

while for a continuous random variable having a density function $f(x)$, it becomes the form

$$E[\alpha(x)] = \int_A^B \alpha(x) f(x) \, dx.$$

In case the distribution function $F(x)$ is a combination of abrupt jumps and segments of continuous growth, we obtain an integral of this second form augmented by a series of terms to be summed of the first form. In such a case, the total probability given by $\int f(x) \, dx$ plus $\Sigma f(x_i)$ must of course equal unity.

PROBLEM. From the profit standpoint of a particular power generating plant, suppose that the profit due to rainfall during the month of July is $\alpha(x) = 5(1 - e^{-x})$ million dollars, where x represents the total number of inches of rainfall during the month. If the probability density of x is $f(x) = e^{-x}$ for $0 \leq x < \infty$, what is the expected profit due to rain for the power company?

SOLUTION. Figure 4.7 shows the profit $\alpha(x)$ as a function of rainfall x. The leveling off of the profit function, as shown in Figure 4.7, is due to the spilling of water over the dam in periods of heavy rainfall. Figure 4.8 shows the probability density of rainfall during July. It is an exponential density function; the probability density decreases exponentially as the rainfall variable x increases. We want to find the expected value of $\alpha(x)$. We have

$$E[\alpha(x)] = \int_0^\infty 5(1 - e^{-x}) e^{-x} \, dx = 5 \int_0^\infty e^{-x} \, dx - 5 \int_0^\infty e^{-2x} \, dx$$

$$= 5(1 - 1/2) = 2^1/2 \qquad \text{million dollars.}$$

The expected value can be interpreted as an average. As we have seen, the expected value of a discrete random variable is a generali-

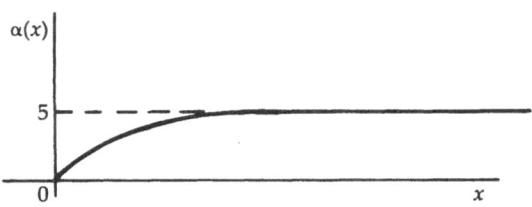

FIGURE 4.7 Profit as a function of rainfall x.

FIGURE 4.8 Probability density of rainfall x.

zation of the simple average. In turn, the expected value of a continuous random variable is the parallel of that for a discrete random variable. By analogy the resulting quantity again may be interpreted as an average. This statement follows from the fact that both the discrete and continuous cases are included in the general definition of the expected value of $\alpha(x)$ in terms of the distribution function of x. In case $\alpha(x) = x$, the expected value of the random variable itself becomes simply

$$E(x) = \sum_i x_i f(x_i)$$

if x is discrete, and

$$E(x) = \int_A^B x f(x)\, dx$$

if x is continuous and has a density function $f(x)$ defined in the range $A \leqslant x \leqslant B$. Furthermore, the interpretation as an average is legitimate for any continuous function $\alpha(x)$, although the connection is indirect. From the definition of $E[\alpha(x)]$, the weight applied to each value of $\alpha(x)$

is the probability increment associated with the corresponding value of x and is, therefore, equal to the probability increment associated with one of the mutually exclusive ways that a particular value of $\alpha(x)$ can occur. If $\alpha(x)$ can have the same value for more than one value of x, these probability increments are automatically added up by the process, and ultimately account for all of the mutually exclusive possibilities.

PROBLEM. Suppose that on an appropriate monetary scale, the distribution of personal income is given to an adequate approximation by $f(x) = e^{-x}$ in the range $0 \leqslant x < \infty$, where x is the number of units of income. Incomes from 0 to 1 are exempt from taxation. A proposed tax rate is $90(1 - e^{1-x})$ percent of income when $x > 1$. What is the tax per capita?

SOLUTION. The tax rate $R(x) = 0.9(1 - e^{1-x})$ and the actual tax $T(x) = xR(x)$ are plotted in Figure 4.9. The tax per capita is simply the expected value of the function $T(x)$. Hence

$$E[T(x)] = \int_0^\infty T(x)f(x)\,dx = \int_1^\infty 0.9x(1 - e^{1-x})e^{-x}\,dx$$

$$= 0.9\int_1^\infty xe^{-x}\,dx - 0.9e\int_1^\infty xe^{-2x}\,dx$$

$$= 0.9\left[-(xe^{-x} + e^{-x}) + \frac{e}{4}(2xe^{-2x} + e^{-2x})\right]_1^\infty$$

$$= 0.9(1.25e^{-1}) \approx 0.414. \qquad Answer.$$

The foregoing problems were given to illustrate the interpretation of expected value as an average. The object of the next three problems is to show how to go about choosing the values of variable quantities that are subject to managerial control in such a way as to maximize the effectiveness of an operation as represented by a suitable measure. For expository purposes, the situations are simplified, but the mathematical principles are of general application.

PROBLEM. In the retailing business, the latest fashions in winter dresses represent a seasonal article which must be ordered in advance and stocked by a department store. Suppose the dresses sell for $100 per unit and cost the store $50 per unit irrespective of disposal. However, any dress not sold during the season must be sold at a sacrifice to a special dealer for $35 per unit. Given that the probability mass function of customer orders for the dresses is given by the Poisson distribution

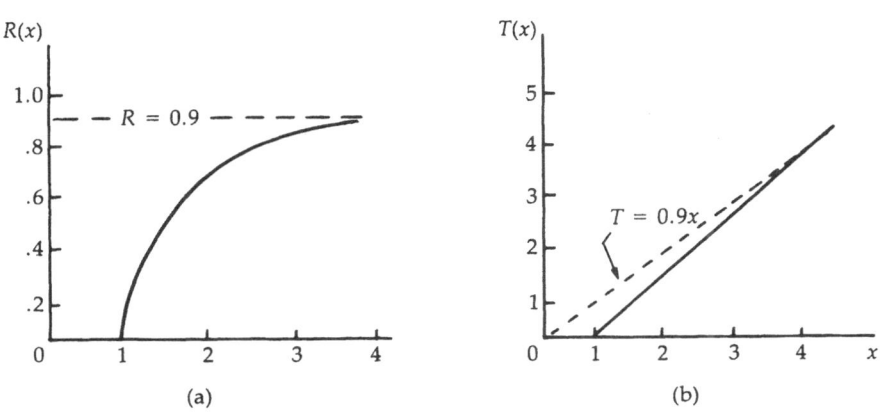

FIGURE 4.9 (*a*) Tax rate (*solid curve*) with asymptote (*dashed line*). (*b*) Actual tax (*solid curve*) with asymptote (*dashed line*).

$$f(x) = e^{-9}9^x/x! \qquad \text{for } x = 0, 1, 2, \cdots$$

with parameter $\mu = 9$. The number x of orders during any season is a random draw from this distribution. How many dresses n should be stocked in a given season in order to maximize the expected value of the profit?

SOLUTION. Let n denote an arbitrary number of dresses stocked. The profit $\alpha(x;n)$ depends upon the random variable x (the number ordered by customers) and the parameter n (the number stocked). We shall express the expected value $E[\alpha(x;n)]$ of the profit as a function of n, and then determine n so as to maximize this function. Depending on the magnitude of the number of orders x relative to n, the profit function assumes two distinct forms. The x-axis is divided by the point n into two regions called L for *left region* and R for *right region*, as shown in Figure 4.10. In region L ($0 \le x \le n$), we sell x dresses at \$100 each, sell the remaining $n - x$ dresses at \$35 each, and at the same time we have to pay for n units at \$50 each. Hence, (neglecting the \$ sign for simplicity) the profit function in the left region is given by

$$\alpha(x;n) = \alpha_L(x;n) = 100x + 35(n - x) - 50n = 65x - 15n.$$

In region R ($x > n$), we have only n dresses available for sale regardless of the actual demand, and since each dress is sold for \$100 but costs the store \$50, the net profit is $100 - 50 =$

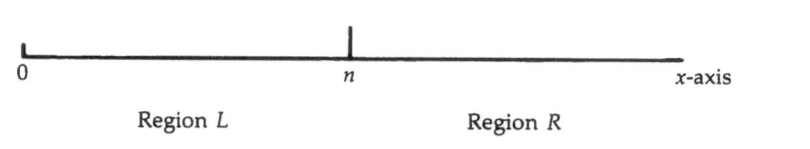

Region L   Region R

FIGURE 4.10 Region L (*left region*) represents overstocking and region R (*right region*) represents understocking.

50. Hence, in the right region the profit function is

$$\alpha(x;n) = \alpha_R(x;n) = 50n$$

The overall expected value of the profit $E[\alpha(x;n)]$ for any value of x is given by

$$E[\alpha(x;n)] = \sum_{x=0}^{\infty} \alpha(x;n)f(x) = \sum_{x=0}^{n} \alpha_L(x;n)f(x) + \sum_{x=n+1}^{\infty} \alpha_R(x;n)f(x)$$

$$= \sum_{x=0}^{n} (65x - 15n)f(x) + \sum_{x=n+1}^{\infty} 50nf(x).$$

Now

$$\sum_{x=n+1}^{\infty} f(x) = 1 - \sum_{x=0}^{n} f(x).$$

Hence

$$E[\alpha(x;n)] = \sum_{x=0}^{n} (65x - 15n)f(x) + 50n\left[1 - \sum_{x=0}^{n} f(x)\right]$$

$$= 50n + 65 \sum_{x=0}^{n} (x - n)f(x).$$

A direct method of arriving at the maximum profit would be to evaluate this function at successive points until the maximum is located. For illustrative purposes we have carried out this calculation, with the results shown in Figure 4.11. The profit function commences at zero when $n = 0$, rises sharply at first and then more slowly; it reaches a maximum at $n = 11$ and then begins to decline; it is very nearly zero at $n = 39$ and is negative when $n \geq 40$.

Let us now give another solution. To determine the optimum value of n, let us investigate whether we would increase or decrease the expected profit by increasing n by one unit. By substitution,

$$E[\alpha(x;n + 1)] = 50(n + 1) + 65 \sum_{x=0}^{n+1} (x - n - 1)f(x)$$

$$= 50(n + 1) + 65 \sum_{x=0}^{n} (x - n - 1)f(x),$$

since $(x - n - 1) = 0$ when $x = n + 1$. Subtracting $E[\alpha(x;n)]$ from $E[\alpha(x;n + 1)]$, we find that the difference is

$$\Delta E[\alpha(x;n)] = E[\alpha(x;n+1)] - E[\alpha(x;n)]$$

$$= 50 + 65 \sum_{x=0}^{n} [(x - n - 1) - (x - n)]f(x)$$

$$= 50 - 65 \sum_{x=0}^{n} f(x) = 50 - 65F(n),$$

where $F(n)$ is the value of the cumulative distribution function of x at $x = n$. From the latter equation, it is clear that $\Delta E[\alpha(x;n)]$ will be positive as long as $F(n) < {}^{50}/_{65}$ and will be negative when $F(n) > {}^{50}/_{65}$, while if $F(n) = {}^{50}/_{65}$, $E[\alpha(x;n)] = E[\alpha(x;n+1)]$. Therefore, the *maximum value* of the *profit* will be obtained if we choose n to be the *smallest* value for which $\Delta E[\alpha(x;n)]$ is zero or negative. The reason is that if $F(n) = {}^{50}/_{65}$ for an integral value of n, then we would get the same expected profit with n as with $n + 1$, and the smaller number has the advantage of smaller investment. Moreover, in case $F(n)$ cannot equal ${}^{50}/_{65}$ for an integral value of n, the smallest value of n such that $\Delta E[\alpha(x;n)]$ is negative has the property that the next larger value of n would yield a smaller profit, in view of this negative difference, while the next smaller value would also yield a smaller profit, because for the latter the difference is positive.

Because the distribution is the well-known Poisson distribution, extensive tables are available. From such tables, we find the following values near ${}^{50}/_{65} \approx 0.769$. Thus the maximum value of the expected profit will be obtained by choosing $n = 11$. This solution, of course, agrees with Figure 4.11.

n	9	10	11
$F(n)$	0.587	0.706	0.803
$\Delta E[\alpha(x;n)]$	11.8	4.1	−2.2

The mean value $E(x)$ of a Poisson random variable is equal to the parameter μ. Thus from the practical standpoint, we know that for the given Poisson distribution the average number of

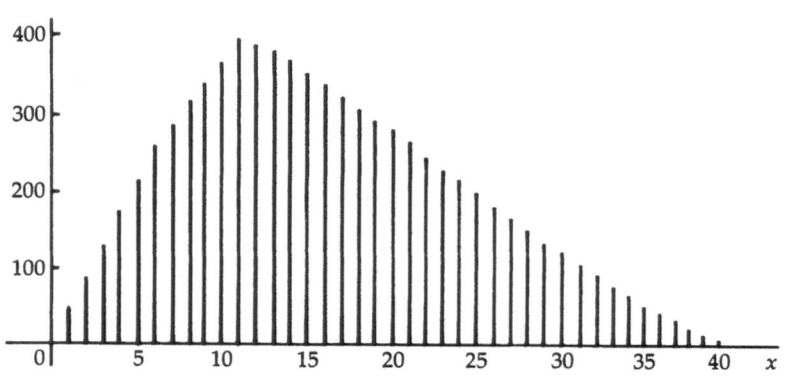

FIGURE 4.11 Expected profit $E[\alpha(x;n)]$ vs. amount stocked n.

orders is 9. From experience the manager will be aware of this fact even though he might not know the form of the distribution. In the absence of this type of analysis, a reasonable decision on his part would be to stock the average number, 9. As it happens, the expected profit would be $373 as opposed to the maximum of $389, and the difference is only about 4 percent of the smaller figure. Of course, there are four values of x (10, 11, 12, 13) which would yield higher profit than $n = 9$, but the curve is so flat near the average-order level that the consequence of a nonoptimum decision is not very serious. This situation is typical of the area in which a qualitative understanding of an operation is sufficient for practical purposes, until conditions become critical. In an enterprise of large scale, of course, a small percentage represents a substantial absolute gain.

PROBLEM. Bathing suits to be sold for the summer season bring a net profit of a dollars for each unit sold and a net loss of b dollars for each unit left unsold when the season ends. If the distribution of order volume x may be approximated by the density function $f(x)$ with admissible range $0 \leqslant x \leqslant B$ as shown in Figure 4.12, determine the optimum number of units n to be stocked in order to maximize the expected value $E[\alpha(x;n)]$ of the profit g.

SOLUTION. There are two forms for the profit function, depending upon whether x is less than n (the case of overstocking) or x is greater than n (the case of understocking). If $x \leqslant n$ the profit function is

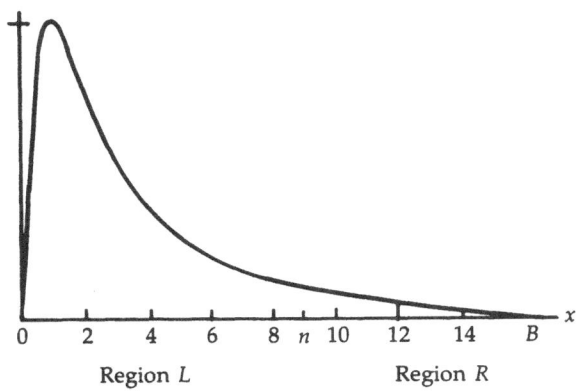

FIGURE 4.12 Region L represents overstocking, and Region R understocking.

$$\alpha(x;n) = \alpha_L(x;n) = ax - b(n - x) = (a + b)x - bn,$$

whereas, if $x > n$

$$\alpha(x;n) = \alpha_R(x;n) = an.$$

Hence the expected value of the profit is

$$E[\alpha(x;n)] = \int_0^n \alpha_L(x;n)f(x)\, dx + \int_n^B \alpha_R(x;n)f(x)\, dx$$

$$= \int_0^n [(a + b)x - bn]f(x)\, dx + \int_n^B anf(x)\, dx.$$

Since

$$\int_n^B anf(x)\, dx = an\left[1 - \int_0^n f(x)\, dx\right],$$

the expected value becomes

$$E[\alpha(x;n)] = an + (a + b)\int_0^n (x - n)f(x)\, dx.$$

We want to differentiate this function with respect to n. Because the quantity within the integral sign as well as the upper limit on the integral sign each depend on n, we can differentiate according to the method given in Appendix B. The result is

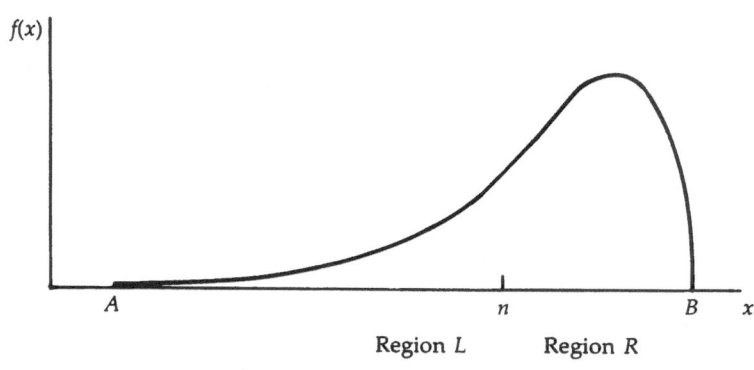

Region L   Region R

FIGURE 4.13 Region L represents overstocking, and Region R understocking.

$$\frac{\partial E[\alpha(x;n)]}{\partial n} = a + (a + b)(n - n)f(n) - (a + b)\int_0^n f(x)\, dx$$

$$= a - (a + b)F(n).$$

Hence the maximum average profit is reached if n is so chosen that

$$F(n) = \frac{a}{a + b} = \frac{\text{unit profit}}{\text{unit profit} + \text{unit loss}}.$$

PROBLEM. On the first working day of August, a manufacturer must order a certain number of units of raw material, which will be worked on in September as orders are received. The manufacturer pays a price of c per unit for them, but if they are not used in September they will be sold for scrap at a fraction of c, say kc, where $0 < k < 1$. The selling price will be determined by adding a margin m per unit to the total cost. The probability density function of sales is known and has the form $f(x)$ where $A \leqslant x \leqslant B$ is the admissible region of the variable x. See Figure 4.13. How many units should be ordered so as to maximize the profit?

SOLUTION. In this problem the manufacturing cost need not be considered, since the manufacturer produces only on order and, therefore, always recoups this money. Let n be the number of units of raw material. If $x \leqslant n$, the manufacturer can sell x units at the price $(c + m)$ each, must dispose of $n - x$ units at the rate kc per unit, and has to pay for n units at the rate c per unit. Hence, in this range, the actual profit for any pair of values of x and n is

(Range L: $x \leqslant n$): $\quad \alpha(x;n) = \alpha_L(x;n) = (c + m)x + kc(n - x) - cn.$

If $x > n$, the manufacturer can sell only the amount for which there was enough raw material, and there is nothing to scrap. Hence in this range, the actual profit is

(Range R: $x > n$): $\quad \alpha(x;n) = \alpha_R(x;n) = (c + m)n - cn = mn.$

Accordingly, the expected value of the profit is

$$E[\alpha(x;n)] = \int_A^n \alpha_L(x;n)f(x)\,dx + \int_n^B \alpha_R(x;n)f(x)\,dx$$

$$= (c + m) \int_A^n xf(x)\,dx + kc \int_A^n (n - x)f(x)\,dx$$

$$- cn \int_A^n f(x)\,dx + mn \int_n^B f(x)\,dx.$$

Rewriting $\int_n^B f(x)\,dx$ as $1 - \int_A^n f(x)\,dx$, we may combine terms, obtaining

$$E[\alpha(x;n)] = mn + (c + m - kc) \int_A^n (x - n)f(x)\,dx.$$

If we use the method of differentiating an integral with respect to a parameter, as given in Appendix B, we find

$$\frac{\partial E[\alpha(x;n)]}{\partial n} = m + (c + m - kc)(n - n)f(n)$$

$$- (c + m - kc) \int_A^n f(x)\,dx$$

$$= m - (c + m - kc)F(n).$$

When this is equated to zero, we find that the value of x that maximizes the profit is given by

$$F(n) = \frac{m}{c + m - kc} = \frac{K}{1 + K - k} \qquad \text{where } K \equiv \frac{m}{c}.$$

This means that the amount n that should be ordered is such that the area to the left of $x = n$ under the probability density of x should equal

$$\frac{K}{1 + K - k}.$$

Apparently then, the successful operation of this business depends upon two operational constants; namely: (1) the ratio of the margin m to the cost c of the raw material, and (2) the fraction k of the initial cost which can be realized for scrap. From common sense it is obvious that if there is no penalty for procuring more raw material than will be needed to meet the actual demand (the situation when $k = 1$), then the number of units of raw material should equal the greatest possible number of orders; that is, $n = B$. This conclusion, of course, is confirmed by the analytic solution; for if $k = 1$, then $F(n) = K/K = 1$ and $n = B$. On the other extreme, if $k = 0$ (which implies that each excess unit of raw material is a total loss), the area should equal

$$\frac{K}{1 + K}.$$

Even in this case, the value of n can still be fairly large if the profit margin is great; for if the margin is very large compared to the cost of the raw material, this last expression also approaches unity and x tends toward its maximum value B. Intermediately, if the margin were \$1.00 per unit and the cost of the raw material \$4.00 per unit and only 25 percent of this is realized on scrap, then the quantity

$$\frac{K}{1 + K - k}$$

would equal 0.25, and thus n should be picked so that only $1/4$ of the area is to the left of n.

This analysis was done on the basis of profit, and no explicit allowance was made for the fact that a certain amount of good will is lost through failure to supply some of the would-be customers. Let us assume that each order lost through insufficient material can be rated as a loss of ϵ dollars per unit. Then when $x > n$ the profit function would be

$$\alpha_R(x;n) = mn - \epsilon(x - n),$$

and the expected profit would become

$$E[\alpha(x;n)] = mn + (c + m - kc) \int_A^n (x - n)f(x)\, dx - \epsilon \int_n^B (x - n)f(x)\, dx.$$

We thus have

$$\frac{\partial E[\alpha(x;n)]}{\partial n} = m - (c + m - kc)\int_A^n f(x)\, dx + \epsilon \int_n^B f(x)\, dx$$

$$= (m + \epsilon) - (c + m + \epsilon - kc)F(n).$$

Therefore, the maximum profit is reached when n is so chosen that

$$F(n) = \frac{m + \epsilon}{c + m + \epsilon - kc} = \frac{K + K_1}{1 + K + K_1 - k} \qquad \text{where } K_1 \equiv \frac{\epsilon}{c}.$$

Obviously, the effect of introducing this factor is to move n to the right and thus increase the outlay of raw material. By assigning various values to K_1, we can determine the extent to which different value ratings of good will affect immediate profit. Reciprocally, we can arrive at a quantitative rating of good will by setting an outside limit on the tolerable decrease in immediate profit and solving for the corresponding values of K_1 and ϵ.

EXERCISES

1. An ice cream vender finds that the demand for ice cream bars is 100 with probability $2/3$ (the day is warm) and 35 with probability $1/3$ (the day is cold). If the vender pays $0.25 per bar, sells bars at $0.50, and must discard unsold bars at the end of the day, what is the expected profit (or loss) if he or she orders 35 bars? 50 bars? 100 bars?

2. Suppose one bus arrives at a given corner each hour. With probability $1/6$ it arrives at 10 minutes past the hour, with probability $1/2$ it arrives at 25 minutes past the hour, and with probability $1/3$ it arrives at 50 minutes past the hour.
   (a) What is the expectation of the waiting time for someone who arrives at the corner on the hour?
   (b) At 15 minutes past the hour?
   (c) Can you find the best time for one to arrive at the corner (best in the sense that it minimizes the expectation of the waiting time)?
   ANSWERS. (a) $185/6$ minutes. (b) $1055/36$ minutes. (c) 25 minutes past the hour.

3. A game played by two persons is said to be "fair" if the expectation of return is zero for each player. Tom and Jean toss a balanced die, and Tom agrees to pay Jean $6 if the score registered is less than or equal to 2. How much should Jean pay Tom when the score is larger than 2 in order that the game be fair?
   ANSWER. $3.

4. A business consists of renting out the use of 10 washing machines. Assume that, under present conditions, any given machine is likely to be out of commission one day out of five and also that break-downs occur at random. What is the probability of more than one machine being out of order on a given day? Assuming the average loss of c dollars per day for a machine out of order, write down an expression for the expected loss due to having machines out of service and evaluate by making the substitution $u = k - 1$ and recognizing the binomial expansion.

ANSWER. $1 - (^{14}/_5)(^4/_5)^9$; $\sum_{k=0}^{10} ck \binom{10}{k} (^4/_5)^{10-k}(^1/_5)^k = 2c$.

5. A lottery has three prizes of $1000, $500, and $100. Five thousand tickets are to be sold and the winning tickets will be drawn as a random sample of size three from them.
   (a) What price should be charged per ticket if this amount is to be the same as the expectation of return on a single ticket?
   (b) Suppose a person bought a ticket at the price determined in (a) and then learned that a total of 4000 tickets were sold. Should that individual complain?
   (c) If 5000 tickets are sold at the price determined in (a), how much profit will the sponsors of the lottery make? Why?
   (d) A woman is approached about buying the last of 5000 tickets at the price determined in (a). Is she justified in thinking that her expectation of return is less than the price of the ticket since all winning tickets may have already been sold?

6. The sales volume of a given concern is given by the following probability density function

$$f(x|\mu) = \frac{4}{\mu^2} xe^{-2x/\mu} \qquad 0 < x < \infty,$$

which has an expected value of μ, that is, $\int_0^\infty xf(x|\mu)\,dx = \mu$. This expected value μ depends upon the selling price s per unit and also upon the total amount of money t spent on advertising. Assume that in the range of application, the relation is

$$\mu = M\frac{1 + bt}{1 + t}e^{-as},$$

where M, a, and b are constant. Also assume that the cost c per item is independent of the volume, since this item is purely one of assembly. The margin of profit m per item is, of course, reduced by the amount spent on production and advertising, so that $m = s - c$

– (t/x). Find the expected value of the product of m times x and then maximize this with respect to t and s in order to determine the best selling price and amount of money to be spent on advertising.

ANSWER.

$$E(mx) = (s - c)\mu - t,$$

$$s = c + 1/a,$$

$$t = -1 + \sqrt{\frac{M(b-1)}{a}}\, e^{-ac-1}.$$

7. A seasonal item which must be stocked in advance by a retailer costs $1 per unit and sells for $2 per unit. Unsold items at the end of the season are a complete loss. The probability of k customer orders is $(1/2)^{k+1}$, $k = 0, 1, \cdots$. Find the expected profit as a function of the number n of items stocked. Note that

$$\sum_{k=0}^{\infty} (1/2)^k = 2,$$

$$\sum_{k=0}^{j} (1/2)^k = 2[1 - (1/2)^{j+1}],$$

$$\sum_{k=0}^{j} k(1/2)^k = 2[1 - (1/2)^j] - j(1/2)^j.$$

ANSWER. $2 - n - (1/2)^{n-1}$.

5

Multivariate Distributions

I value the discovery of a single even insignificant truth more highly than all the argumentation on the highest questions which fails to reach a truth.

GALILEO (1564 - 1642)

Joint Probability Mass Functions

Marginal and Conditional Probability Mass Functions

Joint, Marginal, and Conditional Probability Density Functions

Expectation, Covariance, and Correlation

Regression Curves

Law of Large Numbers

Change of Several Variables

JOINT PROBABILITY MASS
FUNCTIONS

In probability theory the word "experiment" refers to any process that is nondeterministic to some particular observer. The observer's uncertainty may be due to the nature of the process, the state of knowledge of the observer, or both. The study of probability theory is concerned with the analysis of abstractions, or models, of actual physical experiments. The formulation of a model requires a precise statement of an appropriate universe of possible outcomes. Another term for universe is sample space. The universe (or sample space) is a basic (i.e., elemental), mutually exclusive, and collectively exhaustive listing of all possible outcomes of a model of an experiment. That is, a sample space is made up of members that are EEE (which stands for *elemental*, *exclusive*, and *exhaustive*). The members, which are the possible outcomes of the experiment, are also known as the sample points of the sample space. The various events are sets of sample points. Set theory provides the algebra of events.

For the study of experiments whose outcomes may be specified numerically, it is useful to introduce the concept of random variable. As we have seen a random variable is defined by a function that assigns a numerical value of the random variable to each sample point in the sample space of an experiment. Each performance of the experiment generates an experimental value of the random variable. This experimental value of the random variable is equal to the value that the function assigns to the sample point that occurred. Although the full sample space may be required to describe the detailed probabilistic structure of an experiment, our practical interest in each performance of the experiment may relate only to the resulting experimental values of one or more random variables. When this situation obtains, we may prefer to work in an *event space* which distinguishes among outcomes only in terms of the possible experimental values of the random variables of interest. For example, consider the experiment made up of three independent flips of a fair coin. Suppose that our only interest in the performance of this experiment has to do with the resulting experimental value of the random variable X. The random variable X is defined as the total number of heads resulting from the three flips. The event space would consist of the four points:

$$x = 0, \quad x = 1, \quad x = 2, \quad x = 3.$$

These four points form a mutually exclusive, collectively exhaustive listing of all possible experimental outcomes. The event point at any one of these values of x corresponds to the following event: "The experimental value of the random variable X generated on a performance of the experiment is equal to x."

Let us now look at the case of an experiment for which we define two random variables X and Y. Our concern with each performance of the experiment depends only upon the resulting experimental values of these two random variables. As a result, an event space would be a two-dimensional space with orthogonal coordinates x and y. Let us return to our example of the experiment of three independent flips of a fair coin. We define the random variable X as before, but now we also define the random variable Y. Let Y be the length of the longest run resulting from the four flips. (A run is a set of successive flips all of which have the same outcome.) The sample space and the corresponding values of the random variables are:

Sample Space	Probability	X	Y
HHH	$1/8$	3	3
HHT	$1/8$	2	2
HTH	$1/8$	2	1
HTT	$1/8$	1	2
THH	$1/8$	2	2
THT	$1/8$	1	1
TTH	$1/8$	1	2
TTT	$1/8$	0	3

The event space is shown in Figure 5.1. An event point in this space with coordinates (x,y) corresponds to the event: On a performance of this experiment, the random variable (X,Y) takes on the experimental value (x,y). For example, the event point $(2,1)$ represents the event of exactly two heads and no pair of consecutive flips have the same outcome.

The random variables discussed in this example take on only experimental values selected from a set of discrete numbers. Such a random variable is called a *discrete random variable*. In the first part of this chapter, we treat discrete random variables. Later in the chapter, we deal with continuous random variables. A *continuous random variable* takes on experimental values anywhere within continuous ranges. In the case of a discrete random variable, we define a *probability mass function*, whereas in the case of a continuous random variable we define a *probability density function*.

We know that a random variable is defined by a function that assigns a numerical value (i.e., the value of that random variable) to each sample point. These assigned values represent the possible experimental values of the random variable. Each performance of the ex-

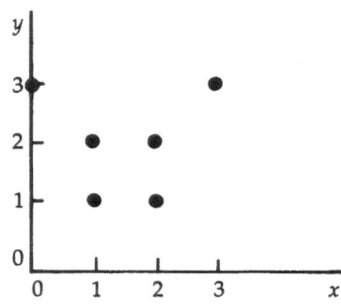

FIGURE 5.1 Event space.

periment generates an experimental value (or observed value) of the random variable. An event space for a single discrete random variable consists of the points on a line corresponding to all possible experimental values of the random variable. We next define a function on this event space that assigns a probability to each event point. This function is known as the probability mass function for the discrete random variable. The *probability mass function* $f(x)$ gives the probability that the experimental value of the random variable X obtained on performance of the experiment is equal to the discrete point x; that is,

$$f(x) = P(X = x) \qquad \text{where } x \text{ is a discrete point.}$$

Often the probability mass function is drawn as a bar graph over the line representing the event space of the random variable. Let us return to the experiment of three independent flips of a fair coin. The random variable X (i.e. the number of heads) has the event space consisting of the discrete event points $x = 0, 1, 2, 3$. The probability mass function is

$$f(0) = {}^1/_8, \quad f(1) = {}^3/_8, \quad f(2) = {}^3/_8, \quad f(3) = {}^1/_8.$$

We obtained these probabilities by adding up the probabilities in the original sample space. The bar graph for the probability mass function is shown in Figure 5.2. We now wish to consider the situation in which values of several random variables are assigned to each point in the sample space of an experiment. We will discuss the case of two random variables. The extension to more than two random variables follows the same lines of reasoning.

In a given performance of an experiment, consider the probability

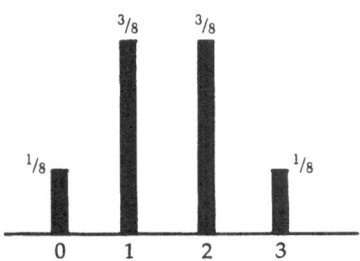

FIGURE 5.2 Bar graph of probability mass function.

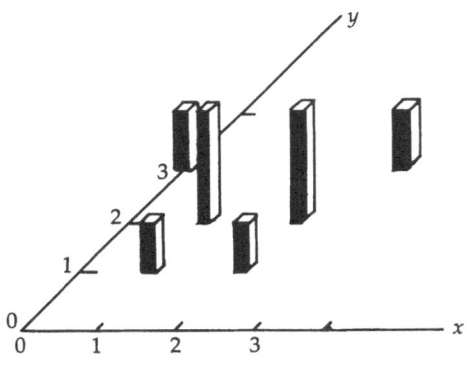

FIGURE 5.3 Joint probability mass function.

that random variable X will take on the experimental value x *and* random variable Y will take on the experimental value y. This probability may be determined in sample space by summing the probabilities of each sample point with this joint attribute (x,y). We thus define the joint probability mass function for two random variables X, Y. The *joint probability mass function* $f(x,y)$ gives the probability that the experimental values of random variables X and Y obtained on a performance of the experiment are equal to the discrete points x and y respectively. A three-dimensional graph of this function would have the possible event points (x,y) marked on a plane with each value of $f(x,y)$ indicated as a bar perpendicular to this plane above each event point. See Figure 5.3. As in the single-variable case, we use the term "event point" because each such event point represents the union of all sample points which yield the particular value of (x,y). For example, in the experiment of three independent flips of a fair coin, the event points

(x,y) are (3,3), (2,2), (2,1), (1,2), (1,1), (0,3). These event points represent the union of sample points as follows:

(3,3) = *HHH*
(2,2) = *HHT* ∪ *THH*
(2,1) = *HTH*
(1,2) = *HTT* ∪ *TTH*
(1,1) = *THT*
(0,3) = *TTT*

Because each sample point has the probability of $1/8$, the joint probability mass function is

$f(3,3) = 1/8$
$f(2,2) = 1/4$
$f(2,1) = 1/8$
$f(1,2) = 1/4$
$f(1,1) = 1/8$
$f(0,3) = 1/8$

We can also represent this joint probability mass function as shown in Figure 5.4. The value of $f(x,y)$ associated with each discrete event point (x,y) is shown beside the point.

The joint probability mass function $f(x,y)$ gives the probability of the joint event $X = x$ and $Y = y$; that is,

$$f(x,y) = P\{(X = x) \cap (Y = y)\}.$$

Alternatively, we may write the joint probability mass function as

$$f(x,y) = P\{(X,Y) = (x,y)\}.$$

Another term for the joint probability mass function is *compound probability mass function*. The joint probability mass function $f(x,y)$ represents the bivariate distribution of the random variables X and Y.

Bivariate probability mass functions follow the same rules as do their single-variate counterparts, namely

$$f(x,y) \geq 0 \qquad \text{for all } (x,y).$$

$$\sum_x \sum_y f(x,y) = 1 \qquad \text{[where the summation is over all } (x,y)\text{]}.$$

$$P\{(x,y) \epsilon A\} = \sum_x \sum_y f(x,y) \qquad \text{[where the summation is over } (x,y) \text{ in the set } A\text{]}.$$

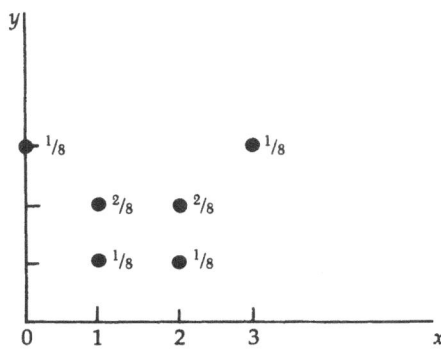

FIGURE 5.4 Joint probability mass function.

The joint probability mass function for two or more random variables represents the multivariate distribution of these random variables. (That is, a bivariate distribution is a multivariate distribution in the case of two random variables.)

For a precise usage of symbols, as we have seen, it is necessary to make a distinction between a random variable X and one of its possible values x. Ordinarily, however, we are not so much interested in a random variable per se as in the properties of its distribution. Therefore, whenever no confusion results, it is convenient and efficient to use the same symbol for the random variable X as for the corresponding real variable x which describes its values; and generally the lower case letter is used. This usage is a common practice, and as the book proceeds we will also make use of it.

EXERCISES

1. An urn contains three balls numbered 1, 2, 3 respectively. We draw two balls at random, with replacement, from the urn. Let X be the number of the first ball drawn and let Y be the number on the second ball drawn. What collection of 2-tuples (i.e., pairs of numbers) would be a reasonable sample space for the experiment, and what are the corresponding probabilities?

   ANSWER. (x,y) for $x = 1, 2, 3$ and $y = 1, 2, 3$; $f(x,y) = \frac{1}{9}$ for each of these 2-tuples.

2. Repeat Exercise 1 for sampling without replacement.

   ANSWER. (x,y) for $x = 1, 2, 3$ and $y = 1, 2, 3$ provided $x \neq y$; $f(x,y) = \frac{1}{6}$ for each of these 2-tuples.

3. In tossing a fair coin, count a head as 1 and a tail as 2. The coin is tossed twice. Let X be the sum of the two numbers that occur and Y be the difference of the two (the first minus the second). Compute the probability mass functions $f_1(x)$ and $f_2(y)$ of X and Y respectively. Also compute the joint probability mass function $f(x,y)$.

ANSWER. $f_1(2) = \frac{1}{4}$, $f_1(3) = \frac{1}{2}$, $f_1(4) = \frac{1}{4}$, $f_2(-1) = \frac{1}{4}$, $f_2(0) = \frac{1}{2}$, $f_2(1) = \frac{1}{4}$, $f(3,-1) = f(2,0) = f(4,0) = f(3,1) = \frac{1}{4}$.

4. The fair coin of Exercise 3 is tossed three times. Let X be the sum of the first two numbers and Y be the sum of the last two numbers. Find the joint probability mass function.

ANSWER. $f(2,2) = f(2,3) = f(3,2) = f(3,4) = f(4,3) = f(4,4) = \frac{1}{8}$, $f(3,3) = \frac{1}{4}$.

5. An urn contains 3 black and 7 red balls. Two balls are selected at random without replacement. Let X and Y be the number of black and red balls drawn respectively. Find $P(X \le Y)$.

ANSWER. $\frac{84}{90}$.

6. Two cards are drawn at random from a standard 52-card deck. Let X be the number of kings that occur and Y the number of hearts. Find $P(X > Y)$.

ANSWER. $\frac{228}{2652}$.

7. Let two random variables X and Y have the joint probability mass function $f(x,y)$ given by the tabulated entries:

	$y = -1$	$y = 1$	$y = 2$	$y = 5$
$x = -1$	$\frac{1}{27}$	$\frac{1}{9}$	$\frac{1}{9}$	$\frac{1}{27}$
$x = 1$	$\frac{1}{9}$	$\frac{2}{9}$	$\frac{1}{9}$	0
$x = 5$	$\frac{4}{27}$	$\frac{1}{9}$	0	0

(a) Find the probabilities of the following events: (i) Y is even, (ii) Y is even and $X^2 < 2$, (iii) XY is odd, (iv) $0 < X < 5$, (v) $0 < Y < 5$.

(b) Find the probability distribution of $X + Y$.

(c) Find the probability distribution of XY.

(d) Find the conditional probability that Y is odd given X is negative.

ANSWERS. (a) $\frac{2}{9}$, $\frac{2}{9}$, $\frac{7}{9}$, $\frac{4}{9}$, $\frac{2}{3}$, (b) $\frac{1}{27}$, $\frac{2}{9}$, $\frac{1}{9}$, $\frac{2}{9}$, $\frac{1}{9}$, $\frac{5}{27}$, $\frac{1}{9}$ for admissible values $x + y = -2, 0, 1, 2, 3, 4, 6$ respectively. (c) $\frac{5}{27}$, $\frac{1}{9}$, $\frac{2}{9}$, $\frac{7}{27}$, $\frac{1}{9}$, $\frac{1}{9}$ for $xy = -5, -2, -1, 1, 2, 5$, respectively. (d) $\frac{5}{8}$.

8. A bin of five transistors contains two that are defective. The transistors are to be tested, one at a time, until the defective ones are identified. Denote by N_1 the number of tests made until the first defective is spotted and by N_2 the number of additional tests until

the second defective is spotted; find the joint probability mass function of N_1 and N_2.

ANSWER. $f(i,j) = {}^1/_{10}$ for $i = 1, 2, 3, 4$ and $j = 1, 2, \cdots, 5 - i$.

9. Consider a sequence of independent Bernoulli trials, each of which is a success with probability p. Let X_1 be the number of failures preceding the first success, and let X_2 be the number of failures between the first two successes. Find the joint mass function of X_1 and X_2.

ANSWER. $f(i,j) = (1 - p)^{i+j}p^2$ for $i = 0, 1, 2, \cdots$ and $j = 0, 1, 2, \cdots$.

MARGINAL AND CONDITIONAL
PROBABILITY MASS FUNCTIONS

By its construction, the event space of two discrete random variables is made up of a mutually exclusive and collectively exhaustive listing of all possible outcomes (x,y) of the random variables (X,Y). Therefore, as we have seen, the joint probability mass function $f(x,y)$ satisfies

$$\sum_x \sum_y f(x,y) = 1,$$

where the double summation is over all the discrete event points (x,y). We now define the *marginal probability mass functions* $f_1(x)$ and $f_2(y)$ as

$$f_1(x) = \sum_y f(x,y) \quad \text{and} \quad f_2(y) = \sum_x f(x,y),$$

where the summation on the y index is over all the discrete event points y and the summation on the x index is over all the discrete event points x. The modifying word "marginal" is attached to the term "probability mass function" in situations where we are concerned with more than one random variable. However, no matter how many random variables are defined on the sample space of the experiment, the function $f_1(x)$ always has the same physical meaning. This physical meaning is that $f_1(x)$ is the univariate probability mass function of the random variable X. Likewise, $f_2(x)$ is the univariate probability mass function of the random variable Y. For example, the probability that the discrete random variable X is equal to the number 3 is given by $f_1(3)$. Thus in general the marginal distributions can be written in terms of probabilities as

$$f_1(x) = P(X = x) \quad \text{and} \quad f_2(y) = P(Y = y);$$

and they satisfy

$$\sum_x f_1(x) = 1 \quad \text{and} \quad \sum_y f_2(y) = 1,$$

where the first summation is over all discrete event points x, and the second is over all discrete event points y.

In the preceding section we treated the example of three tosses of a fair coin, and defined the random variable X as the number of heads, and the random variable Y as the length of the longest run. The marginal probability mass function of X is obtained by collecting the probabilities of the appropriate events in the (x,y) space. We have:

(Event that $X = 0$) = (0,3),
(Event that $X = 1$) = (1,1) $\cup$ (1,2),
(Event that $X = 2$) = (2,1) $\cup$ (2,2),
(Event that $X = 3$) = (3,3).

Thus the marginal probability mass function of X is:

$$f_1(0) = {}^1\!/_8,$$

$$f_1(1) = {}^1\!/_8 + {}^2\!/_8 = {}^3\!/_8,$$

$$f_1(2) = {}^1\!/_8 + {}^2\!/_8 = {}^3\!/_8,$$

$$f_1(3) = {}^1\!/_8.$$

In a similar way, the appropriate events for the marginal probability mass function of Y are

(Event that $Y = 1$) = (1,1) $\cup$ (2,1),
(Event that $Y = 2$) = (1,2) $\cup$ (2,2),
(Event that $Y = 3$) = (0,3) $\cup$ (3,3).

so

$$f_2(1) = {}^1\!/_8 + {}^1\!/_8 = {}^1\!/_4,$$

$$f_2(2) = {}^2\!/_8 + {}^2\!/_8 = {}^1\!/_2,$$

$$f_2(3) = {}^1\!/_8 + {}^1\!/_8 = {}^1\!/_4.$$

Let us now discuss conditional probability. Let us consider a sample space made up of sample points. The sample points, as we know, represent the fine-grain events associated with the probability model in question. The conditioning of the sample space does one or another of two things to the probability of each fine-grain event. If an event does not have the attribute of the conditioning event A, then the conditional probability of that event is set equal to zero. If an event does have the attribute of the conditioning event A, then the conditional probability of that event is set equal to its original probability scaled up by a constant $1/P(A)$. As a result, the sum of the conditional probabilities in the conditional sample space is unity. When we consider a discrete random variable, the notion of conditional probability can be directly applied. Thus we define the *conditional probability mass function* $\phi_1(x|y)$ as the conditional probability that the experimental value of the random variable X is x, given that the experimental values of the random variable Y is y on the same performance of the experiment; that is,

$$\phi_1(x|y) = P(X = x|Y = y).$$

From the definition of conditional probability, it therefore follows that

$$\phi_1(x|y) = \frac{f(x,y)}{f_2(y)}.$$

Likewise, the conditional probability mass function of Y given X is

$$\phi_2(y|x) = \frac{f(x,y)}{f_1(x)}.$$

Let us now return to the example of three flips of a fair coin with X = number of heads and Y = length of the longest run. We recall that the joint probability mass function is

$$f(3,3) = \tfrac{1}{8} \qquad f(2,1) = \tfrac{1}{8} \qquad f(1,1) = \tfrac{1}{8}$$
$$f(2,2) = \tfrac{1}{4} \qquad f(1,2) = \tfrac{1}{4} \qquad f(0,3) = \tfrac{1}{8}$$

Let us find the conditional probability mass function of Y given that $X = 2$. Only the two events $(2,2)$ and $(2,1)$ in the original event space have the attribute of the conditioning event $X = 2$. The relative probability of these points must remain the same in the conditional event space. Thus the required conditional mass function $\phi_2(y|x)$ is

$$\phi_2(1|2) = \frac{f(2,1)}{f_1(2)} = \frac{\tfrac{1}{8}}{\tfrac{3}{8}} = \frac{1}{3},$$

$$\phi_2(2|2) = \frac{f(2,2)}{f_1(2)} = \frac{\tfrac{1}{4}}{\tfrac{3}{8}} = \frac{2}{3}.$$

If the conditioning event is $X = 3$, we see that there is only one event point (x,y) with this conditioning, namely, the event point $(3,3)$. The resulting conditional mass function $\phi_2(y|x)$ is

$$\phi_2(3|3) = \frac{f(3,3)}{f_1(3)} = \frac{1/8}{1/8} = 1.$$

An important aspect of the conditional distribution resides in the fact that by means of it, certain multivariate problems can be reduced virtually to univariate problems and thus they become easier to solve.

Let us now discuss independence. For two random variables to be independent, it is required that no possible experimental value of one random variable can give any new information about the probability of any experimental value of the other random variable. Thus the random variables X and Y are defined to be *independent* if

$$\phi_2(y|x) = f_2(y)$$

for all possible values of x and y. If the conditioning event is of non-zero probability, then from the definition of the conditional distribution we have

$$f(x,y) = f_1(x)\phi_2(y|x) = f_2(y)\phi_1(x|y).$$

If we substitute the above definition of independence into this equation, we see that $\phi_2(y|x) = f_2(y)$ for all x,y requires that

$$f(x,y) = f_1(x)f_2(y) = f_2(y)\phi_1(x|y),$$

so

$$\phi_1(x|y) = f_1(x)$$

for all x,y. Thus an equivalent definition of independence is the statement that the random variables X and Y are independent if

$$f(x,y) = f_1(x)f_2(y)$$

for all x and y. More generally, we define any number of random variables to be *mutually independent* if the joint probability mass function for all the random variables factors into the product of all the marginal probability mass functions for all the arguments.

PROBLEM. A team is made up of two players. Suppose that each player can score 0, 1, or 2 points. The respective probabilities are 0.2,

0.5, and 0.3 for player 1, and 0.1, 0.4, and 0.5 for player 2. If the performances of the players are independent, derive the joint probability mass function of the scores of these two players.

SOLUTION. Let x denote any possible score of player 1 and $f_1(x)$ his probability of gaining that score; also let y denote any possible score of player 2 and $f_2(y)$ the corresponding probability. We next let $f(x,y)$ denote the probability of the joint event that player 1 scores x points while in the same game player 2 scores y points. By the assumption of independence, we have

$$f(x,y) = f_1(x)f_2(y).$$

For example, the probability that both players score 0 is

$$f(0,0) = f_1(0)f_2(0) = (0.2)(0.1) = 0.02.$$

The probability that player 1 scores 0 while player 2 scores 1 is

$$f(0,1) = f_1(0)f_2(1) = (0.2)(0.4) = 0.08$$

and so on. Table 5.1 presents the complete bivariate distribution, comprising nine distinct possibilities together with their associated probabilities. The margins of the table contain the sums of the tabular entries $f(x,y)$ by rows (horizontal) and columns (vertical). Each row total yields precisely the probability $f_1(x)$ that player 1 scores x points, and each column total yields the probability that player 2 scores y points. Because of the fact that the single distributions are displayed in the margins of the bivariate table, single distributions have come to be called *marginal* distributions. Supposing now that the score for the team is the sum of the scores gained by the two players, let us find the distribution of the team score, t. Clearly, the possible values of t are 0, 1, 2, 3, 4; and the probability $g(t)$ of any partic-

TABLE 5.1 Bivariate Distribution of Scores

| | y | | | |
x	0	1	2	Sum $= f_1(x)$
0	0.02	0.08	0.10	0.20
1	0.05	0.20	0.25	0.50
2	0.03	0.12	0.15	0.30
Sum $= f_2(y)$	0.10	0.40	0.50	1.00

ular value is equal to the sum of the probabilities of the distinct ways in which that combined score can be obtained. Hence,

$g(0) = f(0,0) = 0.02,$

$g(1) = f(0,1) + f(1,0) = 0.08 + 0.05 = 0.13,$

$g(2) = f(0,2) + f(1,1) + f(2,0) = 0.10 + 0.20 + 0.03 = 0.33,$

$g(3) = f(1,2) + f(2,1) = 0.25 + 0.12 = 0.37,$

$g(4) = f(2,2) = 0.15.$

In this way, we arrive at the probability mass function of the team score t as shown in Table 5.2.

A joint distribution can be synthesized from the marginal distributions only if the random variables are independent. The joint distribution of dependent random variables can be derived only when there is sufficient mathematical information to determine the joint probabilities either directly or by analytic deduction; without such information, one must resort to empirical estimates based on relative frequencies of joint events. A bivariate distribution with dependent coordinates is illustrated in the next problem.

PROBLEM. General Motors and Ford control 50 percent and 20 percent, respectively, of the American market for automobiles. Two people, each of whom has just purchased an automobile, are selected at random. Let the random variable X be the number of these two people who bought a G.M. car and let the random variable Y be the number of these two people who bought a Ford. For example, if both of these two people bought G.M. cars, we would have $X = 2$ and $Y = 0$: The problem is to derive the bivariate distribution of X and Y.

SOLUTION. The symbol (x,y) represents the event that x of the two people bought G.M. cars, and y of the two people bought Fords. Let $f(x,y)$ be the probability mass function of this event. Since the two companies together sell to only 70 percent of the market, the remaining 30 percent goes to neither. The possible allocations of two buyers are shown in Table 5.3, wherein joint probabilities are computed on the assumption that sampling

TABLE 5.2 Distribution of Team Score

t	0	1	2	3	4	Sum
$g(t)$	0.02	0.13	0.33	0.37	0.15	1.00

TABLE 5.3 Possible Allocations of the Two Buyers

Allocation		Compound Probability of This Allocation	Corresponding Bivariate Event	Total Probability of Bivariate Event
Buyer 1	Buyer 2			
Neither	Neither	(0.3)(0.3) = 0.09	(0,0)	$f(0,0) = 0.09$
Neither	Ford	(0.3)(0.2) = 0.06	(0,1)	
Ford	Neither	(0.2)(0.3) = 0.06	(0,1)	$f(0,1) = 0.06+0.06=0.12$
Ford	Ford	(0.2)(0.2) = 0.04	(0,2)	$f(0,2) = 0.04$
Neither	G.M.	(0.3)(0.5) = 0.15	(1,0)	
G.M.	Neither	(0.5)(0.3) = 0.15	(1,0)	$f(1,0) = 0.15+0.15=0.30$
Ford	G.M.	(0.2)(0.5) = 0.10	(1,1)	
G.M.	Ford	(0.5)(0.2) = 0.10	(1,1)	$f(1,1) = 0.10+0.10=0.20$
G.M.	G.M.	(0.5)(0.5) = 0.25	(2,0)	$f(2,0) = 0.25$

TABLE 5.4 Bivariate Distribution and Marginal Distributions of Buyers

	y			
x	0	1	2	Sum $= f_1(x)$
0	0.09	0.12	0.04	0.25
1	0.30	0.20	0	0.50
2	0.25	0	0	0.25
Sum $= f_2(y)$	0.64	0.32	0.04	1.00

depletions may be ignored. The bivariate distribution itself is presented compactly in Table 5.4. As in the previous problem, the row and column totals yield the marginal distributions, for even though the random variables are not independent, the separate combinations along a row or column represent the exhaustive and mutually exclusive ways in which a fixed value of one of the random variables can occur. The conditional distributions of y for successive choices of x are shown in Table 5.5.

Let us now summarize the main points. The joint distribution of *independent* random variables can be synthesized from the respective marginal distributions, but the joint distribution of *dependent* random variables cannot be synthesized from the marginal distributions. For discrete random variables, the marginal probabilities may be obtained by summing the appropriate joint probabilities; and conditional probabilities may be obtained by dividing joint probabilities by associated marginal probabilities. In turn, the joint distribution can be con-

TABLE 5.5 Conditional Distributions

	Value of y			
	0	1	2	Sum
$\phi_2(y|x=0)$	0.36	0.48	0.16	1.00
$\phi_2(y|x=1)$	0.60	0.40	0	1.00
$\phi_2(y|x=2)$	1.00	0	0	1.00

structed if the marginal and conditional distributions are known. An advantage of the use of the conditional distribution is the comparative ease with which it often can be derived by elementary changes of variable or deduced from physical reasoning.

EXERCISES

1. Let a deck of cards consist of the ace, king, queen, and jack of each of the four suits. If two cards are drawn at random, and X and Y denote the number of clubs and hearts obtained respectively:
   (a) Find the marginal distribution of X
   (b) Find the conditional distribution of Y for $X = 1$.
   ANSWERS. (a) $f_1(0) = {}^{11}/_{20}$, $f_1(1) = {}^{8}/_{20}$, $f_1(2) = {}^{1}/_{20}$. (b) $\phi_2(0|1) = {}^{2}/_{3}$, $\phi_2(1|1) = {}^{1}/_{3}$, $\phi_2(2|1) = 0$.

2. Using the fact that $[(X = x_i) \cap (Y = y_1)] \cup [(X = x_i) \cap (Y = y_2)] \cup \cdots = (X = x_i)$ is a union of mutually disjoint sets, establish the formula

$$\sum_j f(x_i, y_j) = f_1(x_i)$$

   for the marginal probability mass function.

3. Show that the conditional probability mass function is a bona fide probability mass function in that

$$0 \leq \phi_1(x|y) \leq 1 \qquad \phi_1(x_1|y_j) + \phi_1(x_2|y_j) + \cdots = 1.$$

   HINT. For the last assertion, use

$$\sum_i P(X = x_i | Y = y_j) = \sum_i P[(X = x_i) \cap (Y = y_j)]/P(Y = y_j).$$

4. Five balls are randomly distributed among three boxes. Let X be the number of balls in the first box and let Y be the number of balls in the first two boxes. Find the conditional distribution of X given $Y = 4$.
   ANSWER. $\phi_1(x|4) = {}^{1}/_{16}, {}^{4}/_{16}, {}^{6}/_{16}, {}^{4}/_{16}, {}^{1}/_{16}$ for $x = 0, 1, 2, 3, 4$ respectively.

5. If n balls are randomly distributed among m cells, let X be the number in the first cell and Y be the number in the second cell. Show that the conditional distributions of X, given the various values of Y, are

$$\phi_1(x|y) = \binom{y}{x} \left(\frac{1}{2}\right)^y \qquad \text{for } x = 0, 1, \cdots, y.$$

HINT.

$$f(x,y) = \frac{n!}{x!(y-x)!(n-y)!} \left(\frac{1}{m}\right)^y \left(1 - \frac{2}{m}\right)^{n-y}.$$

[Comment on the following interpretation of $\phi_1(x|y)$. If y balls are randomly distributed among two cells, the distribution of the number of balls in the first cell is the same as the conditional distribution $\phi_1(x|y)$.]

6. *Polya Urn Scheme.* An urn contains two white balls and one black ball. A ball is drawn and then replaced along with an additional ball of the same color. The procedure is repeated; that is, a second ball is drawn and is replaced together with an additional ball of the same color as drawn on the second drawing. The procedure is again repeated. Each time a ball is drawn it is replaced together with an additional one of the same color. Let $f_n(x)$ denote the probability mass function of the number x of black balls in the box after n trials have been completed. Verify the following table:

x	$f_1(x)$	$f_2(x)$	$f_3(x)$	$f_4(x)$
0	$^2/_3$	$^3/_6$	$^4/_{10}$	$^5/_{15}$
1	$^1/_3$	$^2/_6$	$^3/_{10}$	$^4/_{15}$
2		$^1/_6$	$^2/_{10}$	$^3/_{15}$
3			$^1/_{10}$	$^2/_{15}$
4				$^1/_{15}$

7. Given that X and Y have the joint probability mass function $f(x,y) = 1/n^2$ for $x = 1, 2, \cdots, n$ and $y = 1, 2, \cdots, n$, verify that X and Y are independent.

8. Given that X and Y have the joint probability mass function $f(x,y) = 2/n(n + 1)$ for $x = 1, 2, \cdots, n$ and $y = 1, 2, \cdots, x$, verify that X and Y are not independent.

9. The team for a certain game of skill is made up of four players, each of whom has a chance of scoring 0, 1, or 2 points, and the score for the team is the sum of the scores made by the individual players.

Assuming independence, derive the distribution of scores for the team of players A, B, C, D whose individual score distributions are shown below:

Score	Players			
	A	B	C	D
0	0.5	0.2	0.3	0.2
1	0.3	0.3	0.5	0.5
2	0.2	0.5	0.2	0.3

From the distribution of team scores find the expected value of the score t for the team and compare this result with the sum of the expected values of the scores of the individual players.

ANSWER.

t	0	1	2	3	4	5	6	7	8	$E(t)$
$f(t)$	.0060	.0376	.1133	.2124	.2614	.2124	.1133	.0376	.0060	4.0000

Player	A	B	C	D	Sum
Expected score	0.7	1.3	0.9	1.1	4.0

JOINT, MARGINAL, AND CONDITIONAL PROBABILITY DENSITY FUNCTIONS

Generally speaking, random phenomena are characterized by several attributes at once. Thus in the human population, all individuals have the attributes of age, height, weight, stature, etc. More specifically, we can often characterize a random phenomenon by the coexistence of quantitative attributes. We are thus led to consider the joint values of two or more random variables. As we know, the corresponding probability distribution is called a *bivariate distribution* in the case of two random variables, or a multivariate distribution in the case of two or more. The probability distribution pertaining to a single random variable is called a *univariate distribution*.

Let us now consider the bivariate case for continuous random variables. We call the two orthogonal axes x_1 and x_2, and let X_1 and X_2 be continuous random variables. Consider now the two inequalities,

$$X_1 \le x_1 \quad \text{and} \quad X_2 \le x_2,$$

each of which represents an event. The joint (or compound) event is given by their intersection

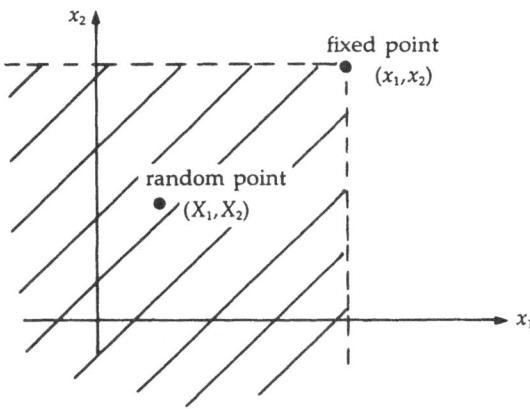

FIGURE 5.5 The joint event $(X_1 \leq x_1) \cap (X_2 \leq x_2)$ is the event that the random point (X_1, X_2) lies within the lower left quadrant (*shaded area*) defined by the fixed point (x_1, x_2).

$$(X_1 \leq x_1) \cap (X_2 \leq x_2).$$

This joint event means that both inequalities must be satisfied simultaneously. That is, this joint event means that the point (X_1, X_2) representing the two random variables must lie within the lower left quadrant defined by the fixed point (x_1, x_2). See Figure 5.5.

We now define the joint distribution. The bivariate joint distribution function $F(x_1, x_2)$ is defined as equal to the probability of the joint event $(X_1 \leq x_1) \cap (X_2 \leq x_2)$. That is, $F(x_1, x_2)$ is defined as

$$F(x_1, x_2) = P\{(X_1 \leq x_1) \cap (X_2 \leq x_2)\}.$$

The joint probability distribution function may be pictured as a surface plotted above the x_1, x_2 plane. The height of the surface at any fixed point gives the probability that a random point will lie within the lower left quadrant defined by the fixed point.

Let us now turn to the multivariate case. Geometrically, a multivariate joint distribution is defined on a space of n dimensions. Each of the n coordinates corresponds to one of the random variables. The multivariate distribution function $F(x_1, x_2, \cdots, x_n)$ is the probability measure of the subset for which the inequalities

$$X_1 \leq x_1, \quad X_2 \leq x_2, \quad \cdots, \quad X_n \leq x_n$$

are satisfied simultaneously. That is, the joint distribution function is defined as

$$F(x_1,x_2,\cdots,x_n) = P\{(X_1 \leq x_1) \cap (X_2 \leq x_2) \cap \cdots \cap (X_n \leq x_n)\}.$$

Thus once the conceptual step is taken from univariate to bivariate distributions, the transition to higher dimensions involves nothing essentially new.

As a direct generalization of the univariate case, a probability differential can be defined for an infinitesimal region about any point in the space. Thus, the symbol $dF(x_1,x_2,\cdots,x_n)$ represents the probability measure of the subset for which the inequalities

$$x_1 \leq X_1 \leq x_1 + dx_1, \quad x_2 \leq X_2 \leq x_2 + dx_2, \quad \cdots, \quad x_n \leq X_n \leq x_n + dx_n$$

are simultaneously satisfied. The density concept introduced for continuous univariate distributions can be extended to any number of dimensions. If the joint distribution function is continuous and differentiable, the joint density function may be defined as its nth partial derivative with respect to each variable taken once:

$$f(x_1,x_2,\cdots,x_n) = \frac{\partial^n}{\partial x_1\, \partial x_2\, \cdots\, \partial x_n}\, [F(x_1,x_2,\cdots,x_n)].$$

Thus, if the first-order partial derivatives of $F(x_1,x_2,\cdots,x_n)$ exist, then the joint probability density function $f(x_1,x_2,\cdots,x_n)$ exists and satisfies

$$dF(x_1,x_2,\cdots,x_n) = f(x_1,x_2,\cdots,x_n)\, dx_1\, dx_2\, \cdots\, dx_n.$$

The meaning of the joint density function is analogous to that of the univariate density function; that is, the expression

$$f(x_1,x_2,\cdots,x_n)\, dx_1\, dx_2\, \cdots\, dx_n$$

represents the probability that a random point X_1, X_2, $\cdots$, X_n will fall within an infinitesimal region around the given point x_1, x_2, $\cdots$, x_n. The total integral of any joint density function over the entire region for which it is defined is necessarily equal to unity.

Random variables representing phenomena of the physical world almost always have natural limits of variation beyond which the random variables cannot extend. Geometric concepts are helpful in discussing multivariate distributions. The simultaneously admissible values of n random variables may be thought of as cutting out a region in n-dimensional space. The limits, of course, depend upon the physical circumstances and may therefore differ from one random variable to another. Because this diversity of natural limits would occasion some analytical inconvenience, the following *range convention* is commonly adopted.

> The limits of all random variables are taken as $-\infty$ to ∞ with the understanding that the probability is zero for that class of values lying beyond the natural limits.

Since this convention exists for the sake of convenience, we shall adhere to it when convenience would thus be served and not otherwise. This range convention is usually taken for granted in connection with bivariate and multivariate distributions, and the probability density functions are automatically assigned the value of zero outside the natural regions of definition.

The region of definition of a joint density function must always be stated to make specification complete. One way of defining the admissible region is to describe its geometric boundaries. It often happens that the admissible region can be defined concisely by stating the extreme limits of one variable and then stating the functional limits of the other for an arbitrary value of the first. For example, suppose the admissible region is the shaded area shown in Figure 5.6. We see that the admissible region of x and y consists of the triangle bounded by the lines $y = 0$, $y = x$, and $x = 1$. Thus we may represent the triangular admissible region of Figure 5.6 by the analytic specification

$$(0 \le x \le 1; \quad 0 \le y \le x).$$

This compact specification is interpreted to mean that the outside limits of the figure are the lines $x = 0$ and $x = 1$; then within these limits, the figure is further bounded by the lines $y = 0$ and $y = x$. For the same region, we could have written equally well the alternative specification,

$$(0 \le y \le 1; \quad y \le x \le 1).$$

In another case, suppose that the admissible region of u and v (Figure 5.7) lies in the first quadrant and consists of that infinite wedge bounded below by the hyperbola $v = 1/u$ and above by the straight line $v = u$. This infinite wedge-shaped region can be defined by the concise specification

$$\left(1 \le u < \infty; \quad \frac{1}{u} \le v \le u\right).$$

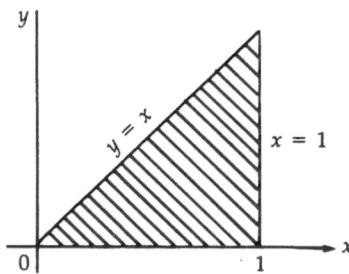

FIGURE 5.6 Admissible region, given by shaded triangle.

Here, however, in order to interchange the roles of u and v, as we did with x and y in the previous example, we would need two statements:

$$\left(0 \leqslant v \leqslant 1; \ \frac{1}{v} \leqslant u < \infty \right) \quad \text{and} \quad (1 \leqslant v < \infty; \ v \leqslant u < \infty).$$

Thus the specification may be simpler from one viewpoint than another, and of course, where possible we shall choose the simpler one. Whenever possible, we shall adopt the convention that constant limits will be given for one variable, followed by corresponding limits for the second when the first assumes an arbitrary value within those constant limits. It is always understood that the density is identically zero outside of the admissible region.

As we have seen, distributions of single random variables are called univariate distributions to distinguish them from bivariate distributions, which involve two random variables simultaneously, or multivariate distributions, which are generally understood to involve three or more random variables simultaneously. The term *multivariate* is sometimes used broadly to designate an arbitrary number of random variables, and in that sense includes univariate and bivariate distributions as special cases. Since natural phenomena are characterized by the coexistence of many attributes and dimensions, it is often necessary to consider multivariate distributions in representing the relevant conditions of practical problems.

Let us now define the marginal density function. The *marginal density function* of each random variable can be obtained from the joint density function by integration with respect to the other variables. In the bivariate case, let the joint density function of x and y be denoted by $f(x,y)$ and the marginal density function of x by $f_1(x)$. Then the marginal density function is given by

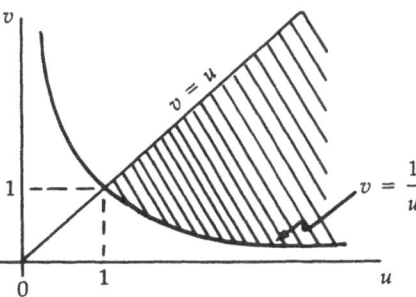

FIGURE 5.7 Admissible region, given by shaded infinite wedge.

$$f_1(x) = \int f(x,y)\, dy,$$

where the integral is over the entire admissible range of y. Let us now explain this integral. The quantity $f_1(x)\, dx$ gives the probability that the random variable X will fall between x and $x + dx$. Also the quantity $f(x,y)\, dx\, dy$ gives the probability that the random point (X,Y) will fall infinitesimally close to the point (x,y). Hence the integral of $f(x,y)\, dx\, dy$ with respect to y for a fixed value of x will yield the sum of the probabilities of all of the mutually exclusive ways of obtaining points with abscissas lying between x and $x + dx$. That is, we have

$$f_1(x)\, dx = P(x \leqslant X \leqslant x + dx) = [\int f(x,y)\, dy]\, dx,$$

which gives the required integral for the marginal density function. This integration is straightforward if the admissible region for $f(x,y)$ is the entire plane. However, in many cases, the admissible region will be bounded by lines or curves, and the integral with respect to y must be taken from the bottom to the top of the vertical strip (Figure 5.8) corresponding to the fixed value of x. The lower limit of y, ordinarily a point on a curve, will be expressed in terms of x as $a_1(x)$ and the upper limit expressed as $b_1(x)$. Consequently, the integral for the marginal density of x is

$$f_1(x) = \int_{a_1(x)}^{b_1(x)} f(x,y)\, dy.$$

By similar reasoning, the marginal density function of y is given by

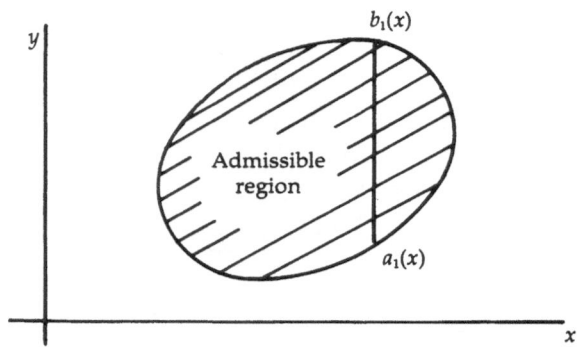

FIGURE 5.8 Path of integration for marginal density of x.

$$f_2(y) = \int_{a_2(y)}^{b_2(y)} f(x,y)\, dx,$$

where $a_2(y)$ and $b_2(y)$ are the functional values of x in terms of y at the two ends of the horizontal strip (Figure 5.9) corresponding to a fixed value of y.

PROBLEM. Find the marginal density functions from the joint density function

$$f(x,y) = 24y(1 - x) \qquad \text{for } 0 \leqslant x \leqslant 1, 0 \leqslant y \leqslant x.$$

SOLUTION. (See Figure 5.10.) The marginal density functions are

$$f_1(x) = \int_0^x f(x,y)\, dy = 24(1 - x) \int_0^x y\, dy = 12x^2(1 - x) \qquad \text{for } 0 \leqslant x \leqslant 1,$$

$$f_2(y) = \int_y^1 f(x,y)\, dx = 24y \int_y^1 (1 - x)\, dx = 12y(1 - y)^2 \qquad \text{for } 0 \leqslant y \leqslant 1.$$

We can verify that the area under these distributions is one by making use of the beta function $\beta(m,n)$, which is defined and illustrated in Chapter 6. The results are as follows:

$$\int_0^1 f_1(x)\, dx = 12 \int_0^1 x^2(1 - x)\, dx = 12\beta(3,2) = \frac{12(2!)(1!)}{4!} = 1,$$

$$\int_0^1 f_2(y)\, dy = 12 \int_0^1 y(1 - y)^2\, dy = 12\beta(2,3) = 1.$$

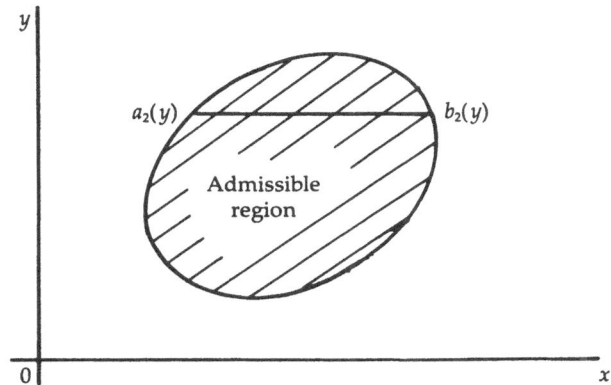

FIGURE 5.9 Path of integration for marginal density of y.

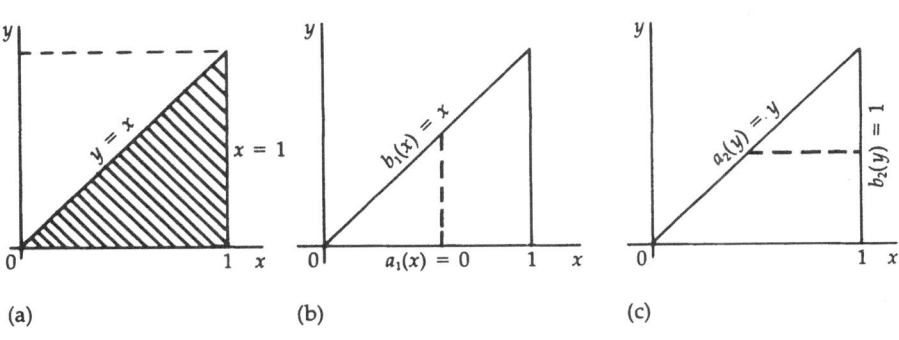

(a)   (b)   (c)

FIGURE 5.10 (a) Admissible region (*shaded area*). (b) Path of integration for marginal distribution of x. (c) Path of integration for marginal distribution of y.

Given the information that the ordinate Y of a random point lies between y and $y + dy$, the conditional probability that its abscissa X will lie between x and $x + dx$ can be obtained in the usual way from the joint and unconditional probabilities:

$$P[(x \leqslant X \leqslant x + dx) \mid (y \leqslant Y \leqslant y + dy)]$$

$$= \frac{P[(x \leqslant X \leqslant x + dx) \cap (y \leqslant Y \leqslant y + dy)]}{P(y \leqslant Y \leqslant y + dy)}.$$

If we define the conditional density function $\phi_1(x|y)$ of x given y as

$$\phi_1(x|y)\, dx = P[(x \leqslant X \leqslant x + dx)|(y \leqslant Y \leqslant y + dy)],$$

then the above equation may be written as

$$\phi_1(x|y)\, dx = \frac{f(x,y)\, dx\, dy}{f_2(y)\, dy}.$$

Thus the *conditional density function* is

$$\phi_1(x|y) = \frac{f(x,y)}{f_2(y)}.$$

The quantity $\phi_1(x|y)\, dx$ gives the conditional probability that X will assume a value arbitrarily close to x when Y is restricted by hypothesis to lie arbitrarily close to y. The conditional density function $\phi_2(y|x)$ of y for a fixed value of x is correspondingly given by

$$\phi_2(y|x) = \frac{f(x,y)}{f_1(x)}$$

and has a similar interpretation.

PROBLEM. From the joint density function given in the previous problem, find the conditional density functions.

SOLUTION. See Figure 5.10. The conditional density functions are

$$\phi_1(x|y) = \frac{24y(1-x)}{12y(1-y)^2} = \frac{2(1-x)}{(1-y)^2} \qquad \text{for } y \leqslant x \leqslant 1,$$

$$\phi_2(y|x) = \frac{24y(1-x)}{12x^2(1-x)} = \frac{2y}{x^2} \qquad \text{for } 0 \leqslant y \leqslant x.$$

As a check we have

$$\int_y^1 \phi_1(x|y)\, dx = \frac{2}{(1-y)^2} \int_y^1 (1-x)\, dx = \frac{2}{(1-y)^2}\, \frac{(1-y)^2}{2} = 1,$$

$$\int_0^x \phi_2(y|x)\, dy = \frac{2}{x^2} \int_0^x y\, dx = \frac{2}{x^2}\, \frac{x^2}{2} = 1.$$

The joint probability density function can be constructed as the product of the marginal density function of one variable and the conditional density function of the other:

$$f(x,y) = f_1(x)\phi_2(y|x) = f_2(y)\phi_1(x|y).$$

By including the range of definition with the equation of the conditional density function, we obtain the complete specification of the conditional distribution. Thus, referring to Figure 5.9, the conditional distribution of x for a fixed value of y would be defined by the equation for $\phi_1(x|y)$ together with the admissible range, $a_2(y) \leqslant x \leqslant b_2(y)$; similarly (Figure 5.8), the conditional distribution of y for a fixed value of x would be given by the equation for $\phi_2(y|x)$ together with the statement that $a_1(x) \leqslant y \leqslant b_1(x)$. As can be verified by applying the appropriate definitions, the conditional density functions have unit area

$$\int_{a_2(y)}^{b_2(y)} \phi_1(x|y)\, dx = 1 \qquad \int_{a_1(x)}^{b_1(x)} \phi_2(y|x)\, dy = 1.$$

Several random variables are *mutually independent* provided that their joint distribution function factors identically into the product of their separate (i.e., marginal) distribution functions, as

$$F(x_1, x_2, \cdots, x_n) = F_1(x_1)F_2(x_2) \cdots F_n(x_n).$$

We shall take this equation as the general definition of independence for random variables. It is, of course, compatible with the definition previously given for qualitative events.

Let us now look at the cases of continuous and discrete random variables. If continuous random variables are independent, their joint probability density function factors identically into the product of their marginal probability density functions:

$$f(x_1, x_2, \cdots, x_n) = f_1(x_1)f_2(x_2) \cdots f_n(x_n).$$

From this fact it follows that the joint density function of independent continuous random variables can be constructed by taking the product of their respective marginal density functions. Likewise, if discrete random variables are independent, their joint probability mass function factors identically into the product of their marginal probability mass functions. The joint mass function of independent discrete random variables can be constructed by taking the product of their respective marginal mass functions.

PROBLEM. The joint density function that appeared in the two foregoing
problems is

$$f(x,y) = 24y(1 - x) \qquad \text{for } 0 \leqslant x \leqslant 1, 0 \leqslant y \leqslant x.$$

This joint density function factors into a product of a pure function of x, namely $(1 - x)$, and pure function of y, namely $24y$.
As a result, are x and y independent?

SOLUTION. This example brings out an important fact. Although
the joint density function of independent random variables factors into pure functions of each, factorization alone is not sufficient for independence; the factors must equal the respective
marginal density functions. Moreover, whenever the conditional distribution of one variable depends either in form or in
range of definition upon another variable, then the variables
are dependent. However, when a conditional distribution does
not involve another variable, either explicitly in the equation
of the density function or implicitly in the range of definition,
the variables are independent. In the present case, it is clear
that $f(x,y) \neq f_1(x)f_2(y)$ despite the fact that $f(x,y)$ is a product of
two pure functions, one in y and one in x. Here the conditional
distributions depend both in form and in range upon the other
variable, and neither agrees with the corresponding marginal
distribution. Thus the two variables are not independent.

PROBLEM. Find the marginal and conditional density functions given the
joint density function

$$f(x,y) = xe^{-x(1+y)} \text{ for } 0 \leqslant x < \infty, 0 \leqslant y < \infty.$$

SOLUTION. The admissible region is the entire first quadrant, and
so the limits of integration will be simply 0 to ∞ in each case.
The total integral of the joint density function is

$$\int_0^\infty \int_0^\infty f(x,y) \, dx \, dy = \int_0^\infty e^{-x} \left(\int_0^\infty xe^{-xy}dy \right)dx = \int_0^\infty e^{-x} \, dx = 1.$$

As already indicated by this integration, the marginal density
of x is given by

$$f_1(x) = e^{-x} \qquad \text{for } 0 \leqslant x < \infty.$$

For the marginal density function of y, the integration with re-
spect to x may be simplified by substituting $u = x(1 + y)$; $x = u/(1 + y)$; and since y is held fixed during the integration, $dx = du/(1 + y)$. The marginal density of y can be evaluated by
the aid of the gamma function $\Gamma(n)$ defined and explained in

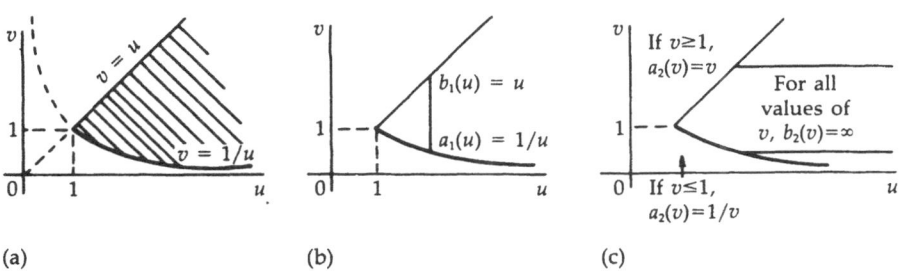

FIGURE 5.11 (a) Admissible region (*shaded area*). (b) Path of integration for marginal distribution of u. (c) Path of integration for marginal distribution of v.

Chapter 6. The result is

$$f_2(y) = \int_0^\infty f(x,y)\, dx = \frac{1}{(1+y)^2} \int_0^\infty ue^{-u}\, du = \frac{\Gamma(2)}{(1+y)^2} = \frac{1}{(1+y)^2}$$

for $0 \leqslant y < \infty$. Note that $\Gamma(2) = 1$. The conditional distributions are

$$\phi_1(x|y) = \frac{f(x,y)}{f_2(y)} = (1+y)^2\, xe^{-x(1+y)} \qquad \text{for } 0 \leqslant x < \infty,$$

$$\phi_2(y|x) = \frac{f(x,y)}{f_1(x)} = xe^{-xy} \qquad \text{for } 0 \leqslant y < \infty.$$

PROBLEM. Find the marginal and conditional density functions given the joint density function (see Figure 5.11)

$$f(u,v) = \frac{1}{2u^2v} \qquad \text{for } 1 \leqslant u < \infty, \frac{1}{u} \leqslant v \leqslant u.$$

SOLUTION. The marginal density function of u is given by

$$f_1(u) = \int_{1/u}^u f(u,v)\, dv = \frac{1}{2u^2} [\ln v]_{1/u}^u = \frac{\ln u}{u^2} \qquad \text{for } 1 \leqslant u < \infty.$$

The marginal distribution of v, however, involves an extra step. For all values of v, the upper limit of integration on u is ∞; but the lower limit assumes two distinct functional forms, depending upon the value of v. When $v \leqslant 1$, the lower limit lies on the hyperbola $u = 1/v$; but when $v \geqslant 1$, the lower limit lies on the straight line $u = v$. When $v = 1$, both functions yield the same value, $u = 1$. Consequently, the marginal density function will have two distinct equations, as follows:

Range I $(0 \leqslant v \leqslant 1)$: $f_2(v) = \int_{1/v}^{\infty} f(u,v)\, du = \frac{1}{2v}\left[-\frac{1}{u}\right]_{1/v}^{\infty} = \frac{1}{2}.$

Range II $(1 \leqslant v < \infty)$: $f_2(v) = \int_{v}^{\infty} f(u,v)\, du = \frac{1}{2v}\left[-\frac{1}{u}\right]_{v}^{\infty} = \frac{1}{2v^2}.$

The marginal density function of v thus requires two statements:

$$f_2(v) = \frac{1}{2} \qquad \text{for } 0 \leqslant v \leqslant 1,$$

$$f_2(v) = \frac{1}{2v^2} \qquad \text{for } 1 \leqslant v < \infty.$$

Corresponding to these two equations, the conditional distribution of u given v assumes two separate forms:

If $0 \leqslant v \leqslant 1$, $\phi_1(u|v) = \frac{1}{u^2 v} \qquad \text{for } \frac{1}{v} \leqslant u < \infty,$

and

If $1 \leqslant v < \infty$, $\phi_1(u|v) = \frac{v}{u^2} \qquad \text{for } v \leqslant u < \infty.$

On the other hand, the conditional distribution of v given u requires only one statement:

$$\phi_2(v|u) = \frac{1}{2 \ln u}\frac{1}{v} \qquad \text{for } \frac{1}{u} \leqslant v \leqslant u.$$

EXERCISES

1. The joint probability density function of X and Y is given by
$$f(x,y) = e^{-(x+y)} \qquad \text{for } 0 \leqslant x < \infty, 0 \leqslant y < \infty..$$
(a) Find $P\{X < Y\}$.
(b) Find $P\{X < a\}$.
ANSWERS. (a) $1/2$. (b) $1 - e^{-a}$.

2. The random variables x and y are jointly distributed as follows (where $\alpha > 0$):

$$f(x,y) = \frac{\sqrt{3}}{6\pi\alpha^2} \exp\left[-\frac{(x^2 - xy + y^2)}{3\alpha^2}\right] \text{ for } -\infty < x < \infty,\ -\infty < y < \infty.$$

Find the marginal distributions of x and y and the conditional distribution of each random variable for a given value of the other. Integrate by completing the square in the exponent.

ANSWER.

$$f_1(x) = \frac{1}{2\alpha \sqrt{\pi}} e^{-x^2/4\alpha^2} \qquad \text{for } -\infty < x < \infty,$$

$$f_2(y) = \frac{1}{2\alpha\sqrt{\pi}} e^{-y^2/4\alpha^2} \qquad \text{for } -\infty < y < \infty,$$

$$\phi_1(x|y) = (1/\alpha \sqrt{3\pi}) \exp\left[-\frac{(x - y/2)^2}{3\alpha^2}\right] \qquad \text{for } -\infty < x < \infty,$$

$$\phi_2(y|x) = (1/\alpha \sqrt{3\pi}) \exp\left[-\frac{(y - x/2)^2}{3\alpha^2}\right] \qquad \text{for } -\infty < y < \infty.$$

3. The random variables x and y are jointly distributed as follows:

$$f(x,y) = 24x^2y(1 - x) \qquad \text{for } 0 \leqslant x \leqslant 1, 0 \leqslant y \leqslant 1.$$

Find the marginal distribution of each random variable and show that they are independent.

ANSWER. $f_1(x) = 12x^2(1 - x)$ for $0 \leqslant x \leqslant 1$; $f_2(y) = 2y$ for $0 \leqslant y \leqslant 1$; $f(x,y) = f_1(x)f_2(y)$.

4. The joint distribution of x and y is

$$f(x,y) = (^1/_8)(x^2 - y^2)e^{-x} \qquad \text{for } 0 \leqslant x < \infty, -x \leqslant y \leqslant x.$$

Find the marginal distribution of each random variable and the conditional distribution of y for a fixed value of x.

ANSWER. $f_1(x) = (^1/_6)x^3e^{-x}$ for $0 \leqslant x < \infty$; $f_2(y) = (^1/_4)e^{-|y|}(1 + |y|)$ for $-\infty < y < \infty$; $\phi_2(y|x) = (^3/_4x^{-3})(x^2 - y^2)$ for $-x \leqslant y \leqslant x$.

5. Find the marginal distribution of y and the conditional distribution of x given y if x and y are jointly distributed as follows:

$$f(x,y) = 4y(x - y)e^{-(x+y)} \qquad \text{for } 0 \leqslant x < \infty, 0 \leqslant y \leqslant x.$$

ANSWER. $f_2(y) = 4ye^{-2y}$ for $0 \leqslant y < \infty$; $\phi_1(x|y) = (x - y)e^{-(x-y)}$ for $y \leqslant x < \infty$.

6. Find the marginal and conditional distributions of x and y that are jointly distributed as follows:

$$f(x,y) = 6xy(2 - x - y) \qquad \text{for } 0 \leqslant x \leqslant 1, 0 \leqslant y \leqslant 1.$$

ANSWER. $f_1(x) = x(4 - 3x)$ for $0 \leqslant x \leqslant 1$; $\phi_1(x|y) = 6x(2 - x - y)/(4 - 3y)$ for $0 \leqslant x \leqslant 1$; $f_2(y) = y(4 - 3y)$ for $0 \leqslant y \leqslant 1$; $\phi_2(y|x) = 6y(2 - x - y)/(4 - 3x)$ for $0 \leqslant y \leqslant 1$.

7. The marginal distribution of x is standard normal (i.e., with parameters $\mu = 0$, $\sigma = 1$); the conditional distribution of y given x is normal with parameters

$$\mu = kx \quad \text{and} \quad \sigma = \sqrt{1 - k^2},$$

and the conditional distribution of z given any pair of values of x and y is normal with parameters

$$\mu = ky \quad \text{and} \quad \sigma = \sqrt{1 - k^2}.$$

(a) State the joint distribution of x, y, z.
(b) Find the marginal distributions of y and z.
(c) Find the conditional distribution of x given y.
(d) State the joint distribution of y and z.
(e) Find the conditional distribution of z given x (but not given y).

HINT. Integration can be carried out by completing the square in the exponent. Make use of formal similarities.

ANSWERS.

(a) $f(x,y,z) = \dfrac{1}{(1 - k^2)(\sqrt{2\pi})^3} e^{-Q/2} \quad$ for $-\infty < x < \infty$, $-\infty < y < \infty$, $-\infty < z < \infty$,

where $Q = \dfrac{(1 - k^2)x^2 + (y - kx)^2 + (z - ky)^2}{1 - k^2}$.

(b) Both y and z are standard normal.
(c) Normal with parameters $\mu = ky$, $\sigma = \sqrt{1 - k^2}$.

(d) $g(y,z) = \dfrac{1}{2\pi\sqrt{1 - k^2}} \exp\left[-\dfrac{(y^2 - 2kyz + z^2)}{2(1 - k^2)}\right]$

for $-\infty < y < \infty$, $-\infty < z < \infty$.

(e) Normal with parameters $\mu = k^2x$, $\sigma = \sqrt{1 - k^4}$.

8. In the joint distribution of two independent standard normal random variables x, y show that the probability that a random point will fall within the square enclosed by the lines $x = -a$, $x = a$, $y = -a$, $y = a$ is less than the probability that it will fall within the circle of equal area defined by $x^2 + y^2 = 4a^2/\pi$. With this fact in mind prove the inequality

$$\frac{1}{\sqrt{2\pi}} \int_{-a}^{a} e^{-x^2/2}\, dx \leqslant \sqrt{1 - e^{-2a^2/\pi}}.$$

HINT. Express the square of the required integral as the product of two integrals of identical form but with one in x and the other in y. Rewrite this as a double integral, and then, changing to polar coordinates, integrate over the circle $r = 2a/\sqrt{\pi}$.

9. The joint probability density function of x and y is given by

$$f(x,y) = c(y^2 - x^2)e^{-y} \qquad \text{for } -y \leqslant x \leqslant y, \, 0 < y < \infty.$$

(a) Find c.
(b) Find the marginal densities of x and y.
ANSWERS. (a) $c = \frac{1}{8}$. (b) $f_1(x) = (\frac{1}{4})e^{-|x|}(1 + |x| + |x|^2)$, $f_2(y) = (\frac{1}{6})y^3e^{-y}$.

10. The joint probability density function of X and Y is given by

$$f(x,y) = \frac{6}{7}\left(x^2 + \frac{xy}{2}\right) \qquad \text{for } 0 < x < 1, \, 0 < y < 2.$$

(a) Verify that this is indeed a joint density function.
(b) Compute the density function of X.
(c) Find $P\{X > Y\}$.
ANSWERS. (b) $f_1(x) = (\frac{6}{7})(2x^2 + x)$. (c) $\frac{15}{56}$.

EXPECTATION, COVARIANCE, AND CORRELATION

The definition of the expected value of a function $\alpha(x)$ can be extended to continuous functions of any number of variables. Thus, if the joint distribution function $F(x_1,x_2,\cdots,x_n)$ of several random variables x_1, x_2, $\cdots$, x_n is defined over the region of space R_X, and $\alpha(x_1,x_2,\cdots,x_n)$ is any continuous function of the x's, then the expected value of this function is defined as

$$E[\alpha(x_1,x_2,\cdots,x_n)] = \int\int_{R_X} \cdots \int \alpha(x_1,x_2,\cdots,x_n)$$
$$\cdot \, dF(x_1,x_2,\cdots,x_n).$$

For a set of discrete random variables, this reduces to a multiple summation, while for continuous random variables having a joint density function, the definition becomes

$$E[\alpha(x_1,x_2,\cdots,x_n)] = \int\int_{R_X} \cdots \int \alpha(x_1,x_2,\cdots,x_n)$$
$$\cdot \, f(x_1,x_2,\cdots,x_n) \, dx_1 \, dx_2 \cdots dx_n.$$

PROBLEM. A die is rolled twice. Let x be the number that shows on the first roll, and y on the second. Find $E(x + y)$ and $E(xy)$.

SOLUTION. Because $f(x,y) = \frac{1}{36}$ for $x = 1, 2, \cdots, 6$ and $y = 1, 2,$ $\cdots, 6$, we have

$$E(x + y) = \sum_{x=1}^{6} \sum_{y=1}^{6} \left(\frac{x + y}{36}\right) = 7.$$

and

$$E(xy) = \sum_{x=1}^{6} \sum_{y=1}^{6} \left(\frac{xy}{36}\right) = 12.25.$$

PROBLEM. Items in a free market fluctuate in both buying price x and selling price y with density $f(x,y) = 2$ for $x < y < 1$ and $0 < x < 1$, and $f(x,y) = 0$ otherwise. If a dealer buys and then sells an item, what is the expected value of his profit $y - x$?

SOLUTION.

$$E(y - x) = \int_{0}^{1} dx \int_{x}^{1} dy \, (y - x)2 = \frac{2}{3} - \frac{1}{3} = \frac{1}{3}.$$

PROBLEM. Find the expected value of the function $z = xy$ given that x and y are jointly distributed as follows:

$$f(x,y) = xe^{-x(1+y)} \qquad \text{for } 0 \leqslant x < \infty \quad \text{and} \quad 0 \leqslant y < \infty.$$

SOLUTION.

$$E(z) = \int_{0}^{\infty} \int_{0}^{\infty} xyf(x,y) \, dx \, dy = \int_{0}^{\infty} e^{-x} \left[\int_{0}^{\infty} x^2 y e^{-xy} \, dy \right] dx = \int_{0}^{\infty} e^{-x} \, dx = 1.$$

As a check, the distribution of z can be found directly from $f(x,y)$. It is found that the probability density function of z is $h(z) = e^{-z}$ for $0 \leqslant z < \infty$. Hence from the distribution of z itself, we have

$$E(z) = \int_{0}^{\infty} ze^{-z} \, dz = 1,$$

which is the same result as before.

As an important special case, the definition of $E[\alpha(x_1, x_2, \cdots, x_n)]$ can yield the expected value of any one of the random variables $x_1, x_2, \cdots, x_n$. The required function is

$$\alpha(x_1, x_2, \cdots, x_n) = x_i,$$

which gives

$$E[\alpha(x_1, x_2, \cdots, x_n)] = E(x_i) = \int_{R_x} \int \cdots \int x_i \, dF(x_1, x_2, \cdots, x_n).$$

For example, suppose that x, y have the density function $f(x, y)$ and we want to find the $E(x)$. We have from the above equation that

$$E(x) = \int_{-\infty}^{\infty} \int_{-\infty}^{\infty} xf(x, y) \, dx \, dy.$$

Let us experiment with this result. We can write this equation as

$$E(x) = \int_{-\infty}^{\infty} x \left[\int_{-\infty}^{\infty} f(x, y) \, dy \right] dx = \int_{-\infty}^{\infty} xf_1(x) \, dx;$$

so we see that $E(x)$ is the mean of the marginal density $f_1(x)$, as we would expect.

The foremost property of expectation is that it is a linear operation. It is this property that is used over and over again in solving problems in probability and statistics.

THEOREM 5.1 (LINEAR PROPERTY OF EXPECTATION). The expectation of a constant times a function is equal to the constant times the expectation of the function; that is,

$$E(c\alpha) = cE(\alpha).$$

The expectation of a sum of functions is equal to the sum of the expectations of each function; that is,

$$E(\alpha_1 + \alpha_2 + \cdots + \alpha_n) = E(\alpha_1) + E(\alpha_2) + \cdots + E(\alpha_n).$$

PROOF. This theorem follows directly from the linear property of the summation in the discrete case, or the integral in the continuous case, in the definition of expected value. For example, in the continuous case for two variables, for the constant c and function $\alpha(x, y)$ we have

$$E(c\alpha) = \iint c\alpha(x, y)f(x, y) \, dx \, dy = c \iint \alpha(x, y)f(x, y) \, dx \, dy = cE(\alpha);$$

and for the two functions $\alpha_1(x, y)$ and $\alpha_2(x, y)$ we have

$$\begin{aligned}
E(\alpha_1, \alpha_2) &= \iint [\alpha_1(x, y) + \alpha_2(x, y)] \, dx \, dy \\
&= \iint \alpha_1(x, y)f(x, y) \, dx \, dy + \iint \alpha_2(x, y)f(x, y) \, dx \, dy \\
&= E(\alpha_1) + E(\alpha_2).
\end{aligned}$$

All of the above integrals are definite integrals over the complete ranges of the variables. Q.E.D.

Two consequences of this theorem are sufficiently important to be stated as corollaries. They concern the *sum* of any number of random variables, or any linear combination of random variables, whether dependent or independent.

COROLLARY 5.1A. If $y = x_1 + x_2 + \cdots + x_n$, then

$$E(y) = E(x_1) + E(x_2) + \cdots + E(x_n).$$

PROOF. In Theorem 5.1, let $\alpha_i = x_i$ for $i = 1, 2, \cdots, n$.

COROLLARY 5.1B. If $z = a_0 + a_1x_1 + a_2x_2 + \cdots + a_nx_n$ where the a_i are constants, then

$$E(z) = a_0 + a_1E(x_1) + a_2E(x_2) + \cdots + a_nE(x_n).$$

The proof is similar to that of Corollary 5.1A.

Let us now give two theorems, one concerning the product of independent random variables, and the other concerning the product of pure functions of independent random variables. The key word in these two theorems is *independent*.

THEOREM 5.2. The expectation of the product of independent random variables is the product of the expectations; that is, if $y = x_1x_2 \cdots x_n$ where $x_1, x_2, \cdots, x_n$ are *independent*, then

$$E(y) = E(x_1)\,E(x_2) \cdots E(x_n).$$

PROOF. If the random variables are independent, their joint probability increment $dF(x_1,x_2,\cdots,x_n)$ factors into the product of the increments of their respective marginal distribution functions:

$$dF(x_1,x_2,\cdots,x_n) = dF_1(x_1)\,dF_2(x_2) \cdots dF_n(x_n).$$

Their respective regions $R_1, R_2, \cdots, R_n$ are also independent. Consequently, the whole expression defining $E(y)$ factors, and the result is

$$E(y) = \int_{R_1} x_1 \, dF_1(x_1) \int_{R_2} x_2 \, dF_2(x_2) \cdots \int_{R_n} x_n \, dF_n(x_n)$$
$$= E(x_1)E(x_2)\cdots E(x_n).$$

THEOREM 5.3. The expectation of the product of pure functions of *independent* random variables is the product of the expectations; that is, if $x_1, x_2, \cdots, x_n$ are independent and

$$y = \alpha_1(x_1)\alpha_2(x_2)\cdots\alpha_n(x_n),$$

then

$$E(y) = E[\alpha_1(x_1)]E[\alpha_2(x_2)]\cdots E[\alpha_n(x_n)].$$

The proof is similar to that of Theorem 5.2.

PROBLEM. Find an expression for the expected value of the function $z = (x + y)^2$.

SOLUTION.

$$E(z) = E[(x + y)^2] = E(x^2 + 2xy + y^2).$$

Hence by Theorem 5.1,

$$E(z) = E(x^2) + 2E(xy) + E(y^2).$$

This result holds for *any* random variables whatever. If the random variables x and y are *independent*, the cross-product term can be factored by virtue of Theorem 5.2 to yield $E(xy) = E(x)E(y)$. Hence, for independent random variables, we have

$$E(z) = E(x^2) + 2E(x)E(y) + E(y^2).$$

For brevity, an expression like $E[(x + y)^2]$ is usually written as $E(x + y)^2$.

Moments for two variables are more complicated than those for a single variable in that one cannot merely speak of the kth moment but must specify which power of each variable is being considered. Moments of two or more variables are called product moments. For two variables, the product moment is defined as

$$\mu'_{pq} = E(x^p y^q) = \int_{-\infty}^{\infty}\int_{-\infty}^{\infty} x^p y^q f(x,y)\, dy\, dx.$$

The corresponding product moment about the mean is defined by

$$\mu_{pq} = E\{(x - \mu_x)^p(y - \mu_y)^q\}$$
$$= \int_{-\infty}^{\infty}\int_{-\infty}^{\infty} (x - \mu_x)^p(y - \mu_y)^q f(x,y)\, dy\, dx.$$

If we choose $p = k$ and $q = 0$, then the product moment becomes

$$\mu'_{k0} = \int_{-\infty}^{\infty} x^k \left[\int_{-\infty}^{\infty} f(x,y) \, dy \right] dx = \int_{-\infty}^{\infty} x^k f_1(x) \, dx = E(x^k).$$

This equation shows that μ'_{k0} is the kth moment of x. In particular, it follows that

$$\mu'_{00} = 1,$$

$$\mu'_{10} = E(x) = \mu_x,$$

$$\mu'_{01} = E(y) = \mu_y,$$

$$\mu_{20} = \text{var } x = \sigma_x^2 \cdot \mu_{02} = \text{var } y = \sigma_y^2.$$

The particular product moment μ_{11} is called the covariance of the two variables. Because the covariance is so important, it has several special notations. The most common are

$$\text{cov } (x,y) = \sigma_{xy} = E\{(x - \mu_x)(y - \mu_y)\}.$$

The covariance is of special interest because the theoretical *correlation coefficient* ρ between the two variables is defined in terms of it as

$$\rho = \frac{\mu_{11}}{\sigma_x \sigma_y} = \frac{\sigma_{xy}}{\sigma_x \sigma_y} = \frac{\text{cov } (x,y)}{\sqrt{\text{var } x \text{ var } y}}.$$

If $\rho = 0$, or equivalently if cov $(x,y) = 0$, then x and y are said to be *uncorrelated*.

A simpler computational formula for the covariance (and hence also for the correlation coefficient) is found as follows:

$$\sigma_{xy} = E\{(x - \mu_x)(y - \mu_y)\} = E\{xy - \mu_x y - x\mu_y + \mu_x \mu_y\}$$
$$= E(xy) - \mu_x E(y) - \mu_y E(x) + \mu_x \mu_y = E(xy) - \mu_x \mu_y.$$

Note the similarity between this equation and the computational formula for the variance

$$\sigma_x^2 = E(x^2) - \mu_x^2.$$

For any random variables x and y, their correlation coefficient ρ satisfies $-1 \le \rho \le 1$. In order to establish this result, we first establish the Schwarz inequality. We introduce two real indeterminants (i.e., dummy variables) a and b and consider the identity

$$E[(ax + by)^2] = E[a^2x^2 + 2\,abxy + b^2y^2]$$
$$= a^2E(x^2) + 2abE(xy) + b^2E(y^2).$$

Because the left member is the expectation of the nonnegative function $(ax + by)^2$, the left member is nonnegative. Thus the right member must be nonnegative also. The right member is a quadratic form, and a well-known mathematical fact says that the coefficients of a non-negative quadratic form must satisfy the inequality $B^2 \leqslant AC$. In the present case, we have $A = E(x^2)$, $B = E(xy)$, and $C = E(y^2)$, so

$$[E(xy)]^2 \leqslant E(x^2)E(y^2),$$

which is the *Schwarz inequality*. If we replace x by $x - \mu_x$ and y by $y - \mu_y$ in the Schwarz inequality, we obtain $\rho^2 \leqslant 1$.

The following theorem gives a very useful general formula for the second moment of a sum of random variables without the assumption of independence.

THEOREM 5.4. If $y = x_1 + x_2 + \cdots + x_n$, then

$$E(y^2) = \sum_{i=1}^{n} E(x_i^2) + 2 \sum_{i=1}^{n} \sum_{j=i+1}^{n} E(x_i x_j).$$

PROOF. This result follows by expanding $(x_1 + x_2 + \cdots + x_n)^2$ and taking expectations.

Observe that for any constant c, the random variable $x + c$ has the same variance as x. Also observe that for any constants c_1 and c_2, the random variables $x + c_1$ and $y + c_2$ have the same covariance as x and y. In Theorem 5.4, let us replace each x_i by $x_i - \mu_i$ where $\mu_i = E(x_i)$. Then we must replace y by $y - \mu_y$ where $\mu_y = \mu_1 + \mu_2 + \cdots + \mu_n$. Thus each random variable becomes centered about its mean, and so the second moments become variances and covariances. Thus we have:

COROLLARY 5.4. If $y = x_1 + x_2 + \cdots + x_n$, then

$$\text{var}\,(y) = \sum_{x=1}^{n} \text{var}\,(x_i) + 2 \sum_{i=1}^{n} \sum_{j=i+1}^{n} \text{cov}\,(x_i, x_j).$$

If, in addition, $x_1, x_2, \cdots, x_n$ are pairwise uncorrelated, then

$$\text{var}\,y = \sum_{i=1}^{n} \text{var}\,(x_i)$$

or, in alternative notation,

$$\sigma_y^2 = \sigma_1^2 + \sigma_2^2 + \cdots + \sigma_n^2.$$

Finally, let us note that if two random variables x and y are independent, then

$$E\{(x - \mu_x)(y - \mu_y)\} = E(x - \mu_x)E(y - \mu_y) = 0;$$

so it follows that x and y are uncorrelated. However, if two variables are uncorrelated it does not necessarily follow that they are independent.

Let us now continue the discussion of random sampling given in Chapter 3 (page 70). We saw there that in order for a set of observations to be a *random sample*, it is necessary that two assumptions be satisfied. One assumption is that the successive trials of the experiment are independent and the other is that the density function of the random variable remains the same from trial to trial. In other words, the sequence of observations in a random sample is a sequence of *independent, identically distributed random variables*. If theory is to be applicable to real experimental data, it is necessary that the data be obtained by a sampling method that possesses these two properties (i.e., independence and identical distribution.) In order to express these properties in a mathematical form, consider the following notation and procedure.

Let $f(x)$ be the density function of the continuous random variable x and let a sample of size n be drawn. The resulting sample values will be denoted by $x_1', x_2', \cdots, x_n'$. If a second sample of size n were drawn, the resulting sample values would be denoted by $x_1'', x_2'', \cdots, x_n''$, and similarly for additional samples. The values can be arranged as follows:

$$\begin{array}{cccc} x_1', & x_2', & \cdots, & x_n' \\ x_1'', & x_2'', & \cdots, & x_n'' \\ x_1''', & x_2''', & \cdots, & x_n''' \\ \vdots & \vdots & & \vdots \end{array}$$

Now consider the values in the first column. These values may be treated as the values of a random variable x_1 with a density function $f_1(x_1)$. In the same manner, the values in the second column may be treated as the values of a random variable x_2 with a density function $f_2(x_2)$, and similarly for the remaining columns. In this notation, the requirement that the density function of the random variable x shall remain constant from trial to trial means that the random variables $x_1, x_2, \cdots, x_n$ must possess the original density function, that is, that

$$f_1(x) = f_2(x) = \cdots = f_n(x) = f(x).$$

In this same notation, the requirement that the trials shall be independent means that the variables x_1, x_2, $\cdots$, x_n must be independent. A method of sampling that possesses these two properties is called *random sampling*. Thus random sampling may be defined mathematically in the following manner:

DEFINITION. Random sampling is a method of sampling for which the joint density function given by the product

$$g(x_1, x_2, \cdots, x_n) = f(x_1)f(x_2)\cdots f(x_n),$$

where $f(x)$, is the density function of the population being sampled and where x_1, x_2, $\cdots$, x_n are random variables corresponding to the n trials of the sample.

Although the variable x in the preceding discussion was treated as a continuous variable, this definition applies to discrete variables as well, with probability mass functions in place of probability density functions.

This definition defines a method of sampling and says nothing about particular samples. It is legitimate to call a sample a random sample only if it has been obtained by a random sampling method. It is frequently not feasible to check many real-life sampling methods for randomness because of the expense or difficulty of obtaining enough data to test the properties required by the definition. Then one must rely on judgment and experience to determine whether the method is sufficiently random to permit the use of models derived on the basis of random sampling.

As an illustration of a continuous random variable for which the sampling method is random, at least to a very good approximation, consider an experiment in which each trial consists of spinning a pointer. Let x be the distance the end of the spinning pointer is from the 0 point, as measured along the circumference, after it comes to rest. If a sample of size six were desired, the pointer would be spun six times and the distances recorded. Now, if a pointer is spun repeatedly, with the resulting values of x marked off into consecutive sets of six, it is usually found that the empirical distributions of the variables x_1, $\cdots$, x_6 will all approach the rectangular distribution $f(x) = 1/c$, where c is the circumference. It also is found that tests of independence, which are studied in statistics, usually substantiate independence of trials here.

The following theorem is useful in sampling theory.

THEOREM 5.5. If we have n independent draws x_1, x_2, $\cdots$, x_n from a population with mean μ and variance σ^2, then the sample mean $\bar{x}$ $= (x_1 + x_2 + \cdots + x_n)/n$ has its mean value equal to the pop-

ulation mean and its variance equal to $(1/n)$ times the population variance, i.e.,

$$E(\bar{x}) = \mu, \qquad \text{var}\,(\bar{x}) = \sigma^2/n.$$

PROOF.

$$E(\bar{x}) = E\{(x_1 + \cdots + x_n)/n\} = [(\mu + \cdots + \mu)/n] = \mu.$$

$$\text{var}\,\bar{x} = \text{var}\,\{(x_1 + \cdots + x_n)/n\} = [\text{var}\,(x_1 + \cdots + x_n)]/n^2$$
$$= (\text{var}\,x_1 + \cdots + \text{var}\,x_n)/n^2 = (\sigma^2 + \cdots + \sigma^2)/n^2 = \sigma^2/n.$$

(Explain why we obtain the factor $1/n^2$ when we remove the
factor $1/n$ from under the variance sign.)

This theorem is a special case of the following theorem, which is
proved in a similar way.

THEOREM 5.6. If $x_1, x_2, \cdots, x_n$ are independent random variables with
means $\mu_1, \mu_2, \cdots, \mu_n$ and variances $\sigma_1^2, \sigma_2^2, \cdots, \sigma_n^2$, respectively,
and if $y = a_1 x_1 + a_2 x_2 + \cdots + a_n x_n$ where the a_i are arbitrary
constants, then

$$\mu_y = a_1 \mu_1 + a_2 \mu_2 + \cdots + a_n \mu_n,$$
$$\sigma_y^2 = a_1^2 \sigma_1^2 + a_2^2 \sigma_2^2 + \cdots + a_n^2 \sigma_n^2.$$

EXAMPLE. In each trial of an experiment, there are only two possible
outcomes, called success and failure. For trial number i, let the
random variable x_i equal 1 for a success and equal 0 for a failure. The probability for a success is given by the same constant
p in each trial, so the probability for a failure is $q = 1 - p$ in
each trial. The trial is called a Bernoulli trial. In a sequence x_1,
$x_2, x_3, \cdots, x_n$ of n independent Bernoulli trials, the random variable $S_n = x_1 + x_2 + \cdots + x_n$ gives the number of successes,
and S_n has a binomial distribution. The expected value and
variance for each trial are respectively

$$E(x_i) = 0 \cdot P(0) + 1 \cdot P(1) = 0 \cdot q + 1 \cdot p = p.$$
$$\text{var}\,(x_i) = (0 - p)^2 P(0) + (1 - p)^2 P(1) = p^2 q + (1 - p)^2 p = pq.$$

Using Theorem 5.6 we see that

$$E(S_n) = p + p + \cdots + p = np.$$
$$\text{var}\,(S_n) = pq + pq + \cdots + pq = npq.$$

If we define $\bar{x} = S_n/n$, then we see that Theorem 5.5 gives

$$E(\bar{x}) = E(S_n)/n = p.$$
$$\text{var}\,(\bar{x}) = \text{var}\,(S_n)/n^2 = pq/n.$$

EXERCISES

1. Iron bars in the shape of slim cylinders are test-measured. Suppose the average length is 10 inches and average area of ends is 1 square inch. The average error made in the measurement of the length is 0.005 inch, that in the measurement of the area is 0.01 square inch. What is the average error made in estimating their weights?

   ANSWER. Since weight is a constant times volume, consider the latter: $V = LA$ where L = length, A = area of ends. Let the errors be ΔL and ΔA respectively; then the error in V is given by

   $$\Delta V = (L + \Delta L)(A + \Delta A) - LA = L\Delta A + A\Delta L + \Delta L\Delta A.$$

   Assume independence between the measurements, so

   $$E(\Delta V) = E(L)E(\Delta A) + E(A)E(\Delta L) + E(\Delta A)E(\Delta L)$$
   $$= 0.105 \quad \text{cubic inch if the last term is ignored.}$$

2. If X and Y are independent, show that $\sigma^2(X + Y) = \sigma^2(X) + \sigma^2(Y)$.

3. Urn 1 contains 5 white and 6 black balls, while urn 2 contains 8 white and 10 black balls. Two balls are randomly selected from urn 1 and are then put in urn 2. If 3 balls are then randomly selected from urn 2, compute the expected number of white balls in the trio.

   HINT. Let $X_i = 1$ if the ith white ball initially in urn 1 is one of the three selected, and let $X_i = 0$ otherwise. Similarly, let $Y_i = 1$ if the ith white ball from urn 2 is one of the three selected, and let $Y_i = 0$ otherwise. The number of white balls in the trio can be written as

   $$\sum_1^5 X_i + \sum_1^8 Y_i.$$

   ANSWER. 1.3364.

4. If 10 married couples are seated at random around a round table, compute (a) the expected number, and (b) the variance of the number of wives that are seated next to their husbands.

   ANSWERS. (a) $20/19$. (b) $[340/(19)^2] + 2\binom{10}{2}[4/(18)(19)^2]$.

5. A die is rolled twice. Let X equal the sum of the outcomes, and let Y equal the first outcome minus the second. Compute $\text{cov}(X,Y)$.

   ANSWER. 0

6. Cards from an ordinary deck are turned face up one at a time. Compute the expected number of cards that need be turned face up in order to obtain (a) 2 aces, (b) 5 spades, and (c) all 13 hearts.
ANSWERS. (a) 21.2. (b) 18.928. (c) 49.214.

7. A prisoner is trapped in a cell containing 3 doors. The first door leads to a tunnel that returns him to his cell after 2 days travel. The second to a tunnel that returns him to his cell after 4 days travel. The third door leads to freedom after 1 day of travel. If it is assumed that the prisoner will always select doors 1, 2, and 3 with respective probabilities 0.5, 0.3, and 0.2, what is the expected number of days until the prisoner reaches freedom?
ANSWER. 12.

8. Calculate the expected sum obtained when 10 independent rolls of a fair die are made.
ANSWER. $E(y) = E(x_1) + E(x_2) + \cdots + E(x_{10}) = 10(^7/_2) = 35$.

9. Expected Number of Matches. A group of N men throw their hats into the center of a room. The hats are mixed up and each man randomly selects one. Find the expected number of men that select their own hats.
ANSWER. $E(y) = E(x_1) + E(x_2) + \cdots + E(x_N) = (1/N)N = 1$.

10. Find the mean of a Pascal-geometric random variable.
ANSWER.

$$E(k) = \sum_{k=1}^{\infty} kpq^{k-1} = p \sum_{k=1}^{\infty} \frac{dq^k}{dk} = p \frac{d}{dk} \sum_{k=1}^{\infty} q^k$$
$$= p \frac{d}{dq} \left(\frac{q}{1-q} \right) = \frac{1}{p}.$$

11. Mean of a Pascal Random Variable. If independent trials, having a constant probability p of being successes, are performed, determine the expected number of trials required to amass a total of r successes.
ANSWER. Let $k = k_1 + k_2 + \cdots + k_r$, where k_1 is the number of trials required to obtain the first success, k_2 the number of additional trials until the second success is obtained, k_3 the number of additional trials until the third success is obtained, and so on. That is, k_i represents the number of additional trials required, after the $(i - 1)$st success, until a total of i successes are amassed. A little thought reveals that each of the variables k_i is a Pascal-geometric random variable with parameter p. Hence, from the results of the above exercise, we have

$$E(k) = E(k_1) + \cdots + E(k_r) = r/p.$$

12. **Mean of a Hypergeometric Random Variable.** If n balls are randomly selected from an urn containing N_1 white and $N - N_1$ black balls, find the expected number of white balls selected.

    ANSWER. Let n_1 be the number of white balls selected, and write $n_1 = x_1 + x_2 + \cdots + x_N$ where $x_i = 1$ if the ith ball selected is white and $x_i = 0$ otherwise. Since the ith ball selected is equally likely to be any of the N, we have $E(x_i) = P(x_i) = N_1/N$. Thus

    $$E(n_1) = E(x_1) + E(x_2) + \cdots + E(x_n) = nN_1/N.$$

13. The following problem was posed and solved in the eighteenth century by Daniel Bernoulli. Suppose that a jar contains $2N$ playing cards, two of them marked 1, two marked 2, two marked 3, and so on. Draw out m cards at random. What is the expected number of pairs that still remain in the jar? (Interestingly enough, Bernoulli proposed the above as a possible probabilistic model for determining the number of marriages that remain intact when there are a total of m deaths among the N married couples.)

    ANSWER. Let $y = x_1 + x_2 + \cdots + x_N$ where $x_i = 1$ if the ith pair remains in the jar and $x_i = 0$ otherwise. Then

$$E(x_i) = P(x_i = 1) = \binom{2N-2}{m} \Big/ \binom{2N}{m}$$

so $\quad E(y) = N\binom{2N-2}{m} \Big/ \binom{2N}{m}.$

14. **Coupon-Collecting Problems.** Suppose that there are N different types of coupons and each time one obtains a coupon it is equally likely to be any one of the N types.
    (a) Find the expected number of different types of coupons that are contained in a set of n coupons.
    (b) Find the expected number of coupons one need amass before obtaining a complete set of at least one of each type.

    ANSWERS.

$$\text{(a)}\, N\left[1 - \left(\frac{N-1}{N}\right)^n\right]. \qquad \text{(b)}\, N\left[1 + \cdots + \frac{1}{N-1} + \frac{1}{N}\right].$$

15. Consider any nonnegative, integer-valued random variable x. If we define, for each $i \geqslant 1$, $x_i = 1$ for $x \geqslant i$ and $x_i = 0$ for $x < i$, then $x = x_1 + x_2 + x_3 + \cdots$. Show that

$$E(x) = \sum_{i=1}^{\infty} P(x \geqslant i).$$

16. Show that for constants a, b, c, d
    (a) cov $(a + bx, c + dy) = bd$ cov (x,y).
    (b) cov $(x + y,z) =$ cov $(x,z) +$ cov (y,z).
    (c) cov $(\Sigma_{i=1}^{n} x_i, \Sigma_{j=1}^{n} y_j) = \Sigma_{i=1}^{n} \Sigma_{j=1}^{n}$ cov (x_i,y_j).
    (d) Let x be the number of 1's and y the number of 2's that occur in n rolls of a fair die. Compute cov (x,y).
    ANSWER. (d) $-n/36$.

17. If X_1, X_2, X_3, X_4 are (pairwise) uncorrelated random variables each having mean 0 and variance 1, compute the correlations of
    (a) $X_1 + X_2$ and $X_2 + X_3$,
    (b) $X_1 + X_2$ and $X_3 + X_4$.
    ANSWER. (a) $1/2$. (b) 0.

18. Independent trials are performed. The ith such trial results in a success with probability P_i.
    (a) Compute the expected value of the number of successes that occur in the first n trials.
    (b) Compute the variance of this number.
    (c) Does independence make a difference in part (a)? In part (b)?
    ANSWERS. (a) $P_1 + P_2 + \cdots + P_n$. (b) $P_1(1 - P_1) + P_2(1 - P_2) + \cdots + P_n(1 - P_n)$.

19. Let $x_1, \cdots, x_n$ be independent and identically distributed random variables having mean μ and variance σ^2, and let $\bar{x} = (x_1 + x_2 + \cdots + x_n)/n$. Show that

$$E\left[\sum_{i=1}^{n} (x_i - \bar{x})^2\right] = (n - 1)\sigma^2.$$

20. Let $x_1, \cdots, x_n$ be independent and identically distributed, continuous random variables. We say that a record value occurs at time j, $j \leq n$, if $x_j \geq x_i$ for all $1 \leq i \leq j$. Show that
    (a) $E[\text{number of record values}] = \Sigma_{j=1}^{n} 1/j$.
    (b) var (number of record values) $= \Sigma_{j=1}^{n} (j - 1)/j^2$.

REGRESSION CURVES

A basic concept associated with a conditional distribution is the curve of regression. Here it is convenient to use the density interpretation of $\phi_2(y|x)$. Let x have the fixed value x_0. Then along the line $x = x_0$ the mean value of y will determine a point whose ordinate will be denoted by $E(y|x_0)$. As different values of x are selected, different mean points along the corresponding vertical lines will be obtained. Thus, the ordinate $E(y|x)$ of the mean point for any such line is a function of the value of x selected. The locus of such mean points will be a curve called the *curve of regression* of y on x. Analytically, the equation of the curve of regression is given by the following formula

$$E(y|x) = \int_{-\infty}^{\infty} y\phi_2(y|x)\, dy.$$

An alternative notation for $E(y|x)$ is $\mu_{y|x}$. This formula may also be expressed in the form

$$E(y|x) = \int_{-\infty}^{\infty} y\, \frac{f(x,y)}{f_1(x)}\, dy.$$

The curve of regression of x on y is defined in the analogous manner:

$$E(x|y) = \int_{-\infty}^{\infty} x\phi_1(x|y)\, dx = \int_{-\infty}^{\infty} x\, \frac{f(x,y)}{f_2(y)}\, dx.$$

The geometrical nature of the two curves of regression can be seen when they are plotted on the x,y plane. However as we will see in the following example, the two curves of regression do not fall on the same geometrical curve.

As a simple example of the curves of regression, consider the joint frequency function

$$f(x,y) = \begin{cases} 2 - x - y & \text{for } 0 < x < 1, 0 < y < 1, \\ 0 & \text{elsewhere.} \end{cases}$$

The marginal distributions are

$$f_1(x) = \int_0^1 (2 - x - y)\, dy = \frac{3}{2} - x \qquad \text{for } 0 < x < 1,$$

$$f_2(y) = \int_0^1 (2 - x - y)\, dx = \frac{3}{2} - y \qquad \text{for } 0 < y < 1.$$

Hence the conditional distributions are

$$\phi_1(x|y) = \frac{f(x,y)}{f_2(y)} = \frac{2 - x - y}{{}^3/_2 - y},$$

$$\phi_2(y|x) = \frac{f(x,y)}{f_1(x)} = \frac{2 - x - y}{{}^3/_2 - x}.$$

Note that there is a complete symmetry between x and y in this example.

From a density point of view, $\phi_2(y|x)$ may be thought of as giving the probability density distribution along the vertical line in the x,y plane corresponding to the fixed value of x. The total mass of this line must be equal to 1. The joint density function $f(x,y)$ as it stands could not be used as a probability density function along such a line because it would not give a total probability mass of one for the entire line unless $f_1(x)$ happened to be equal to 1. The factor $1/f_1(x)$ insures that the total mass of the line will be 1.

The regression curve of y on x is

$$E(y|x) = \int_0^1 y \frac{2 - x - y}{{}^3/_2 - x} \, dy$$

$$= \frac{1}{{}^3/_2 - x} \int_0^1 [(2 - x)y - y^2] \, dy = \frac{3x - 4}{6x - 9}.$$

This is the equation of a hyperbola. By symmetry, the other regression curve is $E(x|y) = (3y - 4)/(6y - 9)$. The graph of these two curves of regression are shown in Figure 5.12.

Let us now continue the simple example given in this section. It follows that

$$E(x) = \int_0^1 x \, f_1(x) \, dx = \int_0^1 x \left(\frac{3}{2} - x \right) dx = \frac{5}{12}.$$

By symmetry, $E(y)$ is also ${}^5/_{12}$. The variance of x is

$$\sigma_x^2 = \mu_{20} = \int_0^1 \left(x - \frac{5}{12} \right)^2 \left(\frac{3}{2} - x \right) dx = \frac{11}{144},$$

and by symmetry the variance of y is also ${}^{11}/_{144}$. The covariance is

$$\mu_{11} = \int_0^1 \int_0^1 \left(x - \frac{5}{12} \right) \left(y - \frac{5}{12} \right) (2 - x - y) \, dy \, dx$$

$$= \int_0^1 \left(x - \frac{5}{12} \right) \int_0^1 \left(y - \frac{5}{12} \right) (2 - x - y) \, dy \, dx$$

$$= \int_0^1 \left(x - \frac{5}{12} \right) \left(\frac{1}{24} - \frac{x}{12} \right) dx = - \frac{1}{144}.$$

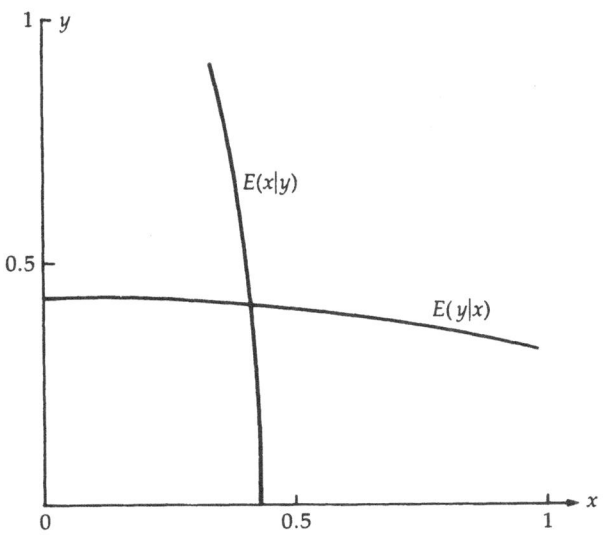

FIGURE 5.12 The two regression curves $E(y|x)$ and $E(x|y)$.

Thus the correlation coefficient is

$$\rho = \frac{-1/144}{11/144} = -\frac{1}{11}.$$

An inspection of Figure 5.12 shows that each regression curve has a slight negative slope throughout its range and therefore it is not surprising that ρ, which measures linear correlation, turned out to be negative.

EXERCISES

1. A fair die is successively rolled. Let x and y denote, respectively, the number of rolls necessary to obtain a 6 and a 5.
   (a) Find $E(x)$.
   (b) Find $E(x|y = 1)$.
   (c) Find $E(x|y = 5)$.
   ANSWERS. (a) 6. (b) 7. (c) 5.81920.

2. If the joint density of x and y is given by

$$f(x,y) = \frac{e^{-y}}{y} \qquad \text{for } 0 < x < y, \text{ and } 0 < y < \infty.$$

Find $E(x^3|y)$.

ANSWER. $y^3/4$.

3. The joint density function of x and y is given by

$$f(x,y) = xe^{-x(y+1)} \qquad \text{for } x > 0, y > 0.$$

Find the conditional density
(a) of x given y,
(b) of y given x.

ANSWERS. (a) $\phi_1(x|y) = (y + 1)^2 xe^{-x(y+1)}$. (b) $\phi_2(y|x) = xe^{-xy}$.

4. The joint density of x and y is

$$f(x,y) = c(x^2 - y^2)e^{-x} \qquad \text{for } 0 \leqslant x < \infty, -x \leqslant y \leqslant x.$$

Find the conditional distribution of y given x.

ANSWER. $\phi_2(y|x) = (1/2) + (3y/4x) - (1/4)(y^3/x^3)$.

5. Suppose the joint density of X and Y is given by

$$f(x,y) = \frac{e^{-x/y}e^{-y}}{y} \qquad \text{for } 0 < x < \infty, 0 < y < \infty.$$

(a) Compute $\phi_1(x|y)$. (b) Compute $E(x|y)$.

ANSWERS. (a) $(1/y)e^{-x/y}$. (This is an exponential distribution with mean y.) (b) y.

6. Discuss the following remark: "Everybody knows that probability and statistics are the same thing, and statistics is nothing but correlation. Now the correlation is just the cosine of an angle. Thus, all is trivial."

7. If $y = a + bx$, show that

$$\rho(x,y) = \begin{cases} +1 & \text{if } b > 0, \\ -1 & \text{if } b < 0. \end{cases}$$

8. If z is a standard normal random variable and if y is defined by $y = a + bz + cz^2$, show that

$$\rho(y,z) = \frac{b}{\sqrt{b^2 + 2c^2}}.$$

9. Show that if x and y are independent, then

$$E(x|y) = E(x) \qquad \text{for all } y$$

(a) in the discrete case.
(b) in the continuous case.

10. Prove that $E[g(x)y|x] = g(x)E(y|x)$.

11. Prove that if $E(y|x) = E(y)$ for all x, then x and y are uncorrelated, and give a counterexample to show that the converse is not true. HINT. Prove and use the fact that $E(xy) = E[xE(y|x)]$.

12. The least-squares linear predictor of y with respect to x_1 and x_2 is equal to $a + bx_1 + cx_2$ where a, b, and c are chosen to minimize

$$E\{[y - (a + bx_1 + cx_2)]^2\}.$$

Find the set of simultaneous equations whose solution yields a, b, and c.

13. The least-squares quadratic predictor of y with respect to x is $a + bx + cx^2$, where a, b, and c are chosen to minimize $E\{[y - (a + bx + cx^2)]^2\}$. Find the set of simultaneous equations whose solution yields a, b, and c.

LAW OF LARGE NUMBERS

We now want to discuss the law of large numbers. Suppose that we toss a fair coin a large number of times n, and let S_n be the number of heads that come up. On intuitive grounds, it seems that the relative frequency S_n/n should be close to $1/2$. In fact, it seems that the larger n becomes, the closer this ratio should be to $1/2$. However, we cannot say with certainty that S_n/n will ever be $1/2$. But we can show that the probability that it differs from $1/2$ by any small amount ϵ tends to zero as n increases. This example is a case of what is called the law of large numbers.

THEOREM 5.7 LAW OF LARGE NUMBERS. Suppose that $x_1, x_2, \cdots, x_k, \cdots$ is a sequence of independent, identically distributed random variables, each with the same mean μ and same variance σ^2. Define a new sequence $\bar{x}_1, \bar{x}_2, \cdots, \bar{x}_k, \cdots$ by

$$\bar{x}_n = (x_1 + x_2 + \cdots + x_n)/n \qquad \text{for } n = 1, 2, 3, \cdots.$$

(Thus $\bar{x}_1 = x_1$, $\bar{x}_2 = (x_1 + x_2)/2$, $\bar{x}_3 = (x_1 + x_2 + x_3)/3$, $\cdots$). Then

$$\lim_{n \to \infty} P(|\bar{x}_n - \mu| > \epsilon) = 0 \qquad \text{for any } \epsilon > 0.$$

PROOF. By Theorem 5.5, we have

$$E(\bar{x}_n) = \mu, \qquad \text{var}(\bar{x}_n) = \sigma^2/n.$$

By Chebychev's inequality, we have

$$P[|\bar{x}_n - \mu| > t\sqrt{\text{var}\,(\bar{x}_n)}] \leqslant 1/t^2$$

for any $t > 0$. If we use the above expression for the variance, and if we choose

$$t = \frac{\epsilon}{\sqrt{\text{var}\,(\bar{x}_n)}} = \frac{\epsilon\sqrt{n}}{\sigma},$$

we see that the inequality becomes

$$P(|\bar{x}_n - \mu| > \epsilon) \leqslant \frac{\sigma^2}{\epsilon^2 n}.$$

For any fixed ϵ, it follows that the right member goes to zero as $n \to \infty$. Since all the terms in this sequence of probabilities must be nonnegative, we clearly have $P(|\bar{x}_n - \mu| > \epsilon) \to 0$ as $n \to \infty$.

The law of large numbers may be equivalently expressed by saying that $P(|\bar{x}_n - \mu| \leqslant \epsilon) \to 1$ as $n \to \infty$. However, we cannot interpret this result as saying that $\bar{x}_n$ is necessarily getting closer and closer to μ as $n \to \infty$. This result does say that the probability tends to one that $\bar{x}_n$ differs from μ by less than ϵ, no matter how small ϵ may be. Thus it is important to realize that the law of large numbers gives the limit of a sequence of probability statements, not a limit of a sequence of random variables $\bar{x}_n$.

Let us consider the application of the law of large numbers to the tossing of a fair coin discussed at the beginning of this section. Conceivably we can toss this coin any number of times we like. Let us define $x_i = 1$ if we get head on the ith flip, and $x_i = 0$ if we get tail on the ith flip, where $i = 1, 2, 3, \cdots$. Thus $x_1, x_2, x_3, \cdots$ is a sequence of independent Bernoulli trials each with probability $p = \frac{1}{2}$. The number of heads in n trials is $S_n = x_1 + x_2 + \cdots + x_n$. We can then define the sequence $\bar{x}_n = S_n/n$ of the proportion of heads that we observe, where $n = 1, 2, 3, \cdots$. For independent Bernoulli trials, we know that $E(\bar{x}) = p$ and var $(\bar{x}) = pq/n$, which in this case are $E(\bar{x}) = \frac{1}{2}$ and var $(\bar{x}) = 1/(4n)$. By the law of large numbers, $P(|\bar{x} - \frac{1}{2}| > \epsilon) \to 0$ as $n \to \infty$. That is, the probability that the proportion of tosses on which we observe heads differs from $\frac{1}{2}$ by more than ϵ tends to zero as the number of tosses increases indefinitely. However, from the law of large numbers we cannot claim that the proportion itself will necessarily get closer and closer to $\frac{1}{2}$ in any particular sequence. For example, one such sequence would be the one in which we get a head on every toss. However, the probability of this latter event happening is shrinking to zero as $n \to \infty$.

For the case in which the sequence $x_1, x_2, x_3, \cdots$ are independent Bernoulli trials, the law of large numbers was first proved by Jakob Bernoulli as a crowning achievement. [Jakob Bernoulli (1654–1705) was a preeminent Swiss mathematician and physicist and the author of the first treatise on probability, *Ars Conjectandi* (1713), which contained this theorem.] His proof depended on direct calculations with binomial coefficients without the benefit of such formulas as Stirling's formula.

For independent Bernoulli trials $x_1, x_2, x_3, \cdots$, with probability p and partial sums $S_n = x_1 + x_2 + \cdots + x_n$, we can state the law of large numbers in the following form reminiscent of the definition of an ordinary limit. For any $\epsilon > 0$, there exists an $n_0(\epsilon)$ such that for all $n \geq n_0(\epsilon)$ we have

$$P\left(\left|\frac{S_n}{n} - p\right| < \epsilon\right) > 1 - \epsilon.$$

If we interpret this as an assertion concerning the proximity of the theoretical probability p to the empirical relative frequency S_n/n, the double hedge (margin of error) implied by the two ϵ's seems inevitable. For, in any experiment, one can neither be 100 percent sure nor 100 percent accurate; otherwise the phenomenon would not be a random one. However, what might not be realized in the empirical world may sometimes be achieved in a purely mathematical sense. Such a possibility was uncovered by Borel who, in 1909, created a new chapter in probability theory by his discovery that in the case of independent Bernoulli trials the following result holds:

$$P\left(\lim_{n \to \infty} \frac{S_n}{n} = p\right) = 1.$$

This is known as a *strong law of large numbers*, which is an essential improvement on Bernoulli's "weak" law of large numbers, namely Theorem 5.7. The strong law asserts the existence of a limiting relative frequency S_n/n equal to the theoretical probability p, for all sample points except possibly a set of probability zero (but not necessarily an empty set). Thus the limit of S_n/n exists, but only for almost all sample points. The empirical theory of relative frequencies used in applied science is justifiable through the strong law of large numbers. The strong law of large numbers is the foundation of a mathematical theory of probability based on the concept of relative frequency. It makes better sense than the weak one and is indispensable for certain theoretical investigations.

EXERCISES

1. Prove the following analogue of Chebyshev's inequality where the absolute first moment is used in place of the second moment:

$$P(|x - \mu| > \alpha) \leq \frac{1}{\alpha} E(|x - \mu|).$$

2. Show that $\lim_{n \to \infty} P(|x_n| > \epsilon) = 0$ for every ϵ if and only if, given any ϵ, there exists $n_0(\epsilon)$ such that $P(|x_n| > \epsilon) < \epsilon$ for $n > n_0(\epsilon)$.

3. Suppose that we have a number n of vacuum tubes, all made by the same manufacturer and quite identical. The time to failure, x_i, for the ith tube is assumed to be an exponential random variable with parameter λ for each i. We put all n of these tubes on test until they fail (independently) and we define $\bar{x} = (x_1 + x_2 + \cdots + x_n)/n$. If we do not know λ, justify why it would be plausible to use $1/\bar{x}$ as a guess of its value.

ANSWER.

$$\lim_{n \to \infty} P\left[\left| \bar{x} - \frac{1}{\lambda} \right| > \epsilon \right] = 0.$$

Thus, if we tested a large number n of tubes and computed the arithmetic average $\bar{x}$ of their failure times, then the probability is high that $|\bar{x} - 1/\lambda|$ is small.

4. If you assume that $x_1, x_2, \cdots, x_n$ (n large) are independent, identically distributed Pascal-geometric random variables with parameter p, what function of $x_1, x_2, \cdots, x_n$ would you guess should be close in value to p?

ANSWER. $1/\bar{x}$.

5. (a) Define the function $g(p) = p(1 - p)$ and show that g has a maximum at $p = \frac{1}{2}$. Thus, $g(p) \leq \frac{1}{4}$ for all p.
   (b) If x is a binomial random variable with parameters n and p, then $\mu_x = np$, $\sigma_x^2 = npq$. Furthermore, if $y = x/n$, then $\mu_y = p$ and $\sigma_y^2 = pq/n$. Show that

$$P(|y - p| < \delta) \geq 1 - pq/n\delta^2.$$

   (c) For y as defined in (b) show that

$$P(|y - p| < \delta) \geq 1 - 1/4n\delta^2.$$

(d) Suppose we wanted to find a value of n such that

$$P(|y - p| < \delta) \geqslant 0.9,$$

where δ is some specified positive constant. Show that if $n \geqslant 2.5/\delta^2$, then the desired probability statement is satisfied.

CHANGE OF SEVERAL VARIABLES

If we know the joint density function of two random variables, then it is possible to obtain the density functions of other random variables which are functions of the two given random variables. For example, if the given random variables are X and Y, then we can find the density functions of the new random variables defined as $X + Y$, XY, and X/Y. The method used requires the techniques of change of variables in multiple integrals. This subject is covered in calculus books, so we will give the main results without proofs. We will restrict our discussion to integrals in two-dimensional space, as it is easy to visualize and it is sufficient for many applications.

First, let us review the 2×2 determinant. A 2×2 matrix C and its determinant $|C|$ are respectively given by

$$C = \begin{bmatrix} c_{11} & c_{12} \\ c_{21} & c_{22} \end{bmatrix}, \qquad \det C = |C| = \begin{vmatrix} c_{11} & c_{12} \\ c_{21} & c_{22} \end{vmatrix} = c_{11}c_{22} - c_{21}c_{12}.$$

It is easy to see that this determinant is equal to zero only if the two rows of the matrix are proportional to each other, or two columns are proportional to each other.

Next, let us consider a multiple integral over some portion A of the (x,y) plane:

$$\int_A \int f(x,y) \, dx \, dy.$$

We want to change the (x,y) variables to new variables (u,v). This change of variables may be written in functional form as

$$u = u(x,y), \qquad v = v(x,y).$$

We suppose that this transformation from the (x,y) plane to the (u,v) plane satisfies the following four conditions:

1. If B is the image of A in the (u,v) plane, then the transformation is a one-to-one transformation from A to B.

2. The transformation has a well-defined inverse

$$x = x(u,v), \qquad y = y(u,v).$$

3. The Jacobian, defined by the determinant

$$\frac{\partial(u,v)}{\partial(x,y)} = \begin{vmatrix} \dfrac{\partial u}{\partial x} & \dfrac{\partial u}{\partial y} \\ \dfrac{\partial v}{\partial x} & \dfrac{\partial v}{\partial y} \end{vmatrix},$$

expressed as a function of x and y is continuous in A and vanishes nowhere there.

4. The Jacobian, defined by the determinant

$$\frac{\partial(x,y)}{\partial(u,v)} = \begin{vmatrix} \dfrac{\partial x}{\partial u} & \dfrac{\partial x}{\partial v} \\ \dfrac{\partial y}{\partial u} & \dfrac{\partial y}{\partial v} \end{vmatrix},$$

expressed as a function of u and v is continuous in B.

If these four conditions are satisfied, then the evaluation of the multiple integral in terms of the new variables (u,v) is given by

$$\int_A \int f(x,y)\, dx\, dy = \int_B \int f[x(u,v),\, y(u,v)] \left| \frac{\partial(x,y)}{\partial(u,v)} \right| du\, dv.$$

Notice that the absolute value of the Jacobian appears here as indicated by the vertical bars. In effect, two sets of vertical bars appear here, one set for the determinant representing the Jacobian and the other set enclosing that Jacobian and specifying its absolute value. The double meaning of vertical bars (determinant and absolute value) is unfortunate here where both uses are required in the same equation. In this equation, the region A on the left is replaced by the region B on the right. In $f(x,y)$, the x and y are replaced by their values in terms of u and v. Finally the differential element $dx\, dy$ is replaced by

$$\left| \frac{\partial(x,y)}{\partial(u,v)} \right| du\, dv.$$

This new differential element is the absolute value of the Jacobian times $du\, dv$ and thus is necessarily positive.

The two Jacobians are reciprocals; that is,

$$\frac{\partial(u,v)}{\partial(x,y)} \frac{\partial(x,y)}{\partial(u,v)} = 1.$$

This property means that we can always compute whichever Jacobian is the simpler, for the other Jacobian can be immediately obtained as the reciprocal of the one computed.

EXAMPLE 5–1. (Transformation to Polar Coordinates). Let the transformation be $u = (x^2 + y^2)^{1/2}$, $v = \tan^{-1}(y/x)$. The inverse transformation is $x = u \cos v$, $y = u \sin v$. Since

$$\frac{\partial(x,y)}{\partial(u,v)} = \begin{vmatrix} \cos v & -u \sin v \\ \sin v & u \cos v \end{vmatrix} = u,$$

it follows that

$$\int_A \int f(x,y) \, dx \, dy = \int_B \int f(u \cos v, u \sin v) u \, du \, dv.$$

Let us now find a general expression for the density function of a product XY of independent nonnegative random variables X and Y having density functions $f(x)$ and $g(y)$ respectively. First, we evaluate the cumulative distribution function of XY, namely,

$$P(XY \leq t) = \iint\limits_{0 \leq xy \leq t} f(x)g(y) \, dx \, dy.$$

This integration is restricted to the first quadrant, since by hypothesis X and Y are nonnegative. Let us make use of the transformation $u = xy$, $v = y$. The inverse of this transformation is $x = u/v$, $y = v$. The Jacobian is

$$\frac{\partial(x,y)}{\partial(u,v)} = \begin{vmatrix} \dfrac{1}{v} & \dfrac{-u}{v^2} \\ 0 & 1 \end{vmatrix} = \frac{1}{v}.$$

The region A is the set of all (x,y) points in the first quadrant such that $xy \leq t$. The region B is thus the set of all (u,v) points where $u \leq t$ and $0 \leq v < \infty$. See Figure 5.13. Because the Jacobian $1/v$ is not continuous for $v = 0$, we must exclude the line $v = 0$ or, equivalently, $y = 0$. However, the probability for such a line is zero, so this exclusion does not matter. Hence for $t > 0$ we have

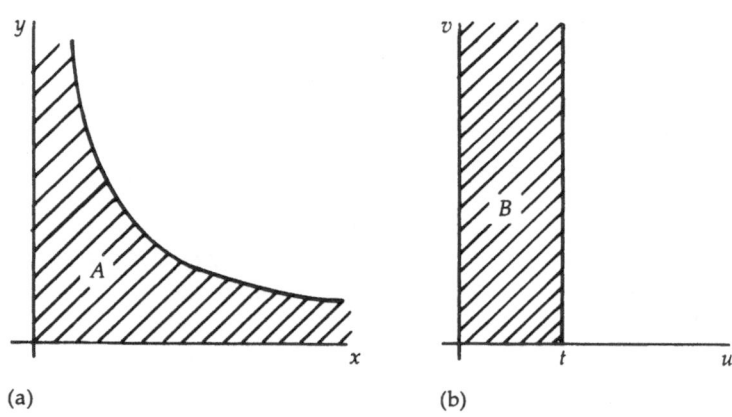

FIGURE 5.13 (a) The set A where $0 \leqslant xy \leqslant t$. (b) The set B where $0 \leqslant u \leqslant t$, $0 \leqslant v < \infty$.

$$P(XY \leqslant t) = P(0 < XY \leqslant t)$$
$$= \int_0^t du \int_0^\infty f\left(\frac{u}{v}\right)g(v)\left(\frac{1}{v}\right) dv = \int_0^t h(u)\, du,$$

where $h(u)$ is the density function of $U = XY$. Thus

$$h(u) = \int_0^\infty f\left(\frac{u}{v}\right)g(v)\left(\frac{1}{v}\right) dv \qquad u > 0.$$

This density function vanishes for negative u.

EXAMPLE 5-2. Suppose X and Y are independent random variables with uniform densities $f(x)$ and $g(y)$ respectively, and each density function is equal to one in the range $(0,1)$ and equal to zero otherwise. Let us find the density function $h(u)$ of the product $U = XY$. Since by hypothesis $f(x) = 1$ for $0 \leqslant x \leqslant 1$ and $f(x) = 0$ otherwise, it follows that $f(u/v) = 1$ for $0 \leqslant u \leqslant v$ and zero otherwise. Since $g(v) = 1$ for $0 \leqslant v \leqslant 1$ and $g(v) = 0$ otherwise, it follows that

$$f(u/v)g(v)(1/v) = \begin{cases} 1/v & \text{for } 0 \leqslant u \leqslant v \leqslant 1, \\ 0 & \text{otherwise.} \end{cases}$$

Hence

$$h(u) = \int_u^1 \left(\frac{1}{v}\right) dv = -\ln u \qquad \text{for } 0 \leqslant u \leqslant 1$$

is the density function of the product $U = XY$.

Following the same method as before, let us now obtain an expression for the density function of the ratio of two independent nonnegative random variables X and Y. Let their density functions be $f(x)$ and $g(y)$, each of which vanishes for negative values of the argument. We first find the cumulative distribution function of $U = X/Y$. The cdf is

$$P\left(\frac{X}{Y} \leqslant t\right) = \iint\limits_{0 \leqslant (x/y) \leqslant t} f(x)g(y) \, dx \, dy.$$

This integral is restricted to the first quadrant because $f(x)g(y)$ is zero elsewhere. We use the transformation $u = x/y$ and $v = u$, whose inverse is $x = uv$ and $y = v$. The Jacobian is

$$\frac{\partial(x,y)}{\partial(u,v)} = \begin{vmatrix} v & u \\ 0 & 1 \end{vmatrix} = v.$$

Hence for $t > 0$, we have

$$P(X/Y \leqslant t) = P(0 < X/Y \leqslant t)$$

$$= \iint\limits_{\substack{0 < u \leqslant t \\ 0 < v < \infty}} f(uv)g(v)v \, du \, dv$$

$$= \int_0^t du \int_0^\infty f(uv)g(v)v \, dv = \int_0^t h(u) \, du.$$

Hence the density function of $U = X/Y$ is

$$h(u) = \int_0^\infty f(uv)g(v)v \, dv \qquad \text{for } u > 0.$$

This density function vanishes for negative u.

EXAMPLE 5-3. Let X and Y be independent and uniform with respective densities $f(x)$ and $g(y)$, each equal to one on $(0,1)$ and vanishing elsewhere. Let us now find the density of the quotient $U = X/Y$. We have

$$f(uv)g(v)v = \begin{cases} v & \text{if } 0 \leqslant v \leqslant 1,\, 0 \leqslant u \leqslant 1, \\[2mm] v & \text{if } 0 \leqslant v \leqslant \dfrac{1}{u},\, u > 1, \\[2mm] 0 & \text{otherwise.} \end{cases}$$

Hence

$$h(u) = \begin{cases} \displaystyle\int_0^1 v\, dv = \frac{1}{2} & \text{if } 0 \leqslant u \leqslant 1, \\[4mm] \displaystyle\int_0^{1/u} v\, dv = \frac{1}{2u^2} & \text{if } u > 1, \\[4mm] 0 & \text{otherwise.} \end{cases}$$

EXAMPLE 5-4. Let X and Y be independent and exponential, with respective densities

$$f(x) = e^{-x}, \quad x > 0; \qquad g(y) = e^{-y}, \quad y > 0.$$

For $u > 0$, the density of the ratio $U = X/Y$ is

$$h(u) = \int_0^\infty e^{-uv} e^{-v} v\, dv = \int_0^\infty e^{-v(1+u)} v\, dv.$$

If we let $z = v(1 + u)$, this integral becomes

$$h(u) = \frac{1}{(1+u)^2} \int_0^\infty z e^{-z}\, dz = \frac{1}{(1+u)^2}, \qquad u \geqslant 0.$$

Thus the ratio X/Y has a Cauchy distribution.

In the foregoing part of this section we have found general expressions for the density functions of products and ratios of two nonnegative independent random variables. Let us now find the density function for the sum of two independent random variables where we do not require the random variables to be nonnegative. As usual, we first find the cumulative distribution function of $X + Y$. It is

$$P(X + Y \leqslant t) = \iint\limits_{x+y \leqslant t} f(x)g(y)\, dx\, dy.$$

We make the change of variables $u = x + y$, $v = y$ which has the inverse $x = u - v$, $y = v$. The Jacobian is

$$\frac{\partial(x,y)}{\partial(u,v)} = \begin{vmatrix} 1 & -1 \\ 0 & 1 \end{vmatrix} = 1.$$

Hence

$$P(X + Y \leq t) = \iint\limits_{\substack{u \leq t \\ -\infty < v < \infty}} f(u - v)g(v) \, du \, dv$$

$$= \int_{-\infty}^{t} du \int_{-\infty}^{\infty} f(u - v)g(v) \, dv = \int_{-\infty}^{t} h(u) \, du.$$

Therefore the density function of the sum $U = X + Y$ is

$$h(u) = \int_{-\infty}^{\infty} f(u - v)g(v) \, dv.$$

If we interchange the roles of X and Y, as we may do, because $X + Y = Y + X$, we see that $h(u)$ may also be written as

$$h(u) = \int_{-\infty}^{\infty} g(u - v)f(v) \, dv.$$

The operation described by either of the above two integrals is an extremely important one. The function h is called the convolution of the functions f and g, and either of the above two integrals is called the *convolutional integral*. An easy way to remember the convolutional integral is to observe that the functions f and g are integrated out over the straight line in the (x,y) plane where $x + y = u$.

EXAMPLE 5-5. Let X and Y be independent and exponential, with $f(x) = e^{-x}$ for $x \geq 0$ and $f(y) = e^{-y}$ for $y \geq 0$. The density function of $U = X + Y$ is

$$h(u) = \int_{-\infty}^{\infty} f(u - v)g(v) \, dv = \int_{0}^{u} e^{-(u-v)}e^{-v} \, dv = ue^{-u} \qquad \text{for } u \geq 0.$$

EXERCISES

1. Let X and Y be independent, where X is exponential with density $f(x) = e^{-x}$ for $x \geq 0$ and Y is uniform with density $g(y) = 1$ for $0 \leq y \leq 1$. What are the density functions of (a) XY and (b) X/Y?

ANSWER.

$$\text{(a)} \quad h(u) = \int_0^1 e^{-u/v} \left(\frac{1}{v}\right) dv$$

$$= \int_u^\infty e^{-v} \left(\frac{1}{v}\right) dv, \qquad u \geq 0$$

$$\text{(b)} \quad h(u) = \int_0^1 e^{-uv} v \, dv, \qquad u \geq 0.$$

2. Suppose that X and Y are independent with densities $f(x) = ax^{a-1}$ for $0 < x \leq 1$ and $g(y) = ay^{a-1}$ for $0 < y \leq 1$, where a > 0. What is the density of XY?

ANSWER. $h(u) = a^2 u^{a-1}(-\ln u)$ for $0 < u \leq 1$.

3. Suppose that X and Y are independent with densities $f(x) = ax^{a-1}$, $0 < x \leq 1$ and $g(y) = a^2 y^{a-1}(-\ln y)$, $0 < y \leq 1$. What is the density of XY?

ANSWER. $h(u) = (a^3/2)u^{a-1} (\ln u)^2$ for $0 \leq u \leq 1$.

4. Let X and Y be independent and uniform, with densities $f(x) = 1$ for $0 \leq x \leq 1$ and $g(y) = 1$ for $0 \leq y \leq 1$. What is the density function of $X + Y$.

ANSWER.

$$h(u) = \begin{cases} u & \text{for } 0 \leq u \leq 1, \\ 2 - u & \text{for } 1 \leq u \leq 2. \end{cases}$$

5. Suppose X and Y are independent and exponential with $f(x) = e^{-x}$ for $x \geq 0$ and $g(y) = e^{-y}$ for $y \geq 0$. Show that the joint density function of $U = X/Y$ and $V = X + Y$ is

$$h(u,v) = (1 + u)^{-2}(ve^{-v}) \qquad \text{for } u \geq 0, v \geq 0,$$

and therefore conclude that U and V are independent.

HINT. Evaluate

$$\iint\limits_{\substack{x/y \leq u \\ x+y \leq v}} e^{-(x+y)} \, dx \, dy$$

by the change of variables $u = x/y$, $v = x + y$. In Example 5–4 we found the marginal density of X/Y and in Example 5–5 we found the marginal density of $X + Y$.

6. Let X and Y be independent where $f(x) = xe^{-x}$ for $x \geqslant 0$ and $g(y) = ye^{-y}$ for $y \geqslant 0$. What is the density of $X + Y$?
   ANSWER. $h(u) = (x^3/3!)e^{-x}$ for $x \geqslant 0$.

7. Let X and Y be independent and exponential, with $f(x) = e^{-x}$ for $x \geqslant 0$ and $g(y) = e^{-y}$ for $y \geqslant 0$. Use the convolution integral to show that the density function of $X - Y$ is $(\frac{1}{2})e^{-|u|}$ for $-\infty < u < \infty$.
   HINT. $W = -Y$ has the density e^w for $w \leqslant 0$.

8. If X is uniformly distributed over $(0,1)$ and Y is exponentially distributed with parameter $\lambda = 1$, find the cumulative distribution function of (a) $Z = X + Y$, and (b) $Z = X/Y$. Assume X and Y are independent.
   HINT. Do not use the Jacobian. In (a), recognize that $P(X + Y \leqslant t)$ is the cdf, so simply evaluate the integral in the equation at the bottom of page 184. In (b), evaluate the integral for $P(X/Y \leqslant t)$ in the second equation on page 183.
   ANSWERS. (a) $z - 1 + e^{-z}$ for $0 \leqslant z \leqslant 1$ and $1 - e^{-z}(e - 1)$ for $z > 1$.
   (b) $z(1 - e^{-1/z})$ for $z \geqslant 0$.

9. If X_1 and X_2 are independent exponential random variables with respective parameters λ_1 and λ_2,
   (a) find the cumulative distribution function of $Z = X_1/X_2$;
   (b) compute $P\{X_1 < X_2\}$.
   HINT. (a) See hint for Exercise 8(b) above.
   ANSWER. (a) $\lambda_1 z/(\lambda_1 z + \lambda_2)$. (b) $\lambda_1/(\lambda_1 + \lambda_2)$.

10. If X and Y have joint density function

$$f(x,y) = \frac{1}{x^2 y^2} \qquad \text{for } x \geqslant 1, y \geqslant 1,$$

   (a) compute the joint density function of $U = XY$, $V = X/Y$;
   (b) what are the marginal densities?
   ANSWER. (a) $f(u,v) = 1/(2u^2 v)$ for $1/u \leqslant v \leqslant u$, $u \geqslant 1$. (b) $f_1(u) = (\ln u)/u^2$ for $u \geqslant 1$, $f_2(v) = \frac{1}{2}$ for $0 \leqslant v \leqslant 1$ and $1/(2v^2)$ for $v > 1$.

11. If X, Y, and Z are independent random variables having identical density functions $f(x) = e^{-x}$, $0 < x < \infty$, derive the joint density of $U = X + Y$, $V = X + Z$, $W = Y + Z$.
   ANSWER. $f(u,v,w) = (\frac{1}{2}) \exp [-(\frac{1}{2})(u + v + w)]$ for $u + v - w > 0$, $u + w - v > 0$, and $w + v - u > 0$.

6

Generating Functions

Simplicity, simplicity, simplicity. I say, let your affairs be as two or three, and not a hundred or a thousand; instead of a million count half a dozen.

HENRY DAVID THOREAU (1817–1862)

Probability Generating Function

Gamma Functions and Beta Functions

Moment Generating Function

Applications of the Moment
Generating Function

PROBABILITY GENERATING
FUNCTION

In this chapter we introduce the concept of a transform of a probability distribution and give applications. One especially important application is in the analysis of sums of independent random variables.

One useful type of transform is known as the generating function. We will introduce this transform only in the case of a probability mass function $f(x)$ for a discrete random variable that can take on only nonnegative integer values. That is, we restrict ourselves to the case of $f(x)$ where $x = 0, 1, 2, 3, \cdots$. The *probability generating function* $\Phi(t)$ is defined as

$$\Phi(t) = \sum_{x=0}^{\infty} t^x f(x).$$

If we wish to indicate the random variable, we may write $\Phi(t)$ more explictly as $\Phi(t;x)$. [The probability generating function is closely related to the z-transform, which is well known in digital signal processing. The z-transform of $f(x)$ is defined as

$$\sum_{x=0}^{\infty} z^{-x} f(x).$$

Thus we see the z-transform is $\Phi(z^{-1})$; that is, by replacing t by z^{-1} in the generating function, we obtain the z-transform.]

We see from its definition that the probability generating function is the expected value of t^x; that is,

$$\Phi(t) = E(t^x).$$

The probability generating function can be shown to be a well-behaved function at least in the region $|t| < 1$. From the definition

$$\Phi(t) = f(0) + tf(1) + t^2 f(2) + t^3 f(3) + \cdots,$$

we see that we can determine the individual terms of the probability mass function $f(x)$ by the formula

$$f(x) = \frac{1}{x!} \left[\frac{d^x}{dt^x} \Phi(t) \right]_{t=0} \qquad \text{where } x = 0, 1, 2, \cdots.$$

Let x and y be nonnegative independent integral-valued random

190

variables with probability mass functions $f(x)$ for $x = 0, 1, 2, \cdots$, and $g(y)$ for $y = 0, 1, 2, \cdots$. The joint event (x,y) has probability $f(x)g(y)$. The sum $s = x + y$ is a new random variable, and the event s is the union of the mutually exclusive events:

$$(x{=}0,\ y{=}s),\ (x{=}1,\ y{=}s{-}1),\ (x{=}2,\ y{=}s{-}2),\ \cdots,\ (x{=}s,\ y{=}0).$$

Therefore the probability mass function $h(s)$ of the sum is given by

$$h(s) = f(0)g(s) + f(1)g(s - 1) + f(2)g(s - 2) + \cdots + f(s)g(0). \quad (6.1)$$

The operation (6.1), leading from the two sequences $f(x)$ and $g(x)$ to a new sequence $h(s)$, occurs so frequently that it is convenient to introduce a special name for it.

DEFINITION. Let $f(x)$ and $g(x)$ for $x = 0, 1, 2, \cdots$ be any two sequences (not necessarily probability mass functions). The new sequence $h(s)$ for $s = 0, 1, 2, \cdots$ defined by (6.1) is called the convolution of $f(x)$ and $g(x)$ and is denoted by $h = f * g$.

In summation notation, the convolution (6.1) is

$$h(s) = \sum_{x=0}^{s} f(x)g(s - x).$$

As an example, suppose $f(x) = g(x) = 1$ for $x = 0, 1, 2, \cdots$. Then their convolution is $h(s) = s + 1$ for $s = 0, 1, 2, \cdots$. As another example, consider $f(x) = x$, $g(x) = 1$. Then $h(s) = 1 + 2 + \cdots + s$. Also if $f(0) = f(1) = \frac{1}{2}$, $f(x) = 0$ for $x = 2, 3, \cdots$, then $h(s) = [g(s) + g(s - 1)]/2$. The sequences $f(x)$ and $g(x)$ have probability generating functions

$$\Phi(t;x) = \sum_{x=0}^{\infty} f(x)t^x \quad \text{and} \quad \Phi(t;y) = \sum_{y=0}^{\infty} g(y)t^y.$$

The product $\Phi(t;x)\Phi(t;y)$ can be obtained by termwise multiplication of the power series for $\Phi(t;x)$ and $\Phi(t;y)$. Collecting terms with equal powers of t, we find that the coefficient $h(s)$ of t^s in the expansion of $\Phi(t;x)\ \Phi(t;y)$ is given by Equation (6.1). We thus have:

THEOREM 6.1. If $f(x)$ for $x = 0, 1, 2, \cdots$ and $g(y)$ for $y = 0, 1, 2, \cdots$ are sequences with generating functions $\Phi(t;x)$ and $\Phi(t;y)$, and $h(s)$ is the convolution of the two sequences, then the generating function

$$\Phi(t;s) = \sum_{s=0}^{\infty} h(s)t^s$$

is the product

$$\Phi(t;s) = \Phi(t;x)\Phi(t;y).$$

If x and y are nonnegative, integral-valued, mutually independent random variables with probability generating functions $\Phi(t;x)$ and $\Phi(t;y)$ respectively, then their sum $x + y$ has the probability generating function $\Phi(t;x)\Phi(t;y)$.

In the study of sums of independent random variables x_n, the special case where the x_n has a common distribution is of particular interest. Let $f(x)$ for $x = 0, 1, 2, \cdots$ be the common probability mass function of the x_n, and let $\Phi(t;x)$ be the probability generating function of $f(x)$. Then the probability mass function of the sum $S_n = x_1 + x_2 + \cdots + x_n$ will be denoted by $[f(x)]^{n^*}$. Thus

$$[f(x)]^{2^*} = f(x) * f(x),$$
$$[f(x)]^{3^*} = [f(x)]^{2^*} * f(x),$$

and so on. In words, $[f(x)]^{n^*}$ is the probability mass function whose generating function is the nth power of $\Phi(t;x)$; that is, $[\Phi(t;x)]^n$.

PROBLEM. Find the probability generating function of the probability mass function (with n a positive integer and with $0 < p < 1$ and $q = 1 - p$):

$$b(x;n,p) = \binom{n}{x} p^x q^{n-x} \qquad \text{for } x = 0, 1, 2, \cdots, n.$$

This probability mass function is that of the binomial distribution.

SOLUTION.

$$\Phi(t) = \sum_{x=0}^{n} \binom{n}{x} (pt)^x q^{n-x} = (q + pt)^n.$$

The fact that this probability generating function is the nth power of $q + pt$ shows that $b(x;n,p)$ is the distribution of a sum $S_n = x_1 + x_2 + \cdots + x_n$ of n independent random variables, each with the common generating function $q + pt$. Thus each variable x_i assumes the value 0 with probability q and the value 1 with probability p. Thus

$$b(x;n,p) = [b(x;1,p)]^{n^*}.$$

The probability generating function may be considered as the *factorial moment generating function*. The reason for this terminology is due to the fact that

$$\Phi(t) = \sum_{x=0}^{\infty} t^x f(x) = E[t^x],$$

$$\frac{d\Phi}{dt} = \sum_{x=1}^{\infty} xt^{x-1}f(x) = E[xt^{x-1}],$$

$$\frac{d^2\Phi}{dt^2} = \sum_{x=2}^{\infty} x(x-1)t^{x-2}f(x) = E[x(x-1)t^{x-2}],$$

$$\frac{d^k\Phi}{dt^k} = \sum_{x=k}^{\infty} x(x-1)\cdots(x-k+1)t^{x-k}f(x) = E[x_{(k)}t^{x-k}].$$

Here we have used the factorial symbol defined as

$$x_{(k)} = x(x-1)(x-2)\cdots(x-k+1).$$

If we let $t = 1$ in the above set of equations we obtain the quantities $E[1] = 1$, $E[x]$, $E[x(x-1)]$, $\cdots$, $E[x_{(k)}]$. These quantities are called the factorial moments of the random variable x, because of their factorial-like structure. It can be shown that knowledge of the factorial moments $E[x_{(k)}]$ for $k = 1, 2, \cdots$ is equivalent to knowledge of the moments $E[x^k]$ for $k = 1, 2, \cdots$, and vice versa.

PROBLEM. Suppose we flip a fair coin until we get a head, and let x be the number of flips required to conclude the experiment. Find the first two moments.

SOLUTION. The probability mass function is $f(x) = 2^{-x}$ for $x = 1$, 2, 3, $\cdots$ and thus the probability generating function is

$$\Phi(t) = E[t^x] = \sum_{x=1}^{\infty} t^x 2^{-x} = \sum_{x=0}^{\infty} \left(\frac{t}{2}\right)^x - 1$$

$$= \frac{1}{1 - t/2} - 1 = \frac{t}{2 - t} \qquad \text{for } |t| < 2.$$

Thus

$$E(x) = \frac{d\Phi}{dt}\bigg]_{t=1} = \frac{2}{(2-t)^2}\bigg]_{t=1} = 2,$$

$$E[x(x-1)] = \frac{d^2\Phi}{dt^2}\bigg]_{t=1} = \frac{4}{(2-t)^3}\bigg]_{t=1} = 4.$$

But

$$E[x(x - 1)] = E[x^2] - E[x],$$

so the second moment is

$$E[x^2] = E[x(x - 1)] + E[x] = 4 + 2 = 6.$$

Probability generating functions were used by the English mathematician Abraham DeMoivre (1667–1754) and by the French mathematician Pierre Simon Laplace (1749–1827). The term "probability generating function" should not be confused with "moment generating function." The probability generating function is $\Phi(t) = E(t^x)$, whereas the moment generating function is $E(e^{\theta x}) = \Phi(e^\theta)$. The moment generating function is related to the Laplace transform, as is discussed in the section on pages 207–212.

EXERCISES

1. Which of the following are valid probability generating functions of a probability mass function for a discrete random variable that can take on only nonnegative integer values?
   (a) $-2 + 2t + t^2$.
   (b) $2 - t$.
   (c) $(2 - t)^{-1}$.

2. Show that convolution is an associative and commutative operation.

3. The quantity $[f(x)]^{1^*}$ is the same as $f(x)$. Show that $[f(x)]^{0^*}$ can be defined as the sequence whose probability generating function is

$$[\Phi(t)]^0 = 1,$$

that is, as the sequence $(1, 0, 0, 0, \cdots)$.

4. Show that the probability generating function for the binomial distribution is $(q + pt)^n$. Show that the multiplicative property

$$(q + pt)^m(q + pt)^n = (q + pt)^{m+n}$$

gives the convolutional relation for the binomial distribution

$$b(x;m,p) * b(x;n,p) = b(x;m+n,p).$$

5. The probability mass function of the Poisson distribution is

$$p(x;\mu) = e^{-\mu}\mu^x/x!,$$

where $\mu > 0$ and $x = 0, 1, 2, \cdots$. Show that the probability generating function is

$$\Phi(t) = e^{-\mu + \mu t}.$$

Also show that the convolutional relation for the Poisson distribution is

$$p(x;\mu_1) * p(x;\mu_2) = p(x;\mu_1 + \mu_2).$$

6. (a) Show that $\Phi(t) = 1$ is the probability generating function of a random variable which is 0 with probability 1. If $P(X = k) = 1$ and $P(X \neq k) = 0$ where k is a nonnegative integer, then the probability generating function is $\Phi(t) = t^k$.

(b) If the probability generating function of X is $\Phi(t)$, show that the generating function of $X + k$ is $t^k\Phi(t)$, where k is a positive integer.

7. (a) A die is thrown repeatedly until a 2 is obtained. Let X be the number of throws required. Show that the probability generating function of X is $t/(6 - 5t)$.

(b) A die is thrown repeatedly until either a 2 or a 3 is obtained. Show that the probability generating function of the number of throws required is $t/(3 - 2t)$.

8. (a) Suppose n balls are randomly distributed among m cells. Let X be the number of unoccupied cells. Show that the probability generating function of X is

$$\sum_{i=0}^{m} \binom{m}{i} \left(\frac{m - i}{m}\right)^n (t - 1)^i.$$

(b) Show that the probability distribution of X is given by

$$P(X = k) = \sum_{i=0}^{m-k} (-1)^i \binom{k+i}{k}\binom{m}{k+i}\left(\frac{m - k - i}{m}\right)^n.$$

9. (a) If Φ is a probability generating function, show that $|\Phi(t)| < 1$ if $|t| < 1$, assuming $P(X = 0) \neq 1$.

(b) Show that $1/[2 - \Phi(t)]$ is a probability generating function.

HINT. Expand $(1/2)[1 - \Phi(t)/2]^{-1}$ as a geometric series in $\Phi(t)/2$.

10. Suppose A throws a die repeatedly until a 2 appears. Independently, B throws a die repeatedly until a 2 or a 3 appears. Let X be the number of throws required by both A and B together. Show that

$$P(X = k) = (^5/_{18})(^5/_6)^{k-2} - (^2/_9)(^2/_3)^{k-2} \qquad \text{for } k = 2, 3, \cdots.$$

HINT. The probability generating function is $[t/(6 - 5t)][t/(3 - 2t)]$. Express this as

$$\frac{5}{18}\frac{t^2}{1 - \frac{5}{6}t} - \frac{2}{9}\frac{t^2}{1 - \frac{2}{3}t}$$

and expand each part as a geometric series.

11. Show that the only way that Φ and $1/\Phi$ can both be probability generating functions is for $\Phi(t)$ to be identically 1.

    HINT. Condition $\Phi(1/\Phi) = 1$ means it is possible for a sum $X + Y$ of two independent, nonnegative, integer-valued random variables to equal 0. Show that it can only happen if $P(X = 0) = P(Y = 0) = 1$.

12. An indicator random variable is defined as a random variable that can take on only two values, either 0 or 1. Suppose that $X_1, \cdots, X_n$ are independent indicator random variables. Define $p_i = P(X_i = 1)$, $i = 1, \cdots, n$. Show that the probability generating function of $X_1 + \cdots + X_n$ is $[p_1 t + (1 - p_1)] \cdots [p_n t + (1 - p_n)]$.

13. (a) Balls are successively distributed among three cells. Let Z be the number of balls that must be distributed to occupy all three cells. Show that the probability generating function of Z is

$$t^3 \frac{\frac{1}{3}}{1 - \frac{2}{3}t}\frac{\frac{2}{3}}{1 - \frac{1}{3}t}.$$

    (b) Notice that each of

$$\Phi_1(t) = \frac{\frac{1}{3}}{[1 - \frac{2}{3}t]} \quad \text{and} \quad \Phi_2(t) = \frac{\frac{2}{3}}{[1 - \frac{1}{3}t]}$$

    is a probability generating function. Conclude that Z is distributed like $X + Y + 3$, where X and Y are independent and have generating functions Φ_1 and Φ_2, respectively.

14. (a) An honest die is thrown six times. Let X be the total number of points. Verify the following entries in the distribution of X:

k	6	7	8	9	10
$P(X = k)$	$1/6^6$	$6/6^6$	$21/6^6$	$56/6^6$	$126/6^6$

    (b) An honest die is thrown n times. Let X be the total number of points. Verify the following entries in the distribution of X.

k	n	$n + 1$	$n + 2$	$n + 3$	$n + 4$
$P(X = k)$	$1/6^n$	$n/6^n$	$\binom{n+1}{2}/6^n$	$\binom{n+2}{3}/6^n$	$\binom{n+3}{4}6^n$

    However this pattern does not continue for all k.

HINT. The probability generating function for X is

$$\frac{(t + t^2 + \cdots + t^6)^n}{6^n} = \frac{t^n (1 - t^6)^n}{6^n (1 - t)^n}$$

$$= \frac{t^n}{6^n} (1 - nt^6 + \cdots) \left[1 + nt + \binom{n+1}{2} t^2 \right.$$

$$\left. + \binom{n+2}{3} t^3 + \cdots \right].$$

15. A box contains four balls, numbered 0, 1, 1, and 2. Suppose n balls are successively drawn, with replacement between drawings. Let X be the sum of the n numbers drawn. Show that

$$P(X = k) = \binom{2n}{k} \frac{1}{2^{2n}} \qquad \text{for } k = 0, 1, \cdots, 2n.$$

HINT. The probability generating function for any one drawing is $(1 + 2t + t^2)/4 = [(1 + t)/2]^2$. The probability generating function of X is $[(1 + t)/2]^{2n}$.

GAMMA FUNCTIONS AND
BETA FUNCTIONS

The *gamma function* is defined as the definite integral

$$\Gamma(n) = \int_0^\infty x^{n-1} e^{-x} \, dx \qquad \text{for } n > 0.$$

The value of the integral is dependent upon the value of n and represents a finite quantity for values of n greater than zero. The gamma function has the other common form

$$\Gamma(n) = 2 \int_0^\infty y^{2n-1} e^{-y^2} \, dy \qquad \text{for } n > 0,$$

which can be derived from the first by making the change of variables $x = y^2$. In the above forms it makes no difference of course whether the dummy variable is called x or y, or by any other symbol inside the integral sign.

Unless n is a positive integer, the value of $\Gamma(n)$ must be obtained by series expansion or some type of numerical integration. For this reason, tables of the gamma function have been worked out for various values of n. Commonly, however, the tables are confined to values of

n between 1 and 2, inasmuch as the function can be determined for other values of n by means of a recurrence formula, which we shall now derive. If the first form above is integrated by parts, we have

$$\Gamma(n) = \int_0^\infty x^{n-1}e^{-x}\,dx = [-x^{n-1}e^{-x}]_0^\infty + (n-1)\int_0^\infty x^{n-2}e^{-x}\,dx$$

$$= (n-1)\Gamma(n-1),$$

since the integrated part vanishes at both limits. Therefore if the gamma function is tabulated for values of n between two consecutive integers, say 1 and 2, the function can be computed for values outside of the table by repeated use of the recurrence formula.

On the other hand, if n is a positive integer, the gamma function of n reduces to the ordinary factorial of $n - 1$ as defined in algebra. To begin with, we determine $\Gamma(1)$ from the defining integral, that is

$$\Gamma(1) = \int_0^\infty e^{-x}\,dx = 1.$$

Then from the recurrence formula we obtain $\Gamma(2)$, $\Gamma(3)$, $\cdots$, $\Gamma(n)$ successively as follows:

$$\Gamma(2) = (1)\Gamma(1) = 1,$$
$$\Gamma(3) = (2)\Gamma(2) = 2 \cdot 1,$$
$$\Gamma(4) = (3)\Gamma(3) = 3 \cdot 2 \cdot 1 \quad \text{etc.}$$

and thus

$$\Gamma(n) = (n-1)(n-2)\cdots3 \cdot 2 \cdot 1 = (n-1)! \qquad \text{for } n = 1, 2, 3, \cdots,$$

where by definition $0! = 1$.

For any positive value of n, we may interpret the gamma function as the area under the gamma curve $y = x^{n-1}e^{-x}$ from $x = 0$ to $x = \infty$. Plots of these gamma curves for various n's are shown in Figure 6.1. The curve in Figure 6.2 shows the areas under the gamma curves and thus is the plot of $\Gamma(n)$ as a function of n.

An important point on Figure 6.2 corresponds to $n = \frac{1}{2}$. To find its value, consider

$$\Gamma\left(\frac{1}{2}\right) = 2\int_0^\infty e^{-x^2}\,dx \quad \text{or} \quad \Gamma\left(\frac{1}{2}\right) = 2\int_0^\infty e^{-y^2}\,dy,$$

where the second form for the Gamma function is being utilized. By

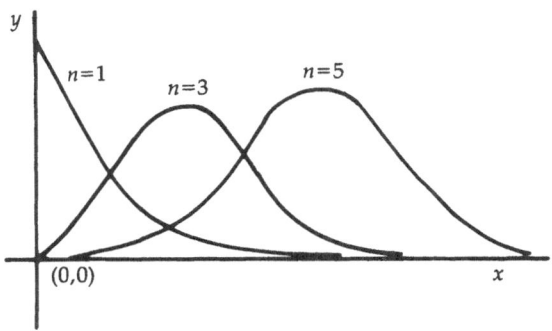

FIGURE 6.1 Gamma curves for $n = 1$, 3, and 5.

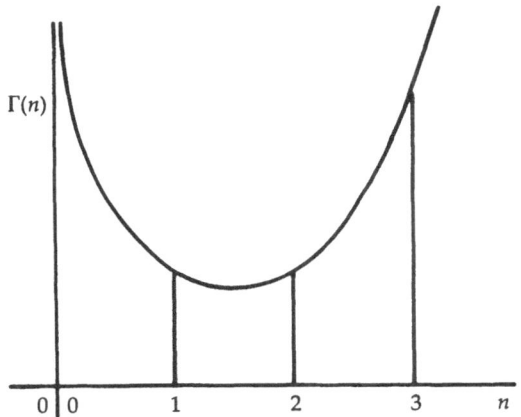

FIGURE 6.2 The gamma function plotted against n.

multiplying these two values of $\Gamma(\frac{1}{2})$ together, we obtain

$$\left[\Gamma\left(\frac{1}{2}\right)\right]^2 = \left[2\int_0^\infty e^{-x^2}\,dx\right]\left[2\int_0^\infty e^{-y^2}\,dy\right]$$

$$= 4\int_0^\infty dx\int_0^\infty e^{-(x^2+y^2)}\,dy.$$

This last expression may be considered as a double integral to be evaluated over the first quadrant in the x,y plane. The above procedure is permitted since both sets of limits are constant and both integrals as well as the product or double integral converge. If polar coordinates are now introduced, we have, since $dx\,dy = r\,d\theta\,dr$,

$$\left[\Gamma\!\left(\frac{1}{2}\right)\right]^2 = 4\int_0^{\pi/2} d\theta \int_0^\infty e^{-r^2} r\,dr = 4\left(\frac{\pi}{2}\right)\left[\frac{e^{-r^2}}{-2}\right]_0^\infty = \pi,$$

where the limits are such that the integral is still evaluated over the first quadrant. Therefore we have shown that $\Gamma(^1\!/_2) = \sqrt{\pi}$.

Many definite integrals can be evaluated by means of the gamma function either because they are of the gamma form or can be transformed into that form. A few illustrations follow, with the appropriate transformations in brackets.

(a) $\displaystyle\int_0^\infty x^{4/5}e^{-x}\,dx = \Gamma\!\left(\frac{9}{5}\right).$

(b) $\displaystyle\int_0^\infty x^2 e^{-2x}\,dx = \frac{1}{8}\int_0^\infty y^2 e^{-y}\,dy = \frac{1}{8}\Gamma(3) = \frac{1}{4},\qquad [2x = y].$

(c) $\displaystyle\int_0^\infty x^9 e^{-x^3}\,dx = \frac{1}{3}\int_0^\infty y^{7/3}\,e^{-y}\,dy = \frac{1}{3}\Gamma\!\left(\frac{10}{3}\right) = \frac{28}{27}\Gamma\!\left(\frac{4}{3}\right),\qquad [x^3 = y.]$

(d) $\displaystyle -\int_0^1 (\ln x)^{1/5}\,dx = -\int_\infty^0 y^{1/5}\,e^{-y}\,dy = \int_0^\infty y^{1/5}\,e^{-y}\,dy = \Gamma\!\left(\frac{6}{5}\right),$

$[\ln x = -y].$

We recall that the nth *moment* μ_n' of a probability density function $f(x)$ is defined as

$$\mu_n' = E(x^n) = \int_{-\infty}^\infty x^n f(x)\,dx.$$

For example, the mean is $\mu = \mu_1' = E(x)$ and the variance is $\sigma^2 = \mu_2' - (\mu_1')^2 = E(x^2) - [E(x)]^2$.

The gamma function is of value in finding the moments of some useful distributions, as we shall now illustrate. The moments of the exponential density function $f(x) = e^{-x}$ for $x \geqslant 0$ are simply

$$E(x^n) = \int_0^\infty x^n e^{-x}\,dx = \Gamma(n + 1) = n!.$$

More generally, we have:

THEOREM 6.2. The moments of the exponential density function $f(x) = \lambda e^{-\lambda x}$ for $x \geq 0$ and positive parameter λ are

$$E(x^n) = \frac{\Gamma(n+1)}{\lambda^n} = \frac{n!}{\lambda^n} \qquad \text{for } n = 0, 1, 2, \cdots.$$

As a result we see that the mean of the exponential distribution is $1/\lambda$ and the variance is $1/\lambda^2$.

THEOREM 6.3. The moments of the gamma density function $f(x) = \lambda^r x^{r-1} e^{-\lambda x} / \Gamma(r)$ for $x \geq 0$ and positive parameters r and λ are

$$E(x^n) = \frac{\Gamma(n+r)}{\lambda^n \Gamma(r)} \qquad \text{for } n = 0, 1, 2, \cdots.$$

As a result, the mean and variance of a gamma distribution are r/λ and r/λ^2. Finally, we will now give two important theorems concerning the standard normal distribution.

THEOREM 6.4. If z has standard normal density $f(z) = (1/\sqrt{2\pi}) \exp(-z^2/2)$, then $x = z^2$ has gamma density with $r = 1/2$ and $\lambda = 1/2$, or equivalently $z^2/2$ has the gamma density with $r = 1/2$ and $\lambda = 1$.

PROOF. We have $z = \sqrt{x}$ and $dz = (1/2)x^{-1/2}dx$. The distribution function of x is

$$P(x \leq t) = P(z^2 \leq t) = P(-\sqrt{t} \leq z \leq \sqrt{t}) = 2P(0 \leq z \leq \sqrt{t}),$$

which is

$$P(x \leq t) = 2 \int_0^{\sqrt{t}} f(z)dz = 2 \int_0^t f(\sqrt{x})(1/2)x^{-1/2}\,dx.$$

Thus the density function of y is

$$2f(\sqrt{x})(1/2)x^{-1/2} = f(\sqrt{x})x^{-1/2} = \frac{1}{\sqrt{2\pi}}x^{-1/2}e^{-x/2} \qquad \text{for } 0 \leq x \leq \infty,$$

which is a gamma density with $r = 1/2$ and $\lambda = 1/2$. The second part of the theorem follows directly from this result.   Q.E.D.

Since, by definition a gamma density with $r = 1/2$ and $\lambda = 1/2$ is a chi-square density with one degree of freedom, it follows that the square of a standard normal variable has a chi-square distribution with one degree of freedom.

THEOREM 6.5. If z has a standard normal distribution, then its absolute moments are

$$E(|z|^n) = 2^{n/2}\pi^{-1/2}\Gamma[(n + 1)/2] \qquad \text{for } n = 0, 1, 2, \cdots$$

and its moments are

$$E(z^n) = \begin{cases} 0 & \text{for } n = 1, 3, 5, \cdots, \\ E(|z|^n) & \text{for } n = 0, 2, 4, 6, \cdots. \end{cases}$$

PROOF. Absolute values are hard to work with, so we convert the absolute moment to its equivalent by the equation

$$E(|z|^n) = E[(z^2)^{n/2}].$$

By Theorem 6.4, $x = z^2$ has a gamma density with $r = \frac{1}{2}$ and $\lambda = \frac{1}{2}$, and by Theorem 6.3 the moments of such a density are

$$E[(z^2)^{n/2}] = E[x^{n/2}] = \frac{\Gamma[(n/2) + r]}{\lambda^{n/2}\Gamma(r)} = \frac{\Gamma[(n/2) + (1/2)]}{(1/2)^{n/2}\Gamma(1/2)},$$

which is the required result. Even moments, of course, are equal to the absolute moments. For an odd moment, we have

$$E(z^n) = \int_{-\infty}^{0} z^n f(z)\, dz + \int_{0}^{\infty} z^n f(z)\, dz = \int_{0}^{\infty} [(-z^n + z^n)f(z)\, dz = 0,$$

because $f(z) = f(-z)$. Thus the odd moments of the standard normal density, like the odd moments of any symmetric density function, vanish.   Q.E.D.

The area under any density function is given by the moment for $n = 0$; that is,

$$E(x^0) = E(1) = \int_{-\infty}^{\infty} f(x)\, dx = 1.$$

Setting $n = 0$ in each of the expressions for $E(x^n)$ in Theorems 6.2, 6.3, and 6.5 we thereby verify that the exponential, gamma, and standard normal density functions each have unit area. Setting $n = 1$ in the same expressions, we find their respective means μ and $1/\lambda$, r/λ, and 0. Setting $n = 2$, we find their respective second moments are $2/\lambda^2$, $(r + 1)r/\lambda^2$, and 1. Using the equation $\sigma^2 = E(x^2) - \mu^2$ we find that their respective variances are $1/\lambda^2$, r/λ^2, and 1.

Another useful type of integral involves two parameters and is called the beta function. It can be defined as follows:

Form (a): $\beta(m,n) = \displaystyle\int_{0}^{1} x^{m-1}(1 - x)^{n-1}\, dx, \qquad m > 0, n > 0.$

Form (b): $\beta(m,n) = 2 \displaystyle\int_0^{\pi/2} \sin^{2m-1}\theta \cos^{2n-1}\theta \, d\theta,$ $m > 0, n > 0.$

The integral has a specific value for any given choice of m and n provided these quantities are greater than zero; otherwise the integral diverges. Form (a) is transformed into form (b) by the change of variable $x = \sin^2\theta$.

It turns out that the beta function is related to the gamma function. If the two integrals defining gamma functions of m and n,

$$\Gamma(m) = 2 \int_0^\infty y^{2m-1} e^{-y^2} \, dy \qquad \text{and} \qquad \Gamma(n) = 2 \int_0^\infty x^{2n-1} e^{-x^2} \, dx,$$

are multiplied together, then upon introducing polar coordinates as before, we obtain

$$\Gamma(m)\Gamma(n) = \left[2 \int_0^{\pi/2} \cos^{2n-1}\theta \sin^{2m-1}\theta \, d\theta \right]\left[2 \int_0^\infty r^{2(m+n)-1} e^{-r^2} \, dr \right].$$

Utilizing our definitions of both beta and gamma functions, this equation is

$$\Gamma(m)\Gamma(n) = \beta(m,n)\Gamma(m+n),$$

and therefore

$$\beta(m,n) = \frac{\Gamma(m)\Gamma(n)}{\Gamma(m+n)}.$$

This equation shows that the beta function is symmetrical in the parameters m and n, i.e.,

$$\beta(m,n) = \beta(n,m).$$

The beta function is the area under the curve $y = x^{m-1}(1-x)^{n-1}$ between the values of $x = 0$ and $x = 1$. Not only is this total area known, but tables are available for the percentage area included between 0 and x where $0 < x < 1$; and these computational aids make the curves very useful for the representation of simple mathematical models. The effect of the parameters m and n on the shapes of these curves may be studied by dividing the ordinates by $\beta(m,n)$ thus mak-

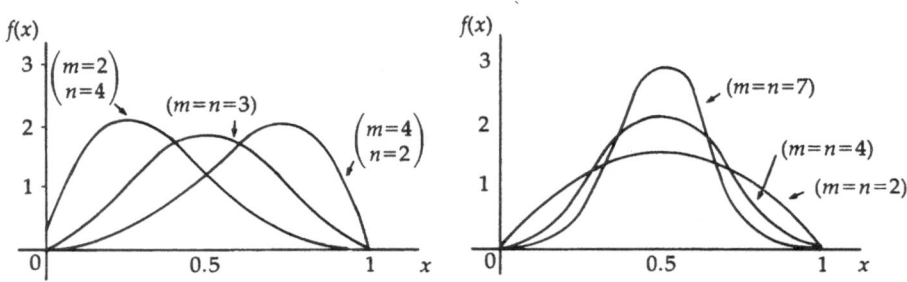

FIGURE 6.3 Graphs of the beta density function for various values of m and n.

ing the total area equal to unity in all cases. The resulting curve is the beta density function. The order of contact of the curves to the x-axis at 0 and 1 is determined by the value of m and n and increases with increasing values of these parameters. When m and n are equal, the curves are symmetrical about $x = \frac{1}{2}$. The effect of assigning a value of m between 0 and 1 is to cause the curve to rise to infinity with $x = 0$ as an asymptote, although the area under the curve is still finite. A similar situation exists at $x = 1$ if n is allowed to take on a value between 0 and 1. Some of these effects are shown in Figure 6.3 where m and n are varied for purposes of illustration. Because each curve is a probability density function, the total area is equal to unity in each case. A wide variety of shapes can thus be obtained with these curves, provided only one maximum is desired. When two or more maxima are required by the mathematical models being considered, several of these curves can be added together in some predetermined proportion in order to give the desired effect.

The range for x from 0 to 1 is not essential, since a simple linear transformation changes the origin and the scale in such a manner that the curve will vanish at any other two points in place of 0 and 1. The following examples illustrate the use of the beta function in evaluating definite integrals.

a. $\displaystyle\int_0^1 x^3(1 - x)^4 \, dx = \beta(4,5) = \frac{\Gamma(4)\Gamma(5)}{\Gamma(9)} = \frac{1}{280}.$

b. $\displaystyle\int_0^{\pi/2} \tan^{1/2} x \, dx = \int_0^{\pi/2} \sin^{1/2} x \cos^{-1/2} x \, dx$

$$= \frac{1}{2} \beta(^3/_4, ^1/_4) = \frac{1}{2} \frac{\Gamma(^3/_4)\Gamma(^1/_4)}{\Gamma(1)}$$

$$= \frac{8}{3} \Gamma(^7/_4)\Gamma(^5/_4).$$

c. $\int_0^1 \frac{x^2 \, dx}{\sqrt{1-x^4}} = \frac{1}{4} \int_0^1 y^{-1/4}(1-y)^{-1/2} \, dy$

$$= \frac{1}{4} \beta(^3/_4, ^1/_2) = \frac{1}{4} \frac{\Gamma(^3/_4)\Gamma(^1/_2)}{\Gamma(^5/_4)}$$

$$= \frac{\sqrt{\pi}}{3} \frac{\Gamma(^7/_4)}{\Gamma(^5/_4)}.$$

EXERCISES

Evaluate each of the following integrals.

1. $\int_0^\infty x^{5/3} e^{-2x} \, dx.$

   ANSWER. $\dfrac{\Gamma(^8/_3)}{2^{8/3}}.$

2. $\int_0^\infty x^4 e^{-5x^2} \, dx.$

   ANSWER. $\dfrac{\Gamma(^5/_2)}{50\sqrt{5}}.$

3. $\int_0^1 x^{-1/5} (1-x)^{-2/5} \, dx.$

   ANSWER. $\beta(0.8, 0.6) = 1.95.$

4. $\int_0^\infty x^n e^{-bx} \, dx.$

   ANSWER. $\dfrac{n!}{b^{n+1}}.$

5. $\int_0^\infty e^{-x^3} \, dx.$

   ANSWER. $\Gamma(^4/_3).$

6. $\displaystyle\int_0^1 \frac{dx}{\sqrt{1-x^3}}.$

ANSWER. $\dfrac{\sqrt{\pi}\Gamma(1/3)}{3\Gamma(5/6)}.$

7. $\displaystyle\int_0^2 x^2(2-x)^{13}\,dx.$

ANSWER. $\dfrac{4096}{105}.$

8. $\displaystyle\int_0^1 \frac{dx}{\sqrt{1-x^4}}.$

ANSWER. $\dfrac{\sqrt{\pi}\Gamma(1/4)}{4\Gamma(3/4)}.$

9. $\displaystyle\int_0^1 \frac{x^6\,dx}{(1-x^5)^{1/3}}.$

ANSWER. $\dfrac{9}{32}\dfrac{\Gamma(7/5)\Gamma(5/3)}{\Gamma(16/15)}.$

10. $\displaystyle\int_2^4 (x-2)^{-1/6}(4-x)^{-5/6}\,dx.$

ANSWER. $\Gamma(5/6)\Gamma(1/6).$

11. $\displaystyle\int_1^4 (x-1)^{-1/4}(4-x)^{-3/4}\,dx.$

ANSWER. $\Gamma(3/4, 1/4).$

12. Show that if a random variable x has the gamma density with parameters r and λ, then $y = \lambda x$ has the gamma density with parameters r and 1. Conversely, show that if y is gamma with parameters r and 1, then y/λ is gamma with parameters r and λ. In particular, complete the proof of Theorem 6.4 by showing that if $x = z^2$ (where z is standard normal) is gamma with $r = 1/2$ and $\lambda = 1/2$, then $y = z^2/2$ is gamma with $r = 1/2$ and $\lambda = 1$.

HINT. Use the fact that if x has the density function $f(x)$, then x/a has the density $af(ax)$ for any positive constant a. This exercise means that most computations concerning the gamma distribution can be obtained by generalizing from the case $\lambda = 1$. We may thus regard the parameter 1 as a scale factor for the gamma density.

MOMENT GENERATING FUNCTION

Another useful transform is the moment generating function, which is often abbreviated as mgf. We recall that the expected value of a random variable is called the first moment of the distribution, and the quantity

$$\mu'_n = E(x^n) = \int_{-\infty}^{\infty} x^n f(x)\, dx \qquad (x \text{ continuous})$$

or, as the case may be,

$$\mu'_n = E(x^n) = \sum_{x=-\infty}^{\infty} x^n f(x) \qquad (x \text{ discrete}),$$

is called the nth moment. The moment generating function incorporates the most valuable properties of the moments themselves. The *moment generating function* (or mgf) $M(\theta;x)$ is defined as

$$M(\theta;x) = \int_{-\infty}^{\infty} e^{\theta x} f(x)\, dx \qquad (x \text{ continuous})$$

or, as the case may be,

$$M(\theta;x) = \sum_{x=-\infty}^{\infty} e^{\theta x} f(x) \qquad (x \text{ discrete}).$$

We see that the moment generating function is equal to the expected value of the function $e^{\theta x}$, where θ is an arbitrary parameter; that is

$$M(\theta;x) = E(e^{\theta x}).$$

Since this expected value is clearly dependent upon both the value of θ and the nature of the distribution of x, we use the symbol $M(\theta;x)$ to represent the moment generating function derived from the distribution of x when the arbitrary parameter is θ. However, $M(\theta;x)$ is not a function of x, as x has been integrated (or summed) out in the calculation of $M(\theta;x)$.

[The moment generating function is closely related to the Laplace transform of $f(x)$. In the case when the random variable can only take on nonnegative values, the Laplace transform is

$$\int_0^\infty e^{-sx} f(x)\, dx.$$

Thus we see that the Laplace transform is $M(-s;x)$; that is, by replacing θ by $-s$ in the moment generating function, we obtain the Laplace transform.]

[It is interesting to note that the probability generating function and moment generating function are mathematically old concepts, as they date back to the work of Laplace in the eighteenth century. However, in the past fifty years they have been more or less replaced by the z-transform and the Laplace transform, respectively, which differ only in the sign of the parameter. This change can be attributed to the mathematical leadership of the Bell Telephone Laboratories.]

If the moment generating function exists for all values of θ in the arbitrarily small interval $(-h,h)$ where $h > 0$, then the distribution of x is uniquely determined by the moment generating function. In that case, all of the moments exist. If we substitute the series expansion

$$e^{\theta x} = \sum_{n=0}^{\infty} \frac{(\theta x)^n}{n!}$$

into

$$M(\theta;x) = E(e^{\theta x}),$$

we obtain

$$M(\theta;x) = 1 + \theta E(x) + \frac{\theta^2}{2!} E(x^2) + \cdots + \frac{\theta^n}{n!} E(x^n) + \cdots$$

$$= 1 + \theta \mu_1' + \frac{\theta^2}{2!} \mu_2' + \cdots + \frac{\theta^n}{n!} \mu_n' + \cdots.$$

Thus we can find the nth moment by differentiating the moment generating function n times, and then setting $\theta = 0$ in the result; that is,

$$\left. \frac{d^n M(\theta;x)}{d\theta^n} \right|_{\theta=0} = E(x^n) = \mu_n'.$$

In mathematical terminology, a function that yields a given system of constants when operated upon in a prescribed manner is called a generating function. It is for this reason that $M(\theta;x)$ is called the moment generating function.

PROBLEM. Find the moment generating function and the mean for the probability mass function $f(x) = (1 - q)q^x$ for $x = 0, 1, 2, \cdots$ and $0 < q < 1$. (This probability mass function is that of geometric distribution.)

SOLUTION.

$$M(\theta;x) = \sum_{x=0}^{\infty} e^{\theta x}f(x) = (1 - q)\sum_{x=0}^{\infty} (qe^\theta)^x = \frac{1 - q}{1 - qe^\theta},$$

$$E(x) = \frac{dM(\theta;x)}{d\theta}\bigg|_{\theta=0} = \frac{(1 - q)qe^\theta}{(1 - qe^\theta)^2}\bigg|_0 = \frac{q}{1 - q}.$$

PROBLEM. Given that $M(\theta;x) = (1 - q)/(1 - qe^\theta)$ and $x = 0, 1, 2, \cdots$, find $f(x)$.

SOLUTION. This problem could be solved immediately by comparison with the above problem. However the information that x is confined to integral values is sufficient to determine $f(x)$ from $M(\theta;x)$ without reference to the previous result. Expand $M(\theta;x)$ in a series

$$M(\theta;x) = (1 - q)\sum_{x=0}^{\infty} q^x e^{x\theta}.$$

But by definition,

$$M(\theta;x) = \sum_{x=0}^{\infty} f(x)e^{x\theta}.$$

Therefore, equating general terms, we have

$$f(x) = (1 - q)q^x \qquad \text{for } x = 0, 1, 2, \cdots,$$

which we recognize as the probability mass function of the geometric distribution.

THEOREM 6.6. The moment generating function of the gamma density $f(x) = \lambda^r x^{r-1} e^{-\lambda x}/\Gamma(r)$ for $x \geqslant 0$, $r > 0$, $\lambda > 0$ is $M(\theta;x) = (1 - \theta/\lambda)^{-r}$.

PROOF.

$$M(\theta;x) = \frac{1}{\Gamma(r)} \int_0^\infty e^{\theta x}\lambda^r x^{r-1}e^{-\lambda x}\, dx$$

$$= \frac{\lambda^r}{\Gamma(r)} \int_0^\infty e^{-(\lambda - \theta)x}x^{r-1}\, dx.$$

Make the change of variables $(\lambda - \theta)x = y$. Then

$$M(\theta;x) = \frac{\lambda^r}{(\lambda - \theta)^r \Gamma(r)} \int_0^\infty e^{-y} y^{r-1} \, dy = \frac{\lambda^r}{(\lambda - \theta)^r} \qquad \text{Q.E.D.}$$

Since the exponential density $f(x) = \lambda e^{-\lambda x}$ is obtained from the gamma density by setting $r = 1$, we see that the moment generating function of the exponential density is $(1 - \theta/\lambda)^{-1}$.

PROBLEM. The length of time t required to accumulate a fixed number of counts r of radiational particles impinging upon the counting meter of a Geiger counter at an average rate λ is distributed as follows:

$$f(t) = \frac{\lambda^r}{\Gamma(r)} t^{r-1} e^{-\lambda t} \qquad \text{for } 0 \leqslant t < \infty, r = 1, 2, \cdots.$$

Find the moment generating function. This density function for t is the Erlang distribution with parameters r and λ. Note that the Erlang distribution is defined as the gamma distribution in the case when the parameter r is a positive integer.

SOLUTION. From Theorem 6.6, we see that the moment generating function is

$$M(\theta;t) = (1 - \theta/\lambda)^{-r}.$$

Thus the mean of the time to accumulate r counts is $\mu = r/\lambda$ and the standard deviation is $\sigma = \sqrt{r}/\lambda$ (see Theorem 6.3). We see that mean time to count one particle (i.e., $1/\lambda$ seconds per particle) is the reciprocal of the average rate (i.e., λ particles per second).

Why does the moment generating function give the moments? Let us look at an alternative explanation. The first derivative of the moment generating function is

$$\frac{dM}{d\theta} = \frac{d}{d\theta} E(e^{\theta x}) = E\left[\frac{d}{d\theta} e^{\theta x}\right] = E(xe^{\theta x}),$$

which, upon setting $\theta = 0$, becomes $E(x)$. Likewise, the second derivative is

$$\frac{d^2 M}{d\theta^2} = \frac{d^2}{d\theta^2} E(e^{\theta x}) = E\left[\frac{d^2}{d\theta^2} e^{\theta x}\right] = E(x^2 e^{\theta x}),$$

which, upon setting $\theta = 0$, becomes $E(x^2)$. Similarly, we see that the nth derivative of $M(\theta;x)$ with respect to θ is $E(x^n e^{\theta x})$, which upon setting $\theta = 0$ gives the nth moment $E(x^n)$.

EXERCISES

1. An indicator random variable u has two admissible values 0, 1 such that $f(0) = q$, $f(1) = p$, and, of course, $q + p = 1$.
   (a) Find the mean and variance of u and thereby obtain the mean and variance of the random variable $x = u_1 + u_2 \cdots + u_n$ where $u_1, \cdots, u_n$ are each independently distributed as u.
   (b) Find the moment generating function of u and thereby derive the moment generating function of x.
   (c) By expanding the latter and equating terms prove that x is binomially distributed.
   ANSWERS. (a) p, pq, np, npq. (b) $q + pe^\theta$, $(q + pe^\theta)^n$.

2. A fair coin is flipped twice. Let x be the number of heads that occur. Find the moment generating function of x, the mean, and the variance.
   ANSWER. $(1/4)(1 + e^\theta)^2$, 1, $1/2$.

3. Suppose that the length of time t a transistor will work in a certain circuit is a random variable with exponential density $f(t) = 100e^{-100t}$ for $t > 0$. Find the moment generating function of t, the mean, and the variance.
   ANSWER. $100/(100 - t)$, $1/100$, $1/(100)^2$.

4. Find the moment generating function and the mean for the uniform density $f(x) = 1$ for $0 \le x \le 1$ and $f(x) = 0$ otherwise.
   ANSWER. $(e^\theta - 1)/\theta$, $1/2$.

5. Find the moment generating function and the moments for the uniform density $f(x) = 1$ for $a \le x \le b$ and $f(x) = 0$ otherwise.
   ANSWER. $(e^{b\theta} - e^{a\theta})/(b - a)\theta$, $E(x^n) = (b^{n+1} - a^{n+1})/(b - a)(n + 1)$.

6. Given that $f(x) = cx$ for $0 \le x \le 1$.
   (a) Find c by integration.
   (b) Find $E(x^n)$ by integration.
   (c) Find $M(\theta;x)$.
   (d) Find $E(x^n)$ from $M(\theta;x)$.
   ANSWERS. (a) 2. (b) $2/(n + 2)$. (c) $2(\theta e^\theta - e^\theta + 1)/\theta^2$. (d) Expand e^θ and simplify.

7. If $f(x) = ce^{-x}$ for $x \ge 0$,
   (a) find c;

(b) find $M(\theta;x)$;
(c) find $E(x^n)$ from $M(\theta;x)$.

ANSWERS. (a) 1. (b) $(1 - \theta)^{-1}$. (c) $n!$.

8. If $f(x) = cx^{\alpha}e^{-x}$ for $x \geq 0$ and α is a positive integer, find the following:

(a) c, using the fact that

$$\int_0^{\infty} x^{\alpha}e^{-x}\, dx = \alpha!.$$

(b) $E(x^n)$ from the definition.
(c) $M(\theta;x)$.
(d) $E(x^n)$ from $M(\theta;x)$.

ANSWERS. (a) $1/\alpha!$. (b) $(\alpha + n)!/\alpha!$. (c) $(1 - \theta)^{-(\alpha+1)}$.

APPLICATIONS OF THE MOMENT GENERATING FUNCTION

An important application of the moment generating function lies in the convenient treatment of functions of random variables. If u is a single-valued continuous function of the random variables $x_1, x_2, \cdots, x_n$, the moment generating function of u can be found from the joint distribution of the x's without first finding the distribution of u itself.

Suppose $g(u)$ is the (unknown) probability density function of u. Its moment generating function is

$$M(\theta;u) = E(e^{\theta u}) = \int_{-\infty}^{\infty} e^{\theta u}g(u)\, du.$$

However, $M(\theta;u)$ can be directly found from the joint density function $f(x_1, x_2, \cdots, x_n)$ by the formula

$$M(\theta;u) = \int_{-\infty}^{\infty}\int_{-\infty}^{\infty} \cdots \int_{-\infty}^{\infty} e^{\theta u}f(x_1, x_2, \cdots, x_n)\, dx_1\, dx_2 \cdots dx_n.$$

Once $M(\theta;u)$ is known, it is often possible to identify the density function $g(u)$ of u by an appeal to the uniqueness property.

PROBLEM. By the method of moment generating functions, find the distribution of $u = xy$ where

$$f(x,y) = xe^{-x(1+y)} \qquad \text{for } 0 \leq x < \infty, 0 \leq y < \infty.$$

SOLUTION. Here the moment generating function is

$$M(\theta;u) = \int_0^\infty \int_0^\infty e^{\theta xy} x e^{-x-xy}\, dx\, dy = \int_0^\infty e^{-x}\left[\int_0^\infty xe^{-xy(1-\theta)}\, dy\right] dx$$

$$= \int_0^\infty \frac{e^{-x}}{1-\theta}\, dx = \frac{1}{1-\theta}\int_0^\infty e^{-x}\, dx = (1-\theta)^{-1}.$$

This is a special case of the moment generating function $(1 - \theta/\lambda)^{-r}$ with $r = 1$, $\lambda = 1$. Hence, making use of Theorem 6.6, we have $f(u) = e^{-u}$ for $0 \le u < \infty$, so the distribution of u is exponential.

By direct application of the above equation for $M(\theta;u)$, we will now deduce three theorems of considerable importance. The first concerns the effect of a linear transformation, the second concerns the sum of independent random variables, and the third concerns the mean values derived from independent random samples of size n from the same infinite population.

THEOREM 6.7. If $u = ax + b$ $(a \ne 0)$, then $M(\theta;u) = e^{b\theta}M(a\theta;x)$.

PROOF.

$$M(\theta;u) = \int_{-\infty}^\infty e^{\theta(ax+b)}f(x)\, dx$$

$$= e^{b\theta}\int_{-\infty}^\infty e^{a\theta x}f(x)\, dx = e^{b\theta}M(a\theta;x).$$

Frequently the moments $\mu_n' = E(x^n)$ are called moments about the origin, to distinguish them from moments about other points. The nth moment about any point $x = c$ is naturally defined as $E[(x - c)^n]$. Of particular interest are the moments about the mean. These moments, called *central moments*, are denoted by $\mu_n = E[(x - \mu)^n]$. Of course, the first central moment μ_1 is zero, and the second central moment μ_2 is the variance σ^2. An interesting special case of Theorem 6.7 is obtained by setting $a = 1$, $b = -\mu$, where $\mu = E(x)$. This yields the moment generating function of the central moments:

$$M(\theta;x-\mu) = e^{-\mu\theta}M(\theta;x) = 1 + \frac{\theta^2}{2!}\mu_2 + \frac{\theta^3}{3!}\mu_3 + \cdots.$$

This relation is often useful. Note that the term in θ to the first power does not occur because $\mu_1 = 0$.

THEOREM 6.8. The moment generating function of the sum of independent random variables is equal to the product of their respective moment generating functions.

PROOF. Let $u = x_1 + x_2 + \cdots + x_n$ where the x's are independent. If $f_i(x_i)$ is the marginal density function of x_i, the quantity $e^{\theta u} f(x_1, x_2, \cdots, x_n)$ factors, yielding

$$M(\theta; u) = \int_{-\infty}^{\infty} e^{\theta x_1} f_1(x_1)\, dx_1 \int_{-\infty}^{\infty} e^{\theta x_2} f_2(x_2)\, dx_2 \cdots \int_{-\infty}^{\infty} e^{\theta x_n} f_n(x_n)\, dx_n$$

$$= M(\theta; x_1) M(\theta; x_2) \cdots M(\theta; x_n).$$

THEOREM 6.9. If $\bar{x} = (x_1 + x_2 + \cdots + x_n)/n$ is the arithmetic average of a random sample $x_1, x_2, \cdots, x_n$ of n observations from a population for which the moment generating function is $M(\theta; x)$, the moment generating function of $\bar{x}$ in repeated independent sampling is

$$M(\theta; \bar{x}) = [M(\theta/n; x)]^n.$$

PROOF. We can regard the independent observations $x_1, x_2, \cdots, x_n$ as independent random variables having the same distribution. Hence each has the same moment generating function, $M(\theta; x)$. Putting $u = x_1 + x_2 + \cdots + x_n$, it follows from Theorem 6.8 that $M(\theta; u) = [M(\theta; u)]^n$. Now $\bar{x} = u/n$ and from Theorem 6.7,

$$M(\theta; \bar{x}) = M(\theta; u/n) = M(\theta/n; u) = [M(\theta/n; x)]^n.$$

PROBLEM. Under similar conditions, n independent tests are made to determine the waiting time to accumulate r counts of radiational particles impinging on the meter at an average rate λ. From the moment generating function, find the distribution of the average length of time $\bar{t}$.

SOLUTION. By the problem on page 210, we know that

$$M(\theta; t) = (1 - \theta/\lambda)^{-r}.$$

Hence by Theorem 6.9, we have

$$M(\theta; \bar{t}) = [M(\theta/n; t)]^n = (1 - \theta/n\lambda)^{-nr}.$$

This is of the same form as $M(\theta; t)$ except that r is replaced by nr and λ by $n\lambda$. Hence the distribution of $\bar{t}$ is Erlang with parameters nr and $n\lambda$:

$$g(\bar{t}) = \frac{(n\lambda)^{nr}}{\Gamma(nr)} (\bar{t})^{nr-1} e^{-n\lambda \bar{t}} \qquad \text{for } 0 \leqslant \bar{t} < \infty.$$

EXERCISES

1. Given that the moment generating function for a standard normal variable z is $M(\theta;z) = \exp(\theta^2/2)$, verify that $E(z) = 0$ and var $(z) = 1$. Find the moment generating function $M(\theta;x)$ of the normal variable $x = \mu + \sigma z$ and verify that $E(x) = \mu$ and var $(x) = \sigma^2$.

   ANSWER. $M(\theta;x) = \exp(\mu\theta + \sigma^2\theta^2/2)$.

2. Given that the moment generating function of a gamma variable with parameters r and λ is $(1 - \theta/\lambda)^{-r}$, find the moment generating function of a chi-square variable with n degrees of freedom.

   ANSWER. A chi-square distribution with n degrees of freedom is defined as a gamma distribution with $r = n/2$ and $\lambda = 1/2$. Thus the required mgf is $(1 - 2\theta)^{-n/2}$.

3. The moment generating function of the random variable x is

$$M(\theta;x) = \cosh\theta = (1/2)(e^\theta + e^{-\theta}).$$

   Find the moment generating function of $\bar{x}$, the sample mean of n independent observations. Also find the mean and variance of $\bar{x}$.

   ANSWER. $M(\theta;\bar{x}) = [\cosh(\theta/n)]^n$; $E(\bar{x}) = 0$, $\sigma_{\bar{x}}^2 = 1/n$.

4. Given that $x = z^2$, where z is a standard normal variable, has a chi-square distribution with one degree of freedom, what is the moment generating function for x?

   ANSWER. In the gamma mgf $(1 - \theta/\lambda)^{-r}$ set $r = 1/2$ and $\lambda = 1/2$ to obtain $M(\theta;x) = (1 - 2\theta)^{-1}$.

5. Given that $y = z_1^2 + z_2^2 + \cdots + z_n^2$ where the $z_1, z_2, \cdots, z_n$ are independent standard normal variables, use the result of the previous exercise to find the mgf of y and identify the distribution of y. What is the mean and variance of y?

   ANSWER. $M(\theta;y) = (1 - 2\theta)^{-n/2}$; chi-square with n degrees of freedom; mean $= n$, variance $= 2n$.

6. Given that z is standard normal, find the moment generating function of $x = z^2$ directly from the density of z.

   ANSWER. $M(\theta;x) = (1 - 2\theta)^{-1/2}$.

7. Show that the mgf of the binomial distribution $b(x;n,p)$ is $(q + pe^\theta)^n$. Hence show that the mean is np and the variance is npq. Show that the sum of two independent binomial variables with parameters n_1,p and n_2,p respectively is a binomial variable with parameters $n_1 + n_2,p$.

8. Show that the mgf of the Poisson distribution $p(x;\mu) = \mu^x e^{-\mu}/x!$ is $\exp[-\mu(1 - \exp\theta)]$. Hence show that the mean and variance are

each equal to μ. Show that the sum of two independent Poisson variables with parameters μ_1 and μ_2 respectively is a Poisson variable with parameter $\mu_1 + \mu_2$.

9. Given that $f(x) = 1$ for $0 \leqslant x \leqslant 1$ and $f(x) = 0$ otherwise, determine the distribution of $u = -2 \ln x$ from its moment generating function.

    ANSWER. $M(\theta;u) = (1 - 2\theta)^{-1}$. Hence u is chi-square with two degrees of freedom.

10. Independent random variables x and y have probability density functions with moment generating functions $M(\theta;x)$ and $M(\theta;y)$. Random variable s is defined by $s = x + y$. Use $M(\theta;s)$ and the moment generating properties of transforms to show that $E(s) = E(x) + E(y)$ and $\sigma_s^2 = \sigma_x^2 + \sigma_y^2$.

11. Let x and y be independent random variables with

$$f(x) = \begin{cases} \lambda e^{-\lambda x} & \text{for } x \geqslant 0, \\ 0 & \text{for } x < 0, \end{cases}$$

$$g(y) = \begin{cases} 0 & \text{for } y > 0, \\ \lambda e^{\lambda y} & \text{for } y \leqslant 0. \end{cases}$$

Random variable s is defined by $s = x + y$. Determine $M(\theta;x)$, $M(\theta;y)$, $M(\theta;s)$, $E(s)$, σ_s^2, $h(s)$. Repeat the exercise in the case when $s = ax + by$.

12. On a South Sea island, the local natives pack souvenir coral rocks into cartons which are then packed into crates. The weight x of a coral rock is an exponential random variable with probability density function

$$f(x) = ae^{-ax} \qquad \text{for } x \geqslant 0.$$

The number k of coral rocks in any carton is a Poisson random variable with probability mass function

$$p(k) = e^{-\mu}\mu_k/k! \qquad \text{for } k = 0, 1, 2, \cdots.$$

The number n of cartons in a crate is a Pascal-geometric random variable with probability mass function

$$g(n) = q^{n-1}(1 - q) \qquad \text{for } n = 1, 2, 3, \cdots.$$

In the above specifications, it is required that $a > 0$, $\mu > 0$, $0 < q < 1$. Assume that random variables x, k, and n are mutually independent. Determine

a. the probability that a crate selected at random contains exactly one coral rock.
b. the conditional probability density functions for the total weight of coral rocks in a carton given that the carton contains less than two coral rocks.
c. the moment generating function of the probability mass function for the total weight of the coral rocks in a crate.
d. the probability that a crate selected at random contains an odd number of coral rocks.

13. Suppose that the number x of customers who shop at the local store on any day is a Poisson variable with probability mass function

$$f_1(x;\alpha) = e^{-\alpha}\alpha^x/x! \qquad \text{for } x = 0, 1, 2, \cdots.$$

Suppose that the number y of items purchased by any customer is a Poisson variable y with probability mass function

$$f_2(y;\beta) = e^{-\beta}\beta^y/y! \qquad \text{for } y = 0, 1, 2, \cdots.$$

Suppose x and y are independent. Two ways that the store can increase by 20 percent the number of items it sells is to either increase α by 20 percent or to increase β by 20 percent. Which of these two changes would lead to a smaller variance of the total items sold per day?

14. The *cumulant function* $K(\theta;x)$ is defined as $\ln M(\theta;x)$; the coefficients κ_n of $\theta^n/n!$ (where $n = 1, 2, \cdots$) in the series expansion of $K(\theta;x)$ are called *cumulants*. Prove that the nth cumulant of the sum of independent random variables is equal to the sum of their respective nth cumulants. Show that all the cumulants of a Poisson distribution with parameter μ have the common value μ. Deduce at once that the sum of independent Poisson variables is Poisson distributed. Why cannot the same be said by any linear combination rather than just the sum? Show that the normal distribution $N(\mu,\sigma^2)$ has only two nonzero cumulants, namely

$$\kappa_1 = \mu, \ \kappa_2 = \sigma^2.$$

15. Prove that the sum of two correlated Poisson variables does not have the Poisson distribution.

HINT. Show that the variance does not equal the mean.

16. The joint distribution of n variables is said to be multivariate normal if the joint density function is of the form

$$f(x_1,x_2,\cdots,x_n) = Ke^{-Q/2},$$

where each variable has admissible region from $-\infty$ to ∞ and

$$Q = \sum_{i=1}^{n} \sum_{j=1}^{n} a_{ij}(x_i - \mu_i)(x_j - \mu_j),$$

in which (1) the μ's are constants, (2) the a's are constants having the property that $Q > 0$ for all selections of the x's other than $x_i = \mu_i$ (all i), and (3) K is a positive constant such that the total integral of the joint density function is unity. Prove that the quadratic function Q is distributed as chi-square with n degrees of freedom by deriving the moment generating function of Q and comparing it with that of chi-square.

HINT. To obtain the moment generating function of Q, substitute

$$y_i = (x_i - \mu_i)\sqrt{1 - 2\theta},$$

remembering the Jacobian, and make use of the condition on K.

17. Given that $M(\theta;x+y) = M(\theta;x)M(\theta;y)$, prove that x and y are independent.

HINT. By subtraction, arrive at the equation $E\{[M(\theta;y) - M(\theta;y|x)]e^{\theta x}\} = 0$ for all θ. Hence $M(\theta;y|x) = M(\theta;y)$, whereupon $\phi(y|x) = g(y)$.

7

Basic Discrete Distributions

*We define the art of
conjecture, or stochastic art,
as the art of evaluating as
exactly as possible the
probabilities of things.*

JAKOB BERNOULLI (1654–1705)

Review of Basic
Combinatorial Methods

Uniform or
Rectangular Distribution

Hypergeometric
Distribution

Bernoulli Process
and Binomial Distribution

Poisson Process
and Poisson Distribution

Geometric and
Pascal Distributions

Negative Binomial
Distribution

REVIEW OF BASIC
COMBINATORIAL METHODS

In this chapter we study some distributions that arise frequently in applications. Two broad areas of application are to be distinguished. One is sampling, performed for the purpose of drawing inferences about the population from which the sample is taken. As examples of such sampling, we would include consumer polls on the one hand, and scientific measurements on the other. In consumer polls, we wish, say, to determine the percentage of people in a certain city who approve of a certain manufactured product. When a complete poll of all persons is prohibitively expensive in time or effort, we sample; and from the percentage observed in the sample, we draw conclusions about the numerical value of the percentage in the entire population. In scientific experiments, we measure the value of an observable variable, usually getting somewhat different numerical values on each measurement. From a few measurements, which in reality constitute a sample, we attempt to establish or estimate the true value. The variation in measured values may be caused by errors of measurement or by inherent randomness in the process being observed.

As a second broad area of application, where a physical or dynamic process does contain inherent randomness, we may want to determine the probability distribution for a certain result of the process. For example, the energies of the molecules of a gas are characterizable by a probability distribution. As a consequence, so is the pressure of the gas. Again, the response of an electrical system to a random input voltage is characterized by a probability distribution for output voltage. Again, in chemical production processes, the output rate can be a random variable whose probability distribution depends upon the probability distributions of the input variables and of the action of the process itself. The problem arises, therefore, of determining the probability distribution of the output by a mathematical analysis of the distributions of the input variables and of the process.

In order to derive the applicable probability mass functions, we will need to count the number of ways in which an event can happen. For this purpose, certain formulas are particularly helpful, so let us review the basic combinatorial methods.

Suppose we have n objects, all different, and suppose that r of these n objects are to be selected, one after another. How many selections of the group of r objects are possible? There are two cases.

1. *Selection with replacement.* If each selection is merely recorded for reference, the object being replaced in the group of n before the next object is selected, then obviously each object can be selected in n ways. Hence there are all-told n^r possible selections if each choice is made

independently of the choices so far. We note that r can be greater than n.

PROBLEM. Suppose a coin is flipped r times, yielding a string of heads and tails, say $HTTHH \cdots TH$. How many such strings are there?

SOLUTION. Here $n = 2$, and the number of strings is 2^r.

PROBLEM. What is the probability, if a coin is flipped four times, that the string $HTHH$ is obtained?

SOLUTION. Here the number of possible strings is 2^4, and since there is only one way of getting the string $HTHH$, the answer is $1/2^4$.

2. *Selection without replacement.* If each selection made is withheld, then the total number of selections is different. In this case, r must be $\leq n$. The first object can be selected in n different ways. However, for each way the first object was selected, the second can be selected in only $n - 1$ ways. Continuing, we find that for each way in which the first $r - 1$ was selected, the last, or rth, can be selected in $n - (r - 1)$ different ways. Hence the total number of ways in which all r objects can be selected is

$$P_r^n = n(n - 1) \cdots (n - r + 1) = \frac{n!}{(n - r)!}. \tag{7.1}$$

This formula is known as the number of *permutations* of n objects taken r at a time. An alternative notation for P_r^n is $n_{(r)}$, which is called the rth *factorial power* of n.

PROBLEM. Suppose we have 4 objects from which 2 are to be selected without replacement. Denote the 4 objects by a, b, c, and d. What are the selections and how many are there?

SOLUTION. The selections can be enumerated as

$$
\begin{array}{ccccc}
ab & ac & ad & \quad ba & bc & bd \\
ca & cb & cd & \quad da & db & dc
\end{array}
$$

Thus from the formula, $P_2^4 = 4.3 = 12$.

PROBLEM. If 5 cards are dealt from an ordinary deck of 52 cards, what is the probability of getting the sequence, card by card, ace, ten, eight, jack, nine, all in spades?

SOLUTION.

$$\frac{1}{P_5^{52}} = \frac{1}{52 \cdot 51 \cdot 50 \cdot 49 \cdot 48}.$$

If, in this last problem, we had been allowed to receive the sequence of specified cards in any order of dealing, then a number of successful deals would qualify as favorable, increasing the numerator in the probability value. In fact, the number of favorable deals would be just the number of ways in which the 5 cards could be rearranged among themselves, or just the number of permutations of 5 things taken 5 at a time. This number is P_5^5 or 5!.

Such a selection, in which the order of arrangement is immaterial, the content being fixed, is called a *combination*. Since each permutation preserves the combination, then the number C_r^n of *combinations* of n objects taken r at a time, is given by

$$C_r^n = \frac{P_r^n}{P_r^r} = \frac{n!}{n!(n-r)!}. \tag{7.2}$$

The notation C_r^n is often written as $C(n,r)$ or as $\binom{n}{r}$.

This formula, it should be noted, gives the coefficients in the expansion of $(a + b)^n$. More precisely,

$$(a + b)^n = \sum_{r=0}^{n} \binom{n}{r} a^r b^{n-r}. \tag{7.3}$$

As a result, $\binom{n}{r}$, or C_r^n, is called a *binomial coefficient*. For example, the number of different hands at bridge is $\binom{52}{13}$, and at poker $\binom{52}{5}$. In general, the number of different ways in which r objects can be drawn from n, when all permutations of the r objects are equally admissible, is $\binom{n}{r}$.

REMARKS

1. Since $C_r^n = \binom{n}{r} = \dfrac{n!}{r!(n-r)!}$, we see that $\binom{n}{r} = \binom{n}{n-r}$.

   Note also that $\binom{n}{n} = 1$.
2. Setting $a = b = 1$ in $(a + b)^n$ above, we get

$$\sum_{r=0}^{n} \binom{n}{r} = 2^n.$$

The binomial coefficient is readily generalizable to the *multinomial coefficient* as follows. When we draw r objects from n in this fashion, we can think of the act of making such a selection as one which finally divides the n objects into two groups. The first group contains the $n_1 = r$ objects selected, the second the $n_2 = n - r$ objects not selected.

More generally, suppose that n objects are to be divided into k groups, where the first group contains n_1 objects, the 2nd, n_2 objects, $\cdots$, the kth, n_k objects. In how many ways can this division be performed? Clearly the answer is

$$\binom{n}{n_1}\binom{n-n_1}{n_2}\binom{n-n_1-n_2}{n_3}\cdots\binom{n-n_1-\cdots-n_{k-2}}{n_{k-1}},$$

which reduces to

$$\frac{n!}{n_1!n_2!\cdots n_k!}, \qquad \text{where } n_1 + n_2 + \cdots + n_k = n.$$

This number is called a *multinomial coefficient*. Two remarks: (1) when $k = 2$, the multinomial coefficient is just the binomial coefficient; (2) while order *within* groups was immaterial, the order of the groups was essential.

EXERCISES

1. What is the probability of drawing a flush (all cards of the same suit) at poker?
   ANSWER. = 0.00198.

2. Find the probability that a string of ten random digits (selected from 0 through 9) contains no digit twice.
   ANSWER. = 0.00036288.

3. What are the number of distinct deals at bridge?
   ANSWER. $52!/(13!)^4$.

4. Prove that $n!/(n_1!\cdots n_k!)$ is the coefficient of the term $a_1{}^{n_1}\cdots a_k{}^{n_k}$ in the expansion of $(a_1 + \cdots + a_k)^n$, thus justifying the name "multinomial coefficient."

5. A random variable X that takes each of values 1, 2, $\cdots$, N with the same probability $1/N$ is said to have a *uniform distribution* on 1, 2, $\cdots$, N. Let $X_1, X_2, \cdots, X_n$ be n mutually independent random variables each with a uniform distribution on 1, 2, $\cdots$, N.
   (a) Find $E(X_i)$ and var (X_i).
   (b) Let $Y = X_1 + X_2 + \cdots + X_n$. Does Y have a uniform distribution on 1, 2, $\cdots$, N? On 1, 2, $\cdots$, nN?
   (c) Find $E(Y)$ and var (Y).
   (d) Let M_n be the largest of $X_1, X_2, \cdots, X_n$ and m_n be the smallest of $X_1, X_2, \cdots, X_n$. Find the probability distribution of M_n and of m_n.

HINT. $M_n \leq k$ if and only if all the random variables X_1, X_2, $\cdots$, X_n take a value less than or equal to k. The event $M_n = k$ is just the event "$M_n \leq k$ and $M_n > k - 1$."

ANSWERS. (a) $E(X_i) = (N + 1)/2$, var $(X_i) = (N + 1)(N - 1)/12$. (d) $P(M_n = k) = (k/N)^n - [(k - 1)/N]^n$ for $1 \leq k \leq N$ and $P(m_n = k) = [(N - k + 1)/N]^n - [(N - k)/N]^n$ for $1 \leq k \leq N$.

6. (a) Show that

$$(-1)^n \binom{-1/2}{n} = \binom{2n}{n} \frac{1}{2^{2n}}.$$

$$\left[\text{For } n = 3, \right.$$

$$\left. \binom{-1/2}{3} = \frac{(-1/2)(-3/2)(-5/2)}{3!} = \frac{1 \cdot 2 \cdot 3 \cdot 4 \cdot 5 \cdot 6}{3!3!2^6} (-1)^3. \right]$$

(b) Show that

$$\sum_{i=0}^{n} \frac{\binom{2n-2i}{n-i}\binom{2i}{i}}{2^{2n}} = 1.$$

HINT. Use $(1 - t)^{-1/2}(1 - t)^{-1/2} = 1 + t + t^2 + \cdots$.

UNIFORM OR RECTANGULAR DISTRIBUTION

Suppose that a box contains N tickets which bear the labels 1, 2, $\cdots$, N. A ticket is chosen at random and the number X written on the ticket is observed. Under the equally likely hypothesis the random variable X has probability $1/N$ of assuming each of its possible values 1, 2, $\cdots$, N, so that its probability mass function is

x	1	2	$\cdots$	N
$P(X = x)$	$\dfrac{1}{N}$	$\dfrac{1}{N}$	$\cdots$	$\dfrac{1}{N}$

The distribution of X is called the uniform distribution (on the integers 1, $\cdots$, N). It is also known as the rectangular distribution since its graph fits into a rectangle.

We have already encountered this distribution when considering the experiment of throwing a fair die. The number X of points showing when a fair die is thrown has the uniform distribution for $N = 6$. As another illustration, if X is a digit produced by a random digit generator then X has a uniform distribution on 0, 1, 2, $\cdots$, 9. The following examples illustrate some other situations in which a uniform distribution may arise.

a. When a calculation is carried to five decimal places and then rounded off to four decimals, the digit X appearing in the fifth place must be one of 0, 1, 2, $\cdots$, 9. In many types of calculation it appears that these ten digits occur about equally often; that is, we may suppose that X has the uniform distribution on the integers 0, $\cdots$, 9. This model is often employed in studying the accumulation of rounding errors in computers.

b. Students of industrial accidents may conjecture that because of a fatigue effect accidents are more likely to occur late than early in the week. A skeptic maintains that no such effect exists, and that an accident is equally likely to occur on Monday, $\cdots$, Friday. If the days of the work week are numbered 1 to 5, he would assume a uniform model for the number X of the day on which an accident occurs. By observing the numbers X for several unrelated accidents, one may investigate the adequacy of the uniform model in comparison with a "fatigue effect" model. A similar approach can be used when studying the possibility of a "birth order effect" for rare maladies that some geneticists think are more likely to occur in later than in earlier children of a family.

The expectation of the uniform random variable X defined on 1, 2, $\cdots$, N is

$$E(X) = 1 \cdot \frac{1}{N} + 2 \cdot \frac{1}{N} + \cdots + N \cdot \frac{1}{N} = \frac{(1 + 2 + \cdots + N)}{N} = \frac{1}{2}(N + 1).$$

This formula shows, for example, that the expected number of points showing when a fair die is thrown, is $7/2$. From the algebraic formula for the sum of squares,

$$1^2 + 2^2 + \cdots + N^2 = (1/6)N(N + 1)(2N + 1),$$

it follows that

$$E(X^2) = (1/6)(N + 1)(2N + 1),$$

and hence that

$$\operatorname{var}(X) = \frac{N^2 - 1}{12}.$$

HYPERGEOMETRIC DISTRIBUTION

An important discrete probability function which we will now derive is the hypergeometric distribution. Suppose we have a population of N objects consisting of N_1 objects of type 1 and $N_2 = N - N_1$ objects of type 2. From the N objects, a sample of n objects is drawn at random, without replacement and with no regard to order. What is the probability, $h(n_1)$, that the sample has exactly n_1 objects of type 1, the rest of the objects in the sample being of type 2?

We see that n_1 objects of type 1 in the sample can be chosen in $\binom{N_1}{n_1}$ ways, and the remaining objects in the sample in $\binom{N-N_1}{n-n_1}$ ways. All n objects in the sample can be chosen in $\binom{N}{n}$ ways. Hence

$$h(n_1) = \frac{\binom{N_1}{n_1}\binom{N-N_1}{n-n_1}}{\binom{N}{n}} \qquad \text{for } n_1 = 0, 1, 2, \cdots, \min[n,N_1] \qquad (7.4)$$

where the symbol $\min[n,N_1]$ used here stands for the smaller of the two numbers n, N_1, if they are different or, of course, their common value if they are the same. Equation (7.4) is the probability mass function of the *hypergeometric distribution*. See Figure 7.1. The hypergeometric random variable is n_1 = number of type 1 objects in the sample, and the parameters are n = number of objects in sample, N_1 = number of type 1 objects in population, and N = number of objects in population. See Figure 7.2. An alternative expression for the hypergeometric probability mass function is

$$h(n_1) = \frac{\binom{N_1}{n_1}\binom{N_2}{n_2}}{\binom{N}{n}} \qquad \begin{array}{l} \text{for } n_1 = 0, 1, 2, \cdots, \min[n,N_1]; \\ \text{for the population } N = N_1 + N_2; \text{ and} \\ \text{for the sample } n = n_1 + n_2. \end{array}$$

We can display the parameters of the hypergeometric distribution by writing its probability mass function more explicitly as $h(n_1;n,N_1,N)$.

Let us now consider an example from the field of industrial sampling inspection. In industrial quality control, it is often convenient to "sample inspect" from a lot. The size of the lot is N. Let N_1 be the true number of "defective" items in the lot. The number N_1 is unknown. Let n_1 be the number of observed defective items in a sample of n items from the lot. The purpose of sampling is to draw inferences about N_1.

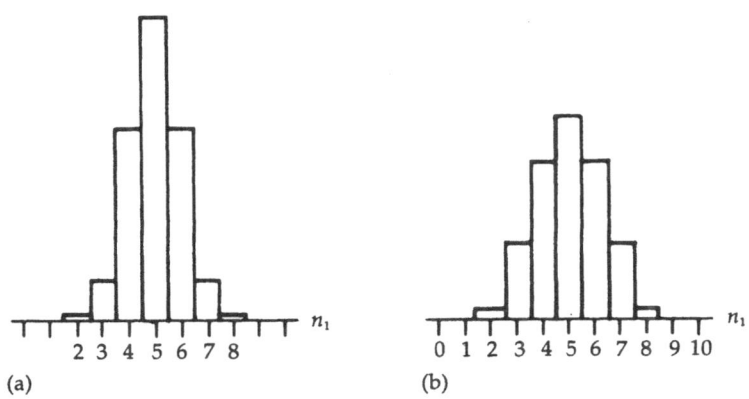

FIGURE 7.1 Probability mass function of hypergeometric random variable n_1: (a) with parameters $n = 10$, $N_1 = 8$, $N = 16$; (b) with parameters $n = 10$, $N_1 = 20$, $N = 40$.

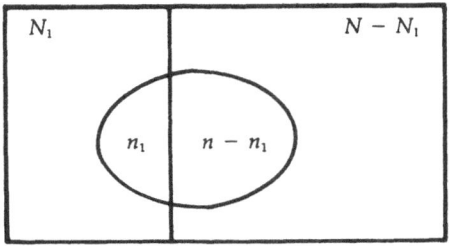

FIGURE 7.2 Partition of objects for the hypergeometric distribution.

In particular, suppose $N = 50$, $n = 5$, and $n_1 = 2$. What is the probability of observing this value of n_1 if $N_1 = 2$? Here

$$h(n_1) = \frac{\binom{2}{2}\binom{48}{3}}{\binom{50}{5}} < 0.01.$$

Since, in customary practice, a probability less than 0.05 is often agreed to make the hypothesis as to the value of N_1 suspect, it follows that $N_1 = 2$ would be suspect in this case.

EXERCISES

1. What is the probability that a bridge hand contains exactly seven spades?

ANSWER.

$$\binom{13}{7}\binom{39}{6}\bigg/\binom{52}{13}.$$

2. The hypergeometric distribution can be easily generalized to the case where the population is composed of more than two types of objects. What will be the formula for such a distribution?

3. A box contains 95 good and 5 defective screws. What is the probability, if 10 screws are used, that none is defective?

ANSWER. 0.5584

4. A group consists of $2n$ boys and $2n$ girls. What is the probability that a group of $2n$ chosen at random will contain exactly n boys and n girls?

ANSWER.

$$\binom{2n}{n}^2\bigg/\binom{4n}{2n}.$$

5. A closet contains n pairs of shoes. If $2r$ shoes are chosen at random, what is the probability that there is no pair among them? (Assume $2r < n$.)

HINT. Recall the *factoral power* notation defined as $n_{(k)} = n(n - 1) \cdots (n - k + 1)$.

ANSWER.

$$\frac{(2n)(2n - 2)\cdots(2n - 4r + 2)}{(2n)_{(2r)}} = \frac{2^{2r}n_{(2r)}}{(2n)_{(2r)}} = \frac{2^{2r}\binom{n}{2r}}{\binom{2n}{2r}}.$$

6. Show that the mean of $h(n_1;n,N_1,N)$ is $E(n_1) = n(N_1/N)$ and the variance is var $(n_1) = n(N_1/N)[1 - (N_1/N)][(N - n)/(N - 1)]$. A lake contains 1000 tagged fish and 5000 untagged fish. If a random sample of size 120 is drawn from the lake without replacement, what is the expectation of the number of tagged fish caught? What is the variance?

7. Let n_1 be a random variable having a hypergeometric distribution with parameters N, N_1, and n. Show that if n is held fixed but N and N_1 tend to infinity in such a way that $N_1/N = p$, then $E(n_1)$

tends to np and var (n_1) tends to $np(1 - p)$. Explain why this phenomenon is not surprising.

HINT. See Exercise 6 and use the fact that $(N - n)/(N - 1) \to 1$ as
$N \to \infty$.

8. Five cards are dealt from a deck of 52 cards; it is assumed that this
is random sampling without replacement. Find
   (a) the probability that exactly two of the cards are kings;
   (b) the probability that no king is drawn on the first two draws;
   (c) the conditional probability that at least one king is drawn given
       that no king is drawn on the first two draws;
   (d) the probability that exactly two aces are drawn and that one of
       these two is the ace of spades.
   ANSWERS. (a) $C(4,2)C(48,3)/C(52,5)$. (b) $(^{48}/_{52})(^{47}/_{51})$. (c) $1 - [C(46,3)/C(50,3)]$. (d) $C(3,1)C(48,3)/C(52,5)$.

9. An experimental colony of fish has 70 fish, of which 20 are salmon
males, 15 are salmon females, 25 are trout males, and the rest are
trout females. A random sample of size 12 is chosen without replacement. Find the probability that:
   (a) There are more salmon than trout in the sample.
   (b) There are exactly seven males in the sample.
   (c) Males are the majority in the sample.
   (d) There are exactly five salmon and exactly three male fish in the
       sample.
   (e) There are exactly five salmon given that there are exactly three
       males in the sample.
   ANSWERS. (a) $\sum_{j=7}^{12} C(35,j)C(35,12-j)/C(70,12)$. (b) $C(45,7)C(25,5)/C(70,12)$. (c) $\sum_{j=7}^{12} C(45,j)C(25,12-j)/C(70,12)$. (d) $\sum_{j=0}^{3} C(20,j)C(25,3-j) C(15,5-j)C(10,4+j)/C(70,12)$. (e) $\sum_{j=0}^{3} C(20,j)C(25,3-j)C(15,5-j) C(10,4+j)/C(45,3)C(25,9)$.

10. An apple grower claims that in each carton of apples he ships there
are 12 large and 8 medium-sized apples. Twenty cartons are inspected by a buyer in the following way: six apples from each carton are chosen randomly without replacement, and, if more than
three of the six are medium size, the carton is rejected. What is the
probability that more than ten of the cartons are rejected, assuming
that the grower is telling the truth? Evaluate this probability by means
of a computer.
   ANSWER.

$$\text{Define } p = \sum_{j=4}^{6} C(8,j)C(12,6-j)/C(20,6).$$

Then the probability that more than 10 cartons are rejected is $\sum_{j=11}^{20} C(20,j)p^j(1-p)^{20-j}$.

11. A supermarket shelf has three brands of canned tuna on it, ten cans of the first brand, five of the second, and three of the third. Suppose six cans are chosen by a customer randomly, without replacement. Find the probability that two cans of each brand will be picked.

ANSWER. $C(10,2)C(5,2)C(3,2)/C(18,6)$.

12. Suppose that x and y have the following joint distribution: (a) X is hypergeometric with parameters $n = 10$, $N_1 = 50$, $N = 200$; and (b) the conditional distribution of y, given x, is hypergeometric with parameters $n = 10$, $N_1 = 50 - x$, $N = 190$. Show that $X + Y$ is hypergeometric with parameters $n = 20$, $N_1 = 50$, $N = 200$. (Intuitive explanation: A first sample of size 10 is drawn, and X is the number of type 1 objects in the sample. Then, without this sample being replaced, a second sample of size 10 is drawn, and Y is the number of type 1 objects in the second sample.)

BERNOULLI PROCESS AND
BINOMIAL DISTRIBUTION

By *Bernoulli trials* we mean identical independent experiments in each of which an event S may occur with probability p with $0 < p < 1$ or fail to occur with probability $q = 1 - p$. Occurrence of the event S is called a success, and nonoccurrence of S (i.e., occurrence of the complementary event F) is called a failure.

A Bernoulli trial then is an event that has only two possible outcomes, which we call "success" and "failure," with probabilities p and $q = 1 - p$, respectively. A sequence of Bernoulli trials is defined as a sequence in which the event is repeated in such a way that p and q are held constant, the repetitions being independent of previous results. A mechanism that generates such a sequence is called a *Bernoulli process*. The classic example is flipping a coin, the coin being called fair if $p = q = 1/2$, biased if $p \neq q$. Any game of chance is an example, if successive games are independent of each other and each game has the same probability p of winning. Obviously p need not be $1/2$, and in fact in most applications p is not $1/2$. It is always assumed that Bernoulli trials are independent, although often the phrase "independent Bernoulli trials" is used to emphasize this fact.

Suppose now that n repetitions, or Bernoulli trials, are performed, with S denoting success and F failure. The result is a string of n events such as $SFSS\cdots FS$. Because of the independence of the trials, we have

$$P(SFSS\cdots FS) = pqpp\cdots qp.$$

We note that there are 2^n such possible strings, since at each new trial a whole new set of combinations is generated by each of the possible outcomes of this trial.

Often we wish the probability $b(x;n,p)$ that in n such trials, exactly x successes occur, where x may be 0, 1, 2, $\cdots$, or n. In such a case, any of the above strings will do so long as the string contains exactly x successes and $n - x$ failures. Since each such string has probability $p^x q^{n-x}$ of occurring, and since the number of such strings is given by the binomial coefficient $\binom{n}{x}$, the probability mass function is

$$b(x;n,p) = \binom{n}{x} p^x q^{n-x} \qquad \text{for } 0 \leqslant x \leqslant n.$$

This is the *binomial distribution*, which in many ways is the most important of all the discrete distributions. For brevity, we will often write $b(x)$ for $b(x;n,p)$ in a given discussion. The random variable x is called a binomial variable with parameters n,p, which explains the use of the notation $b(x;n,p)$.

The first expositions of the binomial distribution were made by Abraham DeMoivre (1667–1754), and by Jakob Bernoulli (1654–1705). Jakob was one of the family of eminent Swiss mathematicians which included his brother Johann (1667–1748) and Johann's son Daniel (1700–1782). Bernoulli studied what we call a sequence of independent and identically distributed indicator random variables . By definition an indicator random variable takes on two values, either 0 or 1. For the ith trial of a Bernoulli sequence, the indicator random variable is 0 for a failure and 1 for a success.

If we think of the binomial distribution as arising from sampling, then we see that

1. the binomial distribution corresponds to sampling with replacement, whereas
2. the hypergeometric distribution corresponds to sampling without replacement.

When the size N, of the population from which sampling is done is large, then the effect of nonreplacement only slightly disturbs the probabilities. Therefore on intuitive grounds, we have for large N that

$$h(n_1;n,N_1,N) \approx b(n_1;n,p) \qquad \text{with} \quad p = N_1/N.$$

To derive the binomial mass function from the hypergeometric mass function, rearrange the hypergeometric as

$$h(n_1) = \frac{\dbinom{N_1}{n_1}\dbinom{N_2}{n_2}}{\dbinom{N}{n}}$$

$$= \binom{n}{n_1} \frac{N_1(N_1 - 1)\cdots(N_1 - n_1 + 1)N_2(N_2 - 1)\cdots(N_2 - n_2 + 1)}{N(N - 1)\cdots(N - n + 1)}.$$

Note that the fraction on the right is the probability of finding the n_1 units of N_1 and the n_2 units of N_2 in any specified order. The numerator of the fraction is a product of polynomials in N_1 and N_2 with leading terms $N_1^{n_1}$ and $N_2^{n_2}$ respectively. The denominator is a polynomial with leading term N^n. Now let N_1, N_2, and N approach infinity such that $N_1/N = p$ and $N_2/N = q$ remain constant. The limit of the fraction is the same as the limit of the leading terms, and thus the limit is $N_1^{n_1}N_2^{n_2}/N^n = p^{n_1}q^{n_2}$. The fraction is multiplied by $\binom{n}{n_1}$. Thus the limit of the hypergeometric is the binomial

$$b(n_1;n,p) = \binom{n}{n_1} p^{n_1}q^{n-n_1}.$$

PROBLEM. If a fair coin is flipped three times, what are the probabilities of 0, 1, 2, 3 heads?

SOLUTION.

$$b(0) = \binom{3}{0}\left(\frac{1}{2}\right)^3 = \frac{1}{8}$$

$$b(1) = \binom{3}{1}\left(\frac{1}{2}\right)^3 = \frac{3}{8}$$

$$b(2) = \binom{3}{2}\left(\frac{1}{2}\right)^3 = \frac{3}{8}$$

$$b(3) = \binom{3}{3}\left(\frac{1}{2}\right)^3 = \frac{1}{8}$$

The probability of at least one head is $1 - 1/8 = 7/8$.

PROBLEM. What is the probability of exactly two "aces" in seven rolls of a die?

SOLUTION. $\binom{7}{2}(1/6)^2(5/6)^5$.

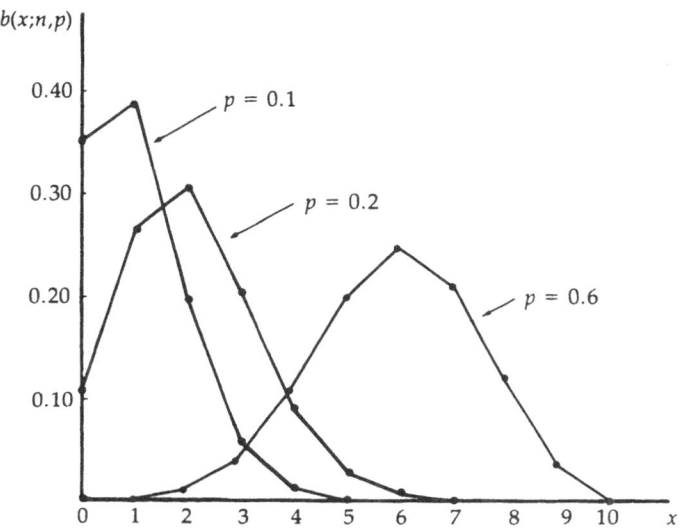

FIGURE 7.3 Binomial distributions for $n = 10$, $p = 0.1, 0.2, 0.6$. The discrete probability values are shown by dots. For clarity, straight lines connect the dots.

Figure 7.3 exhibits three binomial distributions with $n = 10$, thus showing the effect of the value of p on the distribution. The points have been connected in order to facilitate visual comparison.

The name "binomial" comes from the relationship of $b(x)$ to the binomial expansion

$$(q + p)^n = q^n + nq^{n-1} p + \frac{n(n - 1)}{2} q^{n-1} p^2 + \cdots + p^n.$$

The term in this expansion involving p^x, where $x = 0, 1, 2, \cdots, n$, is given by $[n!/x!(n - x)!]q^{n-x}p^x$. But this term is precisely $b(x)$. Thus the binomial expansion is

$$(q + p)^n = b(0) + b(1) + b(2) + \cdots + b(n).$$

Writing now the more explicit notation $b(x;n,p)$ in place of $b(x)$, we see that this equation shows that the sum of the probabilities in a binomial distribution is one:

$$\sum_{x=0}^{n} b(x;n,p) = \sum_{x=0}^{n} \binom{n}{x} p^x q^{n-x} = (p + q)^n = 1.$$

Next, let us compare the value of $b(x;n,p)$ for various x's. We note that

$$\frac{b(x)}{b(x-1)} = \frac{n-x+1}{xq}(1-q) = 1 + \frac{(n+1)p-x}{xq}.$$

This ratio is greater than 1 until $x \geqslant (n+1)p$. If $(n+1)p$ is an integer, then there are two equal largest terms; otherwise, there is one largest term. See Exercise 15 for this section. For example, when $n = 20$, $p = \frac{1}{3}$ then the maximum terms (modes) are at $x = 6, 7$. As another example, when $n = 20$, $p = \frac{1}{2}$, then the maximum term (mode) is at $x = 10$. In any case, the maximum term(s) or mode(s) occur(s) approximately at the point $x = np$.

However, the value of the mode (or modes) which corresponds to the most likely event (or events) may be small. In 100 tossings of a fair coin, the most likely number of heads is 50, but the probability of exactly this event is less than 0.08.

By direct algebraic calculation, one can find the mean and variance of the binomial. The results are

$$E(x) = \mu = \sum_{x=0}^{n} xb(x) = np \qquad \text{and} \qquad \sigma^2 = \sum_{x=0}^{n} (x-\mu)^2 b(x) = npq.$$

We omit this calculation here, in favor of using the moment generating function. The binomial moment generating function is

$$M(\theta;x) = \sum_{x=0}^{n} e^{\theta x} b(x) = \sum_{x=0}^{n} e^{\theta x}\binom{n}{x}p^x q^{n-x}$$

$$= \sum_{x=0}^{n} \binom{n}{x}(pe^\theta)^x q^{n-x}.$$

We recognize the right member as the binomial expansion, but now with $pe^{i\theta}$ in place of p. Hence we have

$$M(\theta;x) = (pe^\theta + q)^n.$$

Its first derivative is

$$\frac{dM(\theta;x)}{d\theta} = n(pe^\theta + q)^{n-1}pe^\theta,$$

from which it follows that the mean is

$$E(x) = \frac{dM(\theta;x)}{d\theta}\bigg|_{\theta=0} = np.$$

Similarly, we can find the variance. The result is

$$\sigma^2 = E(x^2) - [E(x)]^2 = npq.$$

We now want to show that the binomial can be regarded as the distribution of a sum of special random variables, namely, the indicator random variables. As we have seen, the binomial distribution arises from performing n Bernoulli trials. Each such trial, and its result, is independent of the other trials. Let us consider an individual Bernoulli trial, with probability p of success, and define the indicator random variable x_i, corresponding to the ith trial, as follows:

$$x_i = \begin{cases} 1 & \text{if success occurs,} \\ 0 & \text{if failure occurs.} \end{cases}$$

The random variable x_i is also called a zero-one Bernoulli variable. The probability distribution for x_i is

$$b_i(x_i) = \begin{cases} p & \text{for } x_i = 1, \\ q = 1 - p & \text{for } x_i = 0. \end{cases}$$

The indicator moment generating function is then

$$M(\theta;x_i) = b_i(0) + e^\theta b_i(1) = q + pe^\theta.$$

The mean and variance of x_i are then p and pq, respectively.

Now let n Bernoulli trials be performed, so that $i = 1, 2, \cdots, n$, and define

$$x = x_1 + x_2 + \cdots + x_n.$$

We see that x records the number of successes in the n trials. We recall that Theorem 6.8 states that the moment generating function of the sum of independent random variables is equal to the product of their respective moment generating functions. Thus

$$M(\theta;x) = M(\theta;x_1) \cdot M(\theta;x_2)\cdots M(\theta;x_n)$$
$$= (pe^\theta + q)^n$$
$$= \sum_{x=0}^n \binom{n}{x} e^{\theta x} p^x q^{n-x} = \sum_{x=0}^n f(x)e^{\theta x}.$$

Equating coefficients of $e^{\theta x}$ in this expansion, we see that $f(x)$ is the probability mass function

$$f(x) = \binom{n}{x} p^x q^{n-x},$$

i.e., we actually have the binomial distribution for x. Moreover, from Theorem 5.6 the mean is

$$E(x) = \sum_{i=1}^{n} E(x_i) = \sum_{i=1}^{n} p = p + p + \cdots + p = np$$

and the variance is

$$\text{var}(x) = \sum_{i=1}^{n} \sigma^2(x_i) = \sum_{i=1}^{n} pq = pq + pq + \cdots + pq = npq.$$

Thus any binomial distribution is actually the distribution of a sum of independent zero-one Bernoulli variables. Or, in other words, a binomial variable with parameters n and p is the sum of n independent Bernoulli zero-one variables with parameter p.

The binomial distribution is a special case of the multinomial distribution. For suppose that an individual trial has k possible outcomes, $E_1, E_2, \cdots, E_k$, with probabilities $p_1, p_2, \cdots, p_k$, where $p_1 + p_2 + \cdots + p_k = 1$. Let n independent trials be made. Then the probability that E_1 occurs n_1 times, E_2 occurs n_2 times, $\cdots$ E_k occurs n_k times, where $n_1 + n_2 + \cdots + n_k = n$, is just

$$m(n_1, n_2, \cdots, n_k; n, p_1, p_2, \cdots, p_k) = \frac{n!}{n_1! n_2! \cdots n_k!} p_1^{n_1} p_2^{n_2} \cdots p_k^{n_k}.$$

This is the *multinomial distribution*. The binomial distribution is the special case when $k = 2$. The multinomial distribution is of interest in connection with problems of testing the goodness of fit of a theoretical curve to empirical data. It will be noted that the multinomial distribution is a multivariate distribution.

In some applications of the binomial distribution $b(x; n, p)$, n may be large and p small, the mean $\mu = np$ being of moderate size. For example, p may be the probability that an automobile has a flat tire on a certain expressway, and n the daily number of cars using the expressway. In such cases, there is a useful approximation to the binomial distribution first derived by Poisson (1781–1840). We can write

$$b(0;n,p) = q^n = (1 - p)^n = (1 - \mu/n)^n.$$

As $n \to \infty$, this approaches $e^{-\mu}$, and the approximation is even good for finite n if μ^2/n is small, because

$$\log b(0;n,p) = n \log\left(1 - \frac{\mu}{n}\right) = -\mu - \frac{\mu^2}{2n} - \cdots.$$

Further, the general term $b(x;n,p)$ can be written

$$b(x;n,p) = \binom{n}{x} p^x q^{n-x} = \binom{n}{x} p^x q^{-x} q^n = \binom{n}{x} p^x (1 - p)^{-x} b(0;n,p),$$

since $q = 1 - p$ and $q^n = b(0;n,p)$. Thus we have

$$b(x;n,p) = \frac{n(n - 1)\cdots(n - x + 1)}{x!} p^x (1 - p)^{-x} b(0;n,p)$$

$$= \frac{n(n - 1)\cdots(n - x + 1)}{n^x} \frac{(np)^x}{x!} (1 - p)^{-x} b(0;n,p)$$

$$= \left[\left(1 - \frac{1}{n}\right)\left(1 - \frac{2}{n}\right)\cdots\left(1 - \frac{x - 1}{n}\right)(1 - p)^{-x}\right] \frac{\mu^x}{x!} b(0;n,p).$$

Under the assumptions that n is large, p small, and $np = \mu$ is fixed, we see that the factor in the square brackets is approximately one. We have already seen that $b(0;n,p)$ is approximately $e^{-\mu}$. Therefore under the assumptions of large n, small p, and fixed $\mu = np$, we have the approximation

$$b(x;n,p) \approx \frac{\mu^x}{x!} b(0;n,p) \approx \frac{\mu^x}{x!} e^{-\mu}.$$

This equation is called the *Poisson approximation to the binomial*, where the Poisson probability is defined as

$$p(x;\mu) = \frac{\mu^x}{x!} e^{-\mu}.$$

As presented here, $p(x;\mu)$ is represented only as a numerical approximation to the binomial, not as a probability distribution in its own right. (In the next section, we show that the Poisson probability represents in fact a proper probability distribution with the infinite sample space $x = 0, 1, 2, \cdots$; here, x is bounded by n.) We thus have:

THEOREM 7.1. If the probability of success in a single trial p approaches 0 while the number of trials n becomes infinite in such a manner that the mean $\mu = np$ remains fixed, then the binomial distribution approaches the Poisson distribution with mean μ.

As an illustration of the degree of numerical approximation, Table 7.1 compares $b(x;n,p)$ and $p(x;\mu)$ for $n = 100$, $p = {}^1/_{100}$ and $\mu = 1$. Two other comparisons are shown in Figure 7.4. One advantage of the Poisson approximation is that its terms are considerably easier to calculate than the binomial terms.

PROBLEM. If the probability that a fire starts inside a house on any given day is 8×10^{-4}, what is the probability of more than one fire a day in a city with 10^4 houses?

SOLUTION. Here $np = 10^4 \times 8 \times 10^{-4} = 8$. By the Poisson approximation, the probability of more than 1 fire is $1 - p(0;8) - p(1;8) = 1 - e^{-8} - 8e^{-8} = 1 - 0.00306 \approx 0.997$.

PROBLEM. If a book contains an average of 1 misprint every 20 pages, what is the probability of no misprints in 100 consecutive pages?

SOLUTION. Here $n = 100$, $p = 1/20$, $\mu = 5$. Therefore the required probability is approximately $p(0;5) = e^{-5} \approx 0.007$.

PROBLEM. Calculate the probability that at most 5 defective fuses will be found in a box of 200 fuses if experience shows that 2 percent of such fuses are defective.

TABLE 7.1 Comparison of the Binomial and Poisson Distributions for $n = 100$, $p = 0.01$, and $\mu = 1$

x	$b(x;100,0.01)$	$p(x;1)$
0	0.3660	0.3679
1	0.3697	0.3679
2	0.1849	0.1839
3	0.0610	0.0613
4	0.0149	0.0153
5	0.0029	0.0031
6	0.0005	0.0005
7	0.00006	0.00007
8	0.000007	0.000009
9	0.000001	0.000001

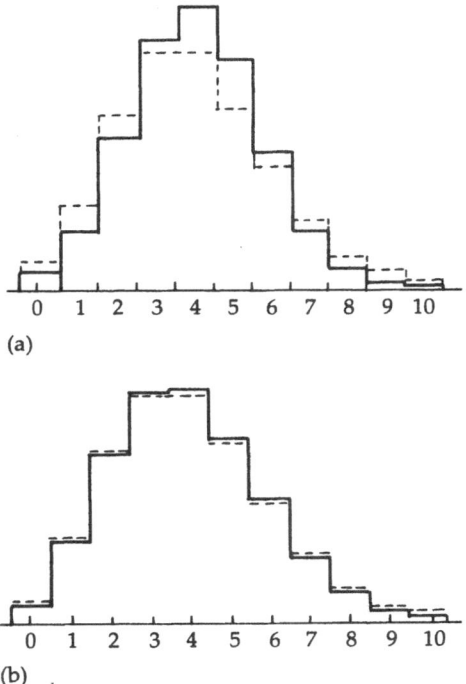

(a)

(b) .

FIGURE 7.4 Comparison of binomial (*solid line*) and Poisson (*dashed line*). (*a*) Case when $n = 12$, $p = \frac{1}{3}$, and $\mu = np = 4$. (*b*) Case when $n = 96$, $p = \frac{1}{24}$, and $\mu = np = 4$. It appears from inspecting these graphs that the Poisson approximation should be sufficiently accurate for most applications if $n \geq 100$ and $p \leq 0.05$.

SOLUTION. Here $\mu = np = 200(0.02) = 4$; hence, the approximate answer will be given by

$$e^{-4}[1 + 4 + (4^2/2) + (4^3/6) + (4^4/24) + (4^5/120)] = 0.785.$$

Lengthy calculations using the binomial yield the answer 0.788; hence, the approximation is very good here.

EXERCISES

1. Assuming that the sexes are equally likely, what is the probability that a family of six children will have at least two girls?

   ANSWER. The distribution of the number of girls is binomial with $p = \frac{1}{2}$, $n = 6$. The required probability is given by

$$1 - [f(0) + f(1)] = 1 - \left[\frac{1}{26} + \frac{6}{26}\right] = 0.89.$$

2. If A, B, and C match pennies and odd player wins, what is the distribution of the number of times that A wins in four matches? Assume that there must be an odd player to constitute a match.

ANSWER. The probability that A wins one particular match is $1/3$. The required distribution, therefore, is binomial with $p = 1/3$, $n = 4$, and is given by the following tabulation:

x	0	1	2	3	4	Sum
$b(x)$	0.198	0.395	0.296	0.099	0.012	1.000

3. For an unbiased coin, the probability of heads is $1/2$. Letting x denote the number of heads in five independent tosses, find the distribution of x and from it, by direct calculation, find μ and σ^2. Thus verify the theoretical results $\mu = np$, $\sigma^2 = npq$.

ANSWER. Here $b(x) = \binom{5}{x}/32$ since $p = q = 1/2$ and $n = 5$. The calculations are as follows.

x	$b(x)$	$xb(x)$	$x^2b(x)$
0	1/32	0	0
1	5/32	5/32	5/32
2	10/32	20/32	40/32
3	10/32	30/32	90/32
4	5/32	20/32	80/32
5	1/32	5/32	25/32
Sum	32/32	80/32	240/32

Hence by direct calculation

$$\sum_0^5 b(x) = {}^{32}/_{32} = 1,$$

$$\mu = \sum_0^5 xb(x) = {}^{80}/_{32} = {}^5/_2 = 2.5,$$

$$\sigma^2 = \sum_0^5 x^2b(x) - \mu^2 = {}^{240}/_{32} - {}^{25}/_4 = {}^5/_4 = 1.25.$$

These results check with the theoretical values, for

$$\mu = np = (5)(1/2) = {}^5/_2,$$
$$\sigma^2 = npq = (5)(1/2)(1/2) = {}^5/_4.$$

4. Show that the binomial probability generating function is $\Phi(t) = E(t^x) = (q + pt)^n$. Derive the mean and variance of the binomial distribution from the probability generating function.

5. In research on drugs to counteract the intoxicating effects of alcohol, 10 subjects were used to test the relative merits of benzedrine and caffeine. Benzedrine brought about more rapid recovery in 9 cases. Basing your analysis on the null hypothesis of a binomial distribution with $p = 0.5$, determine the significance level of this result. Assume that there is no a priori reason to suppose one drug more effective than the other.

ANSWER. Significance level $= b(0) + b(1) + b(9) + f(10) = {}^{22}/_{1024} \approx 0.0215$.

6. The watermark in bond paper is too conspicuous once in 500 times. What is the probability that a ream of bond paper (500 sheets) will contain no more than two sheets with this kind of flaw? Use the Poisson approximation to the binomial.

ANSWER. The true distribution is binomial with $n = 500$, $p = {}^1/_{500}$, and $\mu = np = 1$. Hence the Poisson approximation is $p(x) = e^{-1}/x!$. Since the conditions of the problem are met if $x = 0$, $x = 1$, or $x = 2$, the required probability is

$$P = p(0) + p(1) + p(2) = e^{-1}(1 + 1 + {}^1/_2) \approx 5/[2(2.178)] \approx 0.92.$$

7. Prove that the sum of independent binomial variables having only the same value of p in common is again a binomial variable with this parameter p.

8. An experiment consists of three tosses of a coin. Denoting the outcomes HHT, THT, $\cdots$, and assuming that all eight outcomes are equally likely, find the probability distribution for the total number of heads.

9. Assuming that the sexes are equally likely, what is the probability that a family of six children will have at least two boys?

ANSWER. 0.89.

10. Show that a binomial distribution with $p = 0.50$ is symmetrical. A probability mass function is said to be positively skewed if the long "tail" is on the right and it is said to be negatively skewed if the long "tail" is on the left. Show that the binomial is positively skewed if $p < 0.5$ and negatively skewed if $p > 0.5$.

11. If the annual hurricane season is 20 weeks long, with occurrence each week being a Bernoulli variable with probability 0.05 of success, find the probability of no hurricane

(a) during a whole season;
(b) during two consecutive seasons.

ANSWERS. (a) 0.0903. (b) 0.0083.

12. In successive flips of a fair coin, what is the probability of getting
    (a) 5 heads in 8 trials;
    (b) the 5th head at the 8th trial?

    ANSWERS. (a) 0.219. (b) 0.137.

13. Find the probability of getting 2 sixes, 3 fours, and 4 odd faces in 10 rolls of a die.

    ANSWER. 0.018.

14. If the population (assumed infinite) breakdown is 35 percent male adults, 40 percent female adults, and 25 percent minors, write an expression for the probability that a group of 100 people will consist of 35 men, 40 women, and 25 children.

    ANSWER. $P = \left(\dfrac{100!}{35!40!25!}\right)(0.35)^{35}(0.40)^{40}(0.25)^{25}$.

15. Show that if $(n + 1)p$ is not an integer, then the most probable value for the binomial distribution equals the integral part of the number $(n + 1)p$, and if $(n + 1)p$ is an integer, then the largest value of the probability is attained for the two numbers $(n + 1)p - 1$ and $(n + 1)p$.

16. Let x_i, for $i = 1, \cdots, n$, be the number of successes in the ith trial in n Bernoulli trials with parameter p.
    (a) For any positive integer r, find $E(x_i^r)$.
    (b) Let r, s be any integers. Are x_i^r and x_j^s independent when $i \neq j$? Why?
    (c) Compute $E(x_i^r x_j^s)$ where r, s are positive integers and $i \neq j$.
    (d) Let $x = x_1 + x_2 + \cdots + x_n$, so that x is the total number of successes in the n trials. Using (a)–(c), find $E(x^3)$.
    (e) Write an expression for $E(x^3)$ by using the probability distribution of x. Can you sum this expression?

    ANSWERS. (a) p. (c) p^2. (d) $np + n(n - 1)p^2 + n(n - 1)(n - 2)p^3$.

17. Show that if $n_1, n_2, \cdots, n_k$ are multinomially distributed with parameters n; $p_1, p_2, \cdots, p_k$, then n_1 is binomially distributed with parameters n, p_1.

18. (a) If n independent multinomial trials are performed, and if n_i is the total number of outcomes of the ith kind, (where $i = 1, 2, \cdots, k$), find $E(n_i)$ and var (n_i) in terms of n and the parameters $p_1, p_2, \cdots, p_k$.

(b) Represent n_i as a sum of n independent random variables. What is the distribution of each of these?

(c) By using the representation of n_i as the sum of n independent random variables and by doing the same thing for n_j, find cov (n_i, n_j). Are n_i and n_j independent? Why?

(d) Let $k = 3$. Thus, we have a trinomial distribution with parameters p_1, p_2, p_3. Let n_i be the total number of outcomes of the ith kind (where $i = 1, 2, 3$) when n such independent trinomial experiments are performed. Find $E(n_1)$, var (n_1) and cov (n_1, n_2) by direct computation using the distribution of n_1 and the joint distribution of n_1 and n_2.

ANSWERS. (a) $E(n_i) = np_i$, var $(n_i) = np_i(1 - p_i)$. (d) cov $(n_1, n_2) = -np_1 p_2$.

19. If a machine produces defective items with a probability of $1/20$, what is the expected number of defective items in a random sample of size 500 taken from its output? What is the variance of the number of defective items?

20. Let y be the proportion of successes in n Bernoulli trials with a common probability p of success in a given trial. Find $E(y)$ and var (y).

ANSWER. p, $p(1 - p)/n$.

21. A certain drug is known to have a cure rate of one in four. If 200 patients are given the drug, what is the expectation of the proportion of cures? What is its variance?

22. The random variable x has the binomial distribution with parameters n, p while the independent variable y has the binomial distribution with parameters m, p (that is, p is the same in both). If $z = x + y$ show that the distribution of z is binomial with parameters N, p where $N = n + m$.

ANSWER. $(q + pt)^n(q + pt)^m = (q + pt)^N$ is the probability generating function for $b(z; N, p)$.

23. Suppose we make $n + m$ Bernoulli trials with a common probability of success p. Let x be the number of successes in the first n trials, and let y be the number of successes in the last m trials.

(a) What is the probability distribution of x?

(b) What is the probability distribution of y?

(c) What is the probability distribution of $z = x + y$?

(d) Are x and y independent? Why?

(e) Using your answers to (a) through (d), or otherwise, establish the convolution formula

$$\sum_{x=0}^{n} b(x;n,p)b(z-x;m,p) = b(z,n+m,p).$$

HINT. The right side represents the probability that the total number of successes is z. Consider what values x and y can take if their sum is to be z.

24. Show that the binomial tail probability

$$P(k \geqslant r) = \sum_{k=r}^{n} \binom{n}{k} p^k (1-p)^{n-k}$$

can be expressed by the incomplete beta function

$$P(k \geqslant r) = \frac{1}{\beta(r,n-r+1)} \int_0^p y^{r-1}(1-y)^{n-r} \, dy,$$

for which there exist many tables.

HINT. Define

$$a^k = n\binom{n-1}{k}p^k(1-p)^{n-1-k}.$$

Differentiate term-by-term the first expression for $P(k \geqslant r)$ with respect to p, and thus obtain

$$\frac{dP(k \geqslant r)}{dp} = (a_{r-1} - a_r) + (a_r - a_{r+1}) + \cdots + (a_{n-1} - 0) = a_{r-1}.$$

Using the fact that $1/\beta(r,n-r+1) = n(n-1)!/(r-1)!(n-r)!$, show that the derivative of the second expression for $P(k \geqslant r)$ is also a_{r-1}. Because both of their derivatives are the same, we can use the fundamental theorem of calculus to assert that both of the expressions can differ only by a constant (with respect to p). Set $p = 0$ to see that this constant is actually zero.

25. Let

$$B(k;n,p) = \sum_{r=0}^{k} b(r;n,p).$$

Thus, $B(k;n,p)$ is the probability of at most k successes in n independent Bernoulli trials with a common probability p of success in a given trial. Show that:

$$B(k;n+1,p) = B(k;n,p) - pb(k;n,p) \quad \text{and}$$
$$B(k+1;n+1,p) = B(k;n,p) + qb(k+1;n,p).$$

26. To illustrate the use of the binomial distribution in a problem of

decision making, suppose a manufacturer claims that a particular production process does not turn out more than 10 percent defectives, and that the decision whether to accept or reject this claim has to be made on the basis of a shipment of 20 units among which 5 are defective. To justify the decision one way or the other, find the probability of getting 5 or more defectives in a sample of 20 when the "true" percentage of defectives is 0.10.

ANSWER. Because the probability

$$\sum_{x=5}^{20} b(x;20,0.10) = 1 - B(4;20,0.10) = 0.0432$$

is very small, it would seem reasonable to reject the manufacturer's claim. (Note that this probability would have been even smaller if we had used a value of p less than 0.10.) On the other hand, if the sample had contained only 3 defectives, it would have been much more difficult to reach a decision. The probability of getting 3 or more defectives in a sample of 20, when the probability that any one unit is defective is 0.10, is 0.3230. This can hardly be described as a "rare event" and it certainly does not justify rejecting the manufacturer's claim.

27. The probability that a car driving the entire length of a certain turnpike will have a blowout is 0.05. Find the probability that among 17 cars traveling the length of this turnpike,
(a) exactly one will have a blowout,
(b) at most 3 will have a blowout,
(c) 2 or more will have a blowout.
ANSWERS. (a) 0.3741. (b) 0.9912. (c) 0.2078.

28. Two dice are thrown n times. Show that the number of throws in which the number on the first die exceeds the number on the second die is binomially distributed with parameters n and $p = {}^5/_{12}$.

29. Suppose n balls numbered 1, 2, ..., n are randomly distributed among n cells, also numbered 1, 2, ..., n. Let x be the number of balls that fall in cells marked with the same number as the balls. Show that x is binomially distributed with parameters n and $p = 1/n$.

30. If x is binomially distributed with parameters n,p, show that $n - x$ is binomially distributed with parameters n, $1 - p$.

31. What is more probable in playing against an equal adversary (if the game cannot end in a tie)?
(a) To win three games out of four or five out of eight.
(b) To win at least three games out of four or at least five out of eight.

ANSWERS. (a) To win three games out of four. (b) To win at least five games out of eight.

32. Suppose x is binomially distributed with parameters n,p. Show that if $p = 0$, then x is the constant 0, and if $p = 1$, then x is the constant n.

33. Suppose x is binomially distributed with parameters n,p. Show that x is symmetrically distributed about c if, and only if, $p = \frac{1}{2}$ and $c = n/2$.

    HINT. $p^n = (1 - p)^n$ must hold.

34. Suppose x and y are independent, binomially distributed random variables with parameters n,p, and $n,1-p$, respectively. Show that $x + y$ is distributed like $u_1 + \cdots + u_n$, a sum of n independent random variables, each u_i with the following distribution:

k :	0	1	2
$P(u_i = k)$:	$p(1 - p)$	$(1 - p)^2 + p^2$	$p(1 - p)$.

    HINT. What is the distribution of a sum of two independent indicator random variables that equal one with probabilities p and $1 - p$, respectively?

35. A particle performs a random walk over the positions 0, ± 1, ± 2, $\cdots$ in the following way. The particle starts at 0. It makes successive one-unit steps that are mutually independent; each step is to the right with probability p, or to the left with probability $1 - p$. Let x be the position of the particle after n steps.
    (a) Show that $(x + n)/2$ is binomially distributed with parameters n,p.
    (b) Show that the expected position of the particle after n steps is $n(2p - 1)$.

36. (a) Suppose that a population consists of k types of objects, N_1 of type 1, N_2 of type 2, and so forth. Then $N = N_1 + \cdots + N_k$ is the total population size. Suppose that a sample of size n is drawn without replacement. Let n_1 be the number drawn of type 1, n_2 the number drawn of type 2, and so forth, so $n = n_1 + n_2 + \cdots + n_k$. Show that

$$f(n_1, n_2, \cdots, n_k) = \frac{\binom{N_1}{n_1}\binom{N_2}{n_2}\cdots\binom{N_k}{n_k}}{\binom{N}{n}}$$

$$= \frac{n!}{n_1! n_2! \cdots n_k!} \frac{(N_1)_{(n_1)}(N_2)_{(n_2)} \cdots (N_k)_{(n_k)}}{N_{(n)}}.$$

This is the generalization of the hypergeometric distribution to the multivariate case. [Recall the notation $n_{(k)} = n(n - 1)\cdots(n - k + 1)$.]

(b) Suppose that N, N_1, $\cdots$, N_k all approach infinity in such a way that

$$\frac{N_1}{N} \to p_1, \frac{N_2}{N} \to p_2, \cdots, \frac{N_k}{N} \to p_k \quad \text{with } p_1 + p_2 + \cdots + p_k = 1.$$

Show that the distribution in (a) approaches the multinomial distribution with parameters $n;p_1,p_2,\cdots,p_k$.

POISSON PROCESS AND POISSON DISTRIBUTION

Having observed the Poisson approximation to the binomial, one is led to consider the following reasoning. In Bernoulli trials, the individual trials are made presumably at discrete time intervals, with individual probability p of success. Let us imagine such trials to be performed more and more frequently, approaching a continuous rate of performance. Obviously, then the expected number of successes per unit time will increase without limit if p remains constant. However, if $p \to 0$ as the rate of occurrence of trials $\to\infty$, then the mean rate λ of success per unit time can be finite, so that the way is open to getting a reasonable probability distribution.

Suppose that the events occur randomly in the course of continuous time. For example, we can think in terms of "service calls" (requests for service) arriving randomly at some server (service facility), such as arrivals of motorists at a gas station, telephone calls at an exchange, inquiries at an information desk, etc. Let x be the number of calls that arrive in a time interval from τ to $\tau + t$. Then what is the distribution of the random variable x?

In order to answer this question, we need a model. A useful model is provided by the *Poisson process* which describes a "random flow of events" by the following three properties:

1. The events are independent of one another; more exactly, the random variables x_1, x_2, $\cdots$ are independent if their respective intervals $[\tau_1,\tau_1+t_1]$, $[\tau_2,\tau_2+t_2]$, $\cdots$ are nonoverlapping.
2. The flow of events is stationary; i.e., the distribution of the random

variable x depends only upon the length t of the interval $[\tau,\tau+t]$ and not on the initial time τ.

3. The elementary probability is linear; more exactly, the probability that at least one event occurs in a small time interval $[\tau,\tau+\Delta t]$ is $\lambda\Delta t + o(\Delta t)$, while the probability that more than one event occurs in this small interval is $o(\Delta t)$. Here $o(\Delta t)$ is an infinitesimal of higher order which can be neglected as $\Delta t \to 0$. The constant λ is a positive parameter giving the "mean rate of occurrence" of the events.

Let us now use these properties. Consider the time interval $[0,t]$ and let x be the total number of events occurring in this interval. Divide the interval $[0,t]$ into n equal subintervals $[0,t/n]$, $[t/n,2t/n]$, $\cdots$, $[(n - 1)t/n,t]$, and let $x_1, x_2, \cdots, x_n$ respectively be the number of events occurring in each subinterval. In property (3) we set $\Delta t = t/n$ and neglect higher order terms. Thus

$$P(x_i = 0) = 1 - \frac{\lambda t}{n}, \; P(x_i = 1) = \frac{\lambda t}{n} \qquad \text{for } i = 1, 2, \cdots, n.$$

Hence the probability generating function for each x_i is

$$\Phi_n(z) = P(x_i = 0) + zP(x_i = 1) = \left(1 - \frac{\lambda t}{n}\right) + \frac{\lambda t}{n}z,$$

where we have used the symbol z in the pgf instead of the customary t, as t is now being used as the symbol for time. Because the x_i are independent, it follows that the probability generating function of their sum $x = x_1 + x_2 + \cdots + x_n$ is

$$\Phi(z) = [\Phi_n(z)]^n = [1 + \lambda t(z - 1)/n]^n,$$

which is the limit as $n \to \infty$ becomes

$$\Phi(z) = \exp[\lambda t(z - 1)] = \exp[\lambda tz] \exp[-\lambda t].$$

We now expand the exponential, and obtain

$$\Phi(z) = e^{-\lambda t} \sum_{x=0}^{\infty} \frac{(\lambda t)^x}{x!} z^x = \sum_{x=0}^{\infty} p(x)z^x.$$

Thus the probability mass function of x is

$$p(x) = \frac{(\lambda t)^x}{x!} e^{-\lambda t} \qquad \text{for } x = 0, 1, 2, \cdots.$$

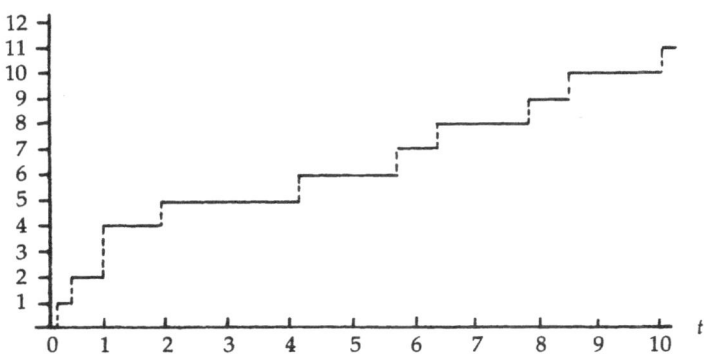

FIGURE 7.5 The number x of events plotted against arrival time t with a mean rate of $\lambda = 1$ count per unit time. For this as well as any other Poisson process the random variable x is discrete but time t is continuous.

The function $p(x)$ is the *Poisson distribution* for parameter λt, and is often written as $p(x;\lambda t)$. As we have just seen, the *Poisson probability generating function* is exp $[\lambda t(z - 1)]$. One realization of a Poisson process is shown in Figure 7.5.

The Bernoulli process is defined in terms of a probabilistic description of the "arrivals" of successes in a series of independent discrete trials. Its continuous analog, namely the Poisson process, is defined in terms of a probabilistic description of the behavior of arrivals at points on a continuous line. The following two theorems give the important results we have derived concerning these two processes.

THEOREM 7.2. In a Bernoulli process with parameter p, the probability of x successes in n trials is given by the binomial mass function $b(x;n,p)$.

THEOREM 7.3. In a Poisson process with parameter λ, the probability of x successes in time t is given by the Poisson mass function $p(x;\lambda t)$.

Applied research has turned up many examples of the Poisson process and thus of the Poisson distribution. These examples range all the way from radioactive emissions in physics, to rate of customer arrivals in business, to frequency of accidental machine breakdown in industrial engineering. In the case of radioactive disintegration, the number of particles reaching a counter per unit time has been observed to be Poisson in distribution. Table 7.2 compares the observed

TABLE 7.2 Comparison of Observed Frequency
with the Poisson Approximation

x	Observed Frequency	Poisson Probability (approximation)
0	57	55
1	203	211
2	383	407
3	525	524
4	532	509
5	408	394
6	273	253
7	139	141
8	45	68
9	27	29
≥10	16	17

Source: Ernest Rutherford, James Chadwick, and C. D. Ellis, *Radiation from Radioactive Substances* (Cambridge: Cambridge University Press, 1920), p. 172.

frequencies and the corresponding numbers predicted by the Poisson distribution.

Let us now look at the properties of the Poisson distribution. In general, we will denote the parameter by μ, so the Poisson probability mass function is

$$p(n;\mu) = \frac{\mu^n}{n!} e^{-\mu} \qquad \text{for } \mu > 0,\ n = 0, 1, 2, \cdots.$$

It is consistent as a probability distribution, as can be seen by summing the terms. Since

$$e^{-\mu} \sum_{n=0}^{\infty} \frac{\mu^n}{n!} = e^{-\mu} e^{\mu} = 1,$$

we get 1 as we should. As for the maximum term, the ratio of $p(n+1;\mu)$ to $p(n;\mu)$ is just $\mu/(n + 1)$. Thus the maximum occurs at the first value of n for which $n + 1 > \mu$, roughly in the neighborhood of μ.

Let us now compute the Poisson moment generating function. We have

$$M(\theta;x) = \sum_{x=0}^{\infty} e^{\theta x} p(x;\mu) = \sum_{x=0}^{\infty} e^{\theta x} \frac{\mu^x}{x!} e^{-\mu}$$

$$= e^{-\mu} \sum_{x=0}^{\infty} \frac{(e^{\theta}\mu)^x}{x!} = e^{-\mu} e^{e^{\theta}\mu}.$$

$$= e^{-\mu(1-e^{\theta})}.$$

From this we find that $E(x) = \mu$ and var $(x) = \mu$. Thus, the mean and variance of the Poisson distribution are each equal to the parameter μ. The following problem, however, finds the mean by direct summation.

PROBLEM. The discrete variable x has the Poisson distribution:

$$p(x;\mu) = e^{-\mu}\frac{\mu^x}{x!} \qquad \text{for } \mu > 0, x = 0, 1, 2, \cdots.$$

Find the mean value of x.

SOLUTION. Since $0p(0;\mu) = 0$, we may write

$$E(x) = \sum_0^\infty xp(x;\mu) = \sum_1^\infty xp(x;\mu).$$

Now if $x > 0$ we have

$$xp(x;\mu) = e^{-\mu}\frac{\mu^x}{(x-1)!} = \mu e^{-\mu}\frac{\mu^{x-1}}{(x-1)!}.$$

Setting $y = x - 1$ and noting that the Maclaurin expansion for e^μ is

$$e^\mu = \sum_0^\infty \frac{\mu^y}{y!},$$

we obtain

$$E(x) = \mu e^{-\mu}\sum_0^\infty \frac{\mu^y}{y!} = \mu e^{-\mu}e^\mu = \mu.$$

In Figure 7.6, Poisson distributions are plotted for $\mu = 1$ and $\mu = 10$.

In the inspection and quality control of manufactured goods the proportion of defective articles in a large lot should be small. Consequently the number of defectives in the lot might be expected to follow a Poisson distribution. For this reason, the Poisson distribution plays an important role in the development of plans for the quality control of manufactured articles.

A much quoted example of a good fit of a Poisson distribution is the number of men in the Prussian Army who were killed during a year by the kick of a horse. At any given time, the men are exposed to a small probability of being kicked. Thus we may consider the event of being kicked as a success in a Poisson process with average success rate μ. For the sequence of such kicks, we may form a new process

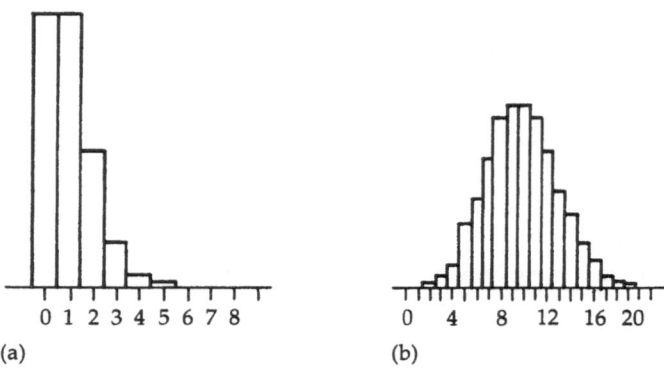

FIGURE 7.6 Poisson distribution for (a) $\mu = 1$ and (b) $\mu = 10$.

by performing an independent Bernoulli trial for each success in the Poisson process. Let us assume that the kicked man has probability p of living. Thus, with probability p, any success (man kicked) in the original Poisson process is also considered a success (kicked man lives) at the same time in the new process. With probability $1 - p$, any particular success (man kicked) in the original process does not appear in the new process (i.e., the kicked man dies). The new process formed in this manner (i.e., by independent random erasures) can be shown to be a Poisson process with an average success rate of μp. See following Exercise 9.

The Poisson distribution was first published in 1837 by Simeon Denis Poisson (1781–1840), French mathematician, in his book *Recherches sur la probabilite des jugements en matiere criminelle et en matiere civile*. For many years, however, the Poisson distribution was regarded as more of an esoteric mathematical result than as a useful scientific tool. Applications were limited to such things as the number of Prussian soldiers who lived after the kick of a horse. A turning point occurred in 1907 when the famous statistician, W. S. Gosset (who wrote under the pseudonym "Student" and who discovered the t distribution) deduced that the number of minute corpuscles found in a drop of a liquid (under the assumption that the corpuscles are distributed at random throughout the liquid) have the Poisson distribution. Then, as we have seen in this section, Ernest Rutherford (1871–1937), the British physicist who theorized the existence of the atomic nucleus, and his coworkers showed in 1920 that the Poisson distribution fitted the distribution of the number of particles discharged per unit time from a radioactive substance. Since then there have been many applications of the Poisson distribution, and now it stands with the binomial and normal distributions as one of the most useful distributions in probability theory.

EXERCISES

1. Graphically compare the binomial probability mass function for $n = 10$, $p = 0.05$ with the Poisson approximation to the binomial. Compare also the variances.

2. A book of 500 pages contains 500 misprints. Selecting pages at random and using the Poisson distribution, find the probability:
   (a) of 3 misprints on a given page.
   (b) of no misprints in 10 pages.
   (c) of more than 2 misprints on a page.
   ANSWERS. (a) 0.0613. (b) 0.000222. (c) 0.264.

3. Independent variables x_1, x_2, x_3 have Poisson distributions with respective parameters $\mu_1 = 0.27$, $\mu_2 = 0.32$, $\mu_3 = 0.41$. Determine the probability that $y \geq 4$, where $y = x_1 + x_2 + x_3$.
   ANSWER. $P = 1 - 8/(3e) \approx 0.019$.

4. The average number of rainy days for a specific period in a certain region is five, and the probability of any number of rainy days for the same period is given by the Poisson distribution. The monetary value of rain as affecting crops in this area increases exponentially with the number of rainy days x according to the expression $(e^{2x} - 1)$. What is the expected monetary value of the rain?
   HINT. Same idea as moment generating function.
   ANSWER. $e^{5(e^2-1)} - 1$.

5. Show that, if two random variables each having the Poisson distribution are correlated, their sum does not have the Poisson distribution.
   HINT. It is sufficient to show that the variance of the sum does not equal the mean of the sum.

6. Suppose $p = 0.2$ is the probability of success in a single trial of a Bernoulli process. Estimate the total probability of obtaining less than 6 or more than 15 successes in 50 trials.
   ANSWER. 0.082.

7. (a) Independent random variables x_1, x_2 have the Poisson distribution with respective parameters μ_1, μ_2. Show that the variable $y = x_1 + x_2$ has the Poisson distribution with parameters $\mu = \mu_1 + \mu_2$. Extend by induction to any number of independent Poisson random variables.

HINT. Set up the joint distribution of x_1, x_2; replace x_2 by $y - x_1$; multiply and divide by $y!$, then sum over x_1 from $x_1 = 0$ to $x_1 = y$ using the binomial theorem.

(b) Using the same approach as in (a), show analytically that the sum of two independent binomial random variables with respective parameters (n_1, p) and (n_2, p) is binomially distributed with parameters (n, p) where $n = n_1 + n_2$. Extend this result to any number of independent binomial variables.

(c) Independent random variables x_1, x_2, x_3 have the Poisson distribution with respective parameters $\mu_1 = 0.27$, $\mu_2 = 0.32$, $\mu_3 = 0.41$. Using the results of (a), determine the probability that $y \geq 4$, where $y = x_1 + x_2 + x_3$.

ANSWER. 0.019.

8. Show that the probability generating function for the Poisson distribution with parameter μ is $\Phi(t) = E(t^x) = \exp[\mu(t - 1)]$. (Do not confuse the symbol t used here in the definition of the probability generating function with the symbol t used in the text to denote time.) Also show that the probability generating function for the binomial with parameters n, p, $q = 1 - p$ is $(q + pt)^n$. Using probability generating functions, do again Exercise 7(a) and 7(b).

9. The variable x has the Poisson distribution with parameter μ and the conditional distribution of y given x is binomial with parameters (x, p), that is, $\phi(y|x) = C(x, y)p^y q^{x-y}$ for $y = 0, 1, 2, \cdots, x$. Show that the marginal distribution of y is Poisson with parameter μp.

HINT. Set up the joint distribution of x, y. Introduce a new variable $u = x - y$ and sum over u from 0 to ∞ by recognizing the series expansion.

10. (a) Suppose that x has possible values 0, 1, 2, $\cdots$, and the probability generating function of x is $\Phi(t)$. If k is a positive integer, show that the probability generating function of kx is $\Phi(t^k)$. (The possible values of kx are 0, k, $2k$, $3k$, $\cdots$).

(b) If x is Poisson distributed with parameter μ and k is a positive integer, show that the probability generating function of kx is $\exp[\mu(t^k - 1)]$.

11. Define the distribution of x and y as $f(x, y) = f_2(y)\phi(x|y)$ such that $f_2(y)$ is Poisson with parameter μ and the conditional distribution $\phi(x|y)$ is that of $x = x_1 + x_2 + \cdots + x_y$ where the x_i's are independent nonnegative integer-valued random variables, each having the same distribution. The distribution $f(x, y)$ so defined is called the *compound Poisson distribution*. Suppose that the common distribution of the x_i's has possible values 0, 1, 2, $\cdots$, n with probabilities p_0, p_1,

$\cdots$, p_n respectively. Let the probability generating function of this common distribution be $\Phi(t) = p_0 + p_1 t + \cdots + p_n t^n$.

(a) Show that the probability generating function of x is

$$\Psi(t) = \exp\{\mu[\Phi(t) - 1]\}.$$

HINT. From the definition of the compound Poisson distribution, the probability generating function of x given y is $E(t^x|y) = [\Phi(t)]^y$ for $y = 0, 1, 2, \cdots$. Thus the probability generating function of x is

$$\Psi(t) = E(t^x) = \sum_{y=0}^{\infty} E(t^x|y) f_2(y) = \sum_{y=0}^{\infty} [\Phi(t)]^y \mu^y e^{-\mu}/y!.$$

(b) Suppose that $n = 1$ so the $x_1, x_2, \cdots, x_y$ are indicator random variables (Bernoulli trials) with $\Phi(t) = q + pt$. Then x given y is a binomial variable with parameters y, p. Show that x is a Poisson variable with parameter μp.

HINT. $\Psi(t) = \exp[\mu(q + pt - 1) = \exp[\mu p(t - 1)]$.

(c) Suppose that $n = 2$ with $\Phi(t) = (1 + t + t^2)/3$. Show that x has the generating function

$$\Psi(t) = \exp[\mu(1 + t + t^2 - 3)/3]$$
$$= \exp[\mu(t - 1)/3] \exp[\mu(t^2 - 1)/3].$$

Thus show that x is distributed like $u_1 + 2u_2$ where u_1 and u_2 are independent and each is Poisson with parameter $\mu/3$.

HINT. Use Exercise 10(b).

(d) In general suppose $\Phi(t) = p_0 + p_1 t + \cdots + p_n t^n$. Show that x is distributed like $u_1 + 2u_2 + \cdots + nu_n$, where $u_1, u_2, \cdots, u_n$ are mutually independent, Poisson distributed, with parameters μp_1, $\mu p_2, \cdots, \mu p_n$, respectively.

HINT.

$$E(t^x) = \exp \mu \left[\sum_{i=1}^{n} p_i t^i - 1 \right] = \exp[\mu p_1(t - 1) + \cdots + \mu p_n(t^n - 1)].$$

12. (a) Suppose that x and y are independent, Poisson distributed with parameters λ_1 and λ_2. Show that the conditional distribution of x, given that $x + y = n$, is binomial with parameters n and $p = \lambda_1/(\lambda_1 + \lambda_2)$ for $n > 0$.

(b) Suppose that $x_1, \cdots, x_m$ are mutually independent, Poisson distributed variables with parameters $\lambda_1, \cdots, \lambda_m$. Let $\lambda = \lambda_1 + \cdots + \lambda_m$. Show that the joint conditional distribution of $x_1, \cdots, x_m$, given that $x_1 + \cdots + x_m = n$, is multinomial with parameters n; $\lambda_1/\lambda, \cdots, \lambda_m/\lambda$ for $n > 0$.

13. Why should we expect that the variance of the Poisson distribution is equal to its mean μ?

ANSWER. This would be expected since the binomial variance npq tends to its mean np when q tends to one.

GEOMETRIC AND PASCAL
DISTRIBUTIONS

In the hypergeometric and binomial distributions, we essentially consider situations in which a definite number of trials is performed, and ask for the probability of x successes among these trials. Of equal importance, in many applications, is a different random variable, namely the number k of trials required to obtain the first success. Two important distributions arise as a result.

First, returning to the process from which the hypergeometric distribution arose, suppose that in a population of N objects, there are N_1 favorable to a certain event, and $N_2 = N - N_1$ unfavorable. Let objects be selected one by one, each choice being random, but let any object selected be permanently withheld. (Sampling without replacement.) What is the probability $f(k)$ that a favorable object is first drawn on the kth trial?

Here $k = 1, 2, \cdots, N_2 + 1$, and

$$f(1) = \frac{N_1}{N},$$

$$f(2) = \frac{N_2}{N} \cdot \frac{N_1}{N-1} = \frac{(N-N_1)N_1}{N(N-1)},$$

$$\vdots$$

$$f(N_2 + 1) = \frac{N_2}{N} \frac{N_2 - 1}{N-1} \cdots \frac{1}{N_1 + 1} \cdot \frac{N_1}{N_1} = \frac{(N-N_1)!N_1!}{N!}.$$

The general term is seen to be

$$f(k) = \frac{\binom{N-N_1}{k-1}}{\binom{N}{k-1}} \cdot \frac{N_1}{N-k+1}.$$

This equation of course simply says that the first $k - 1$ objects must be selected from among the N_2 unfavorable ones, according to the hypergeometric probability

$$\frac{\binom{N_1}{0}\binom{N_2}{k-1}}{\binom{N}{k-1}} = \frac{\binom{N-N_1}{k-1}}{\binom{N}{k-1}};$$

and then the last object must be selected from among the N_1 favorable ones with probability $N_1/(N - k + 1)$.

The simplest case of this distribution occurs when $N_1 = 1$. In that case

$$f(k) = \frac{1}{N} \qquad \text{for } k = 1, 2, \cdots, n,$$

which is a discrete *rectangular distribution*. Its mean is $(N + 1)/2$.

Suppose instead that any unfavorable object selected is replaced. The trials now become Bernoulli trials with $p = N_1/N$ for success, N now becoming irrelevant. If k is the number of the trials on which the first success occurs, then k can be any positive integer and the probability $f(k)$ is $f(k) = (1 - p)^{k-1}p$ for $k = 1, 2, 3, \cdots$. According to our usual notation, we define $q = 1 - p$, so

$$f(k) = q^{k-1}p \qquad \text{for } k = 1, 2, 3, \cdots.$$

This is the probability mass function of the *Pascal-geometric distribution*. It has found many applications (due essentially to the usefulness of the concept of Bernoulli trials). We note some of its properties:

First, the mode occurs at $k = 1$, so that the most likely number of trials to success is 1. Next, the moment generating function is

$$M(\theta;k) = \sum_{k=1}^{\infty} e^{\theta k}f(k)$$

$$= \frac{pe^{\theta}}{1 - qe^{\theta}}.$$

From this we find by successive differentiations that $E(k)$, the mean of the distribution, is $1/p$, and that the variance is $\text{var}(k) = q/p^2$. We note that the sample space for k is infinite.

A typical Pascal-geometric distribution, with $p = 0.6$, is given in Figure 7.7.

We have just described the Pascal-geometric distribution in which

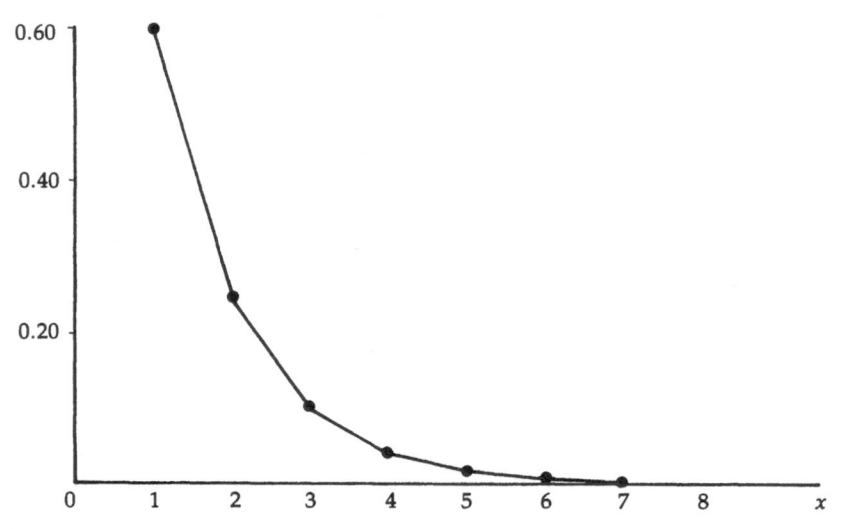

FIGURE 7.7 Pascal-geometric distribution with $p = 0.6$.

the random variable is k = number of trials for first success. In contrast, the regular *geometric distribution* has the random variable s = number of failures before the first success, so that the range of s is $s = 0, 1, 2, \cdots$. Since $s = k - 1$, the geometric probability mass function is

$$g(s) = q^s p, \qquad s = 0, 1, 2, \cdots.$$

As would be expected, this distribution has the mean $E(s) = (1/p) - 1 = q/p$. The variance, as can be verified, is the same as before, namely q/p^2.

In the Pascal-geometric distribution, we found the probability distribution for the number k of the trial at which success first occurs. Let us now generalize this to the number k of the trials at which the rth success occurs (assuming Bernoulli trials as before).

The required process can be described as follows. Let us imagine a succession of r independent experiments in which a new set of Bernoulli trials is begun as soon as each success is obtained. It follows that the total number k of trials needed for r successes (where k is called the Pascal variable) is precisely the sum of the lengths k_i of the r individual sequences. Thus the *Pascal variable* is $k = k_1 + k_2 + \cdots + k_r$, where each k_i is an independent identically distributed Pascal-geometric variable. Since the mean of the common Pascal-geometric variable is $1/p$ and its variance is q/p^2, it follows that the mean and variance of the Pascal variable are

$$E(k) = r/p \quad \text{and} \quad \text{var}(k) = rq/p^2.$$

It also follows that the Pascal probability mass function is the convolution of r identical Pascal-geometric probability mass functions $q^{k_i-1}p$. Also the Pascal moment generating function is the product of r Pascal-geometric moment generating functions:

$$M(\theta;k) = (pe^\theta)^r(1 - qe^\theta)^{-r}.$$

There are several ways we might attempt to take the inverse of this transform to obtain the Pascal probability mass function $f(k;r,p)$. However, instead it is easier to derive the required result directly.

Because $f(k;r,p)$ is the probability that the rth success in a Bernoulli process arrives on the kth trial, we have

$$f(k;r,p) = P(A)P(B|A),$$

where A is the event of having exactly $r - 1$ successes in the first $k - 1$ trials, and B is the event of having the rth success on the kth trial. Thus $P(A)$ is the binomial probability mass function for $r - 1$ successes in $k - 1$ trials. Since each trial is independent, it follows that $P(B|A) = P(B) = p$. Thus the above equation is

$$f(k;r,p) = b(r-1;k-1,p)p,$$

which reduces to

$$f(k;r,p) = \binom{k-1}{r-1} p^r q^{k-r}, \quad k = r, r + 1, r + 2, \cdots.$$

This expression is the required probability mass function of the *Pascal distribution*.

Pierre Fermat (1601–1665) and Blaise Pascal (1623–1662) were founders of the mathematical theory of probability. The specific questions that stimulated these great mathematicians to think about the matter came from requests of noblemen gambling in dice or cards. In the words of Poisson in 1827: "A problem concerning games of chance, proposed by a man of the world to an austere Jansenist, was the origin of the calculus of probabilities." This man of the world was the Chevalier de Mere who approached Pascal with a question concerning the so-called problem of points. Pascal began a correspondence with Fermat on the problem and on related questions, thereby establishing some of the foundations of probability theory (1654). Blaise Pascal at the age of 16 had discovered "Pascal's theorem" concerning a hexagon inscribed in a circle. It was published in 1641 on a single sheet of paper. A few years later Pascal invented a computing machine. At the age of 25 he decided to live the ascetic life of a Jansenist at the convent of Port Royal, but continued to devote time to science and literature. In addition to the name of the Pascal distribution, a modern computing language is named after him.

EXERCISES

[In our terminology a Pascal-geometric distribution is geometric with possible values 1, 2, 3, $\cdots$ whereas a (regular or ordinary) geometric distribution is geometric with possible values 0, 1, 2, $\cdots$.]

1. Derive the mean of the Pascal-geometric distribution by direct summation.

2. The distribution of the variable x is Pascal-geometric with parameter p. Given that x_1, x_2, $\cdots$, x_r are random observations of x and that $y = x_1 + x_2 + \cdots + x_r$. Show that y has the Pascal distribution with the same parameter p.

3. (a) A pair of coins are thrown together. What is the distribution of the number of throws required for both to show heads simultaneously?

   ANSWER. Pascal-geometric with $p = 1/4$.

   (b) What is the distribution of the number of throws required for at least one of the coins to show a head?

   ANSWER. Pascal-geometric with $p = 3/4$.

4. Suppose that x and y are independent random variables that have Pascal-geometric distributions with parameters p_1, $q_1 = 1 - p_1$ and p_2, $q_2 = 1 - p_2$ respectively, where $p_1 \neq p_2$. Show that

$$P(x + y = k) = \frac{p_1 p_2}{p_2 - p_1} [q_1^{k-1} - q_2^{k-1}] \qquad \text{for } k = 2, 3, \cdots.$$

   HINT. The probability generating function of $x + y$ is

$$\frac{p_1 t}{1 - q_1 t} \frac{p_2 t}{1 - q_2 t} = \frac{p_1 p_2 t^2}{p_2 - p_1} \left[\frac{q_1}{1 - q_1 t} - \frac{q_2}{1 - q_2 t} \right].$$

5. For the Pascal-geometric distribution $p(k) = q^{k-1} p$ for $k = 1, 2, 3,$ $\cdots$, show that the *tail probability* defined as $P(k > n)$ is equal to q^n

for $n = 0, 1, 2, \cdots$. [Because $P(k > n) \to 0$ as $n \to \infty$ if $p > 0$, it follows that $f(1) + f(2) + \cdots = 1$. Thus no matter how small the probability p of a success may be, as long as $p > 0$, a success will eventually occur within a finite waiting time.

HINT. The event that the first n trials are failures is the same as the event $k > n$.

6. Show that the probability generating function of a Pascal geometric variable is $\Phi(t) = pt/(1 - qt)$. Since $\Phi^{(k)}(1) = E[x_{(k)}]$ show that the factorial moments are $E[x_{(k)}] = k!q^{k-1}/p_k$ for $k = 1, 2, \cdots$.

7. A coin is tossed until a head is obtained. What is the expected number of tosses?

ANSWER. 2.

8. In a dichotomous population of size N there are N_1 objects of type 1 and $N_2 = N - N_1$ objects of type 2. Random drawings are made with replacement until a type 1 object is obtained. What is the expected number of drawings?

ANSWER. N/N_1.

9. Consider a continuing one-line history of generations of an organism (A begets B, B begets C, etc.). In any organism a certain genetic mutation takes place with probability $1/10000$. What is the expected number of generations that must evolve before a mutation takes place?

ANSWER. 10000.

10. The most interesting property of a Pascal-geometric variable k is that its excess over a boundary is itself a Pascal-geometric variable. That is, if it is known that k is greater than n, then the amount by which k exceeds n still has a Pascal-geometric distribution. Show that this property, called the "lack of memory" or "agelessness" property, is true.

HINT.

$$P(k = n + m | k > n) = \frac{P(k = n + m)}{P(k > n)} = \frac{q^{n+m-1}p}{q^n}$$
$$= q^{m-1}p \quad \text{for} \quad m = 1, 2, \cdots.$$

11. Suppose that one claims that the life expectancy in years for people is given by a Pascal-geometric variable k, with $k = 1$ corresponding to the first year of life (time period from birth to first birthday), $k = 2$ to the second year of life (from first to second birthday), etc. Does this model make sense in view of the agelessness property (Exercise 10) of the postulated distribution?

ANSWER. No, because if the model were true then a person in his kth year would have the same future life span as a baby in his first year. In other words, the life expectancy tables for octogenarians would be the same as that of newborn babies.

12. Suppose that m has possible values 1, 2, $\cdots$, and $P(m = 1) > 0$. Suppose also that the variable k has the following agelessness property:

$$P(k = n + 1 | k > 1) = P(k = n) \qquad \text{for } k = 1, 2, \cdots.$$

Show that k must have a Pascal-geometric distribution with parameter $p = P(k = 1)$.

HINT. The following is typical of the computations that are needed: For $n = 1$,

$$P(k = 2 | k > 1) = \frac{P(k = 2)}{P(k > 1)} = P(k = 1).$$

Hence, $P(k = 2) = p(1 - p)$. Similarly, for $n = 2$, $P(k = 3) = P(k = 2)P(k > 1) = p(1 - p)^2$.

13. Show that the minimum y of a set of independent variables $x_1, \cdots, x_n$ each with a Pascal-geometric distribution with parameters $p_1, \cdots, p_n$ respectively has a Pascal-geometric distribution with parameter $1 - q_1 q_2 \cdots q_n$, where $q_1 = 1 - p_1, \cdots, q_n = 1 - p_n$.

HINT. If $P(y > k) = q^k$ for some positive q, then y must have a Pascal-geometric distribution with parameter $p = 1 - q$, because

$$P(y = k) = P(y > k - 1) - P(y > k) = q^{k-1} - q^k = q^{k-1}p.$$

The result now follows from:

$$P(y > k) = P[(x_1 > k) \cap \cdots \cap (x_n > k)] = q_1^k q_2^k \cdots q_n^k = (q_1 q_2 \cdots q_n)^k.$$

14. The Pascal-geometric distribution is sometimes used as a model to describe the life span of a nuclear particle. If p is the parameter of the distribution, then $1/p$ is the expected lifetime. If the expected life spans of n different particles are $1/p_1, 1/p_2, \cdots, 1/p_n$, what is the expected waiting time for the first extinction?

ANSWER. $[1 - (1 - p_1)(1 - p_2) \cdots (1 - p_n)]^{-1}$.

15. (a) Suppose that x and y are independent, each with Pascal-geometric distributions over 1, 2, $\cdots$ with parameter p. Show that

$$P[\max (x, y) = k] = 2p(1 - p)^{k-1} - p(2 - p)(1 - p)^{2k-2}, \qquad k = 1, 2, \cdots.$$

(b) Each of two persons independently throws a coin until he obtains a head. Show that the maximum number of throws has the following probability mass function:

$$f(k) = \left(\frac{1}{2}\right)^{k-1} - \frac{3}{4}\left(\frac{1}{4}\right)^{k-1}, \qquad k = 1, 2, \cdots.$$

16. Two independent Bernoulli processes have parameters p_1 and p_2 respectively. A new process is defined to have a success on its kth trial ($k = 1, 2, 3, \cdots$) only if exactly one of the other two processes has a success on its kth trial. Determine the probability mass function for the number of trials at which the rth success occurs in the new process. Is the new process a Bernoulli process?

NEGATIVE BINOMIAL DISTRIBUTION

Let us now change the variable in the Pascal distribution. We will let s denote the total number of failures in the sequence before the rth success. Thus $s + r$ is equal to the number k of trials necessary to produce r successes, i.e., $s + r = k$. As before, r is a fixed positive integer. We now substitute $s + r$ for the k occurring in the Pascal distribution $f(k;r,p)$ and thus obtain the probability mass function defined as

$$b^-(s;r,p) = f(s+r;r,p) = \binom{s+r-1}{r-1}p^r q^s, \qquad s = 0, 1, 2, \cdots.$$

This is the probability that the rth success occurs at trial number $s + r$. It is also the probability that exactly s failures precede the rth success. It is called the *negative binomial distribution* (which explains the use of the symbol b^-). This probability mass function consists of an infinite sequence of terms, of which the first three are

$$b^-(0;r,p) = f(r;r,p) = p^r,$$
$$b^-(1;r,p) = f(r+1;r,p) = rp^r q,$$
$$b^-(2;r,p) = f(r+2;r,p) = \binom{r+1}{2}p^r q^2.$$

The mean of the negative binomial distribution is

$$E(s) = E(k - r) = E(k) - r = (r/p) - r = rq/p,$$

and the variance is the same as that of the Pascal distribution, namely rq/p^2.

In a sense, the negative binomial distribution is the converse of the binomial distribution. In the binomial distribution, $b(x;n,p)$, the number of trials is fixed at n, and the number x of successes is a random variable. In the negative binomial distribution, however, the number of successes is fixed at r and the number of trials required to generate the r successes is the random variable k. With some ingenuity, simpler problems involving the negative binomial can be translated into problems solvable by the binomial. For example, the probability that 30 throws of a die will suffice to generate $r = 5$ aces is the same as the probability that at least five aces will appear in 30 throws; that is, $P(n \leqslant 30 | r = 5) = P(r \geqslant 5 | n = 30)$.

We have used the binomial coefficients $\binom{x}{i}$ only when x is an integer, but it is convenient to extend their definition. The *factorial power*

$$x_{(i)} = x(x - 1)(x - 2)\cdots(x - i + 1)$$

is well defined for all real x provided only that i is a positive integer. For $i = 0$ we put $x_{(0)} = 1$. Then

$$\binom{x}{i} = \frac{x_{(i)}}{i!} = \frac{x(x-1)(x-2)\cdots(x-i+1)}{i(i-1)\cdots\cdots3\cdot2\cdot1}$$

defines the binomial coefficients for all values of x and all positive integers i. For $i = 0$, we put $\binom{x}{0} = 1$ and $0! = 1$. For negative integers i, we define $\binom{x}{i} = 0$. As an application, we see that for $r > 0$ we have

$$(-1)^k\binom{-r}{s} = \binom{s+r-1}{s} = \binom{s+r-1}{r-1}.$$

Thus the negative binomial may be written as

$$b^-(s;r,p) = \binom{-r}{s}p^r(-q)^s, \qquad s = 0, 1, 2, \cdots.$$

The binomial theorem due to Sir Isaac Newton states that

$$(1 - q)^{-r} = \sum_{s=0}^{\infty} \binom{-r}{s}(-q)^s.$$

Thus

$$\sum_{s=0}^{\infty} b^-(s;r,p) = p^r(1 - q)^{-r} = 1.$$

The negative binomial thus derives its name from the fact that $b^-(s;r,p)$ is a general term in the expansion of $p^r(1 - q)^{-r}$. The moment generating function of this distribution can be calculated from Newton's binomial formula as

$$M(\theta;s) = \sum_{j=0}^{\infty} b^-(s;r,p)e^{\theta s} = p^r \sum_{s=0}^{\infty} \binom{-r}{s}(-qe^{\theta})^s = p^r(1 - qe^{\theta})^{-r}.$$

Similarly, the probability generating function of the negative binomial is $\Phi(t) = p^r(1 - qt)^{-r}$.

If $r = 1$, then the negative binomial becomes the *geometric distribution*

$$b^-(s;1,p) = pq^s, \qquad s = 0, 1, 2, \cdots,$$

with moment generating function $M(\theta;s) = p(1 - qe^{\theta})^{-1}$. Because the mgf of the negative binomial distribution is the rth power of the mgf of the geometric distribution, it follows that the negative binomial variable is the sum of r independent identical geometric variables. This result, of course, is apparent from the way the negative binomial variable is defined, as well as from the previous result that the Pascal variable is the sum of r independent identical Pascal-geometric variables. We recall that Theorem 6.7 in Chapter 6 states that if $u = ax + b$, then $M(\theta;u) = e^{b\theta}M(a\theta;x)$. Since $k = s + r$, we see that the mgf $M(\theta;k)$ of the Pascal distribution can be found from the mgf $M(\theta;s)$ of the negative binomial as

$$M(\theta;k) = e^{r\theta}M(\theta;s) = e^{r\theta}p^r(1 - qe^{\theta})^{-r}.$$

This result checks with $M(\theta;k)$ as given in the previous section.

EXERCISES

[In our terminology, the geometric distribution is geometric with range $0, 1, 2, \cdots$, whereas the Pascal-geometric distribution is geometric with range $1, 2, 3, \cdots$.]

1. Derive the mean of the geometric distribution by direct summation.

2. Let $x_1, \cdots, x_m$ be independent random variables. Show that:
   (a) If x_i has the binomial distribution with parameters n_i and p, then $x_1 + \cdots + x_m$ has the binomial distribution with parameters $n_1 + \cdots + n_m$ and p.

   ANSWER. The probability generating function of $x_1 + \cdots + x_m$ is $(pt + q)^{n_1 + \cdots + n_m}$, which is the same as that of a random variable having a binomial distribution with parameters $n_1 + \cdots + n_m$ and p.

(b) If x_i has the negative binomial distribution with parameters r_i and p, then $x_1 + \cdots + x_m$ has the negative binomial distribution with parameters $r_1 + \cdots + r_m$ and p.

ANSWER. The probability generating function of $x_1 + \cdots + x_m$ is $[p/(1 - qt)]^{r_1 + \cdots + r_m}$, which is the same as that of a random variable having a negative binomial distribution with parameters $r_1 + \cdots + r_m$ and p.

3. Suppose s has a geometric distribution with parameter p. Compute the probabilities of the following events. (As usual, $q = 1 - p$.)
   (a) $s > 3$.
   (b) $4 \leq s \leq 7$ or $s > 9$.
   (c) $3 \leq s \leq 5$ or $7 \leq s \leq 10$.
   ANSWERS. (a) q^4. (b) $q^4 - q^8 + q^{10}$. (c) $q^3 - q^6 + q^7 - q^{11}$.

4. Let s be geometric with parameter p, and $q = 1 - p$. Let $y = s$ if $s < m$ and $y = m$ if $s \geq m$; that is, $y = \min[s, m]$. Find the probability mass function of y.
   ANSWER. $f(y) = pq^y$ for $y = 0, 1, \cdots, m - 1$ and $f(y) = q^m$ for $y = m$.

5. Let s be geometrically distributed with parameter p, and $q = 1 - p$. Find the probability mass functions of
   (a) $i = s^2$.
   (b) $j = s + 3$.
   ANSWER. (a) $f(i) = pq^{\sqrt{i}}$ for $i = 0, 1, 4, 9, 16, \cdots$. (b) $g(j) = pq^{j-3}$ for $j = 3, 4, 5, \cdots$.

6. A discrete random variable x has a geometric distribution. Let y be an integer. Given that $x > y$, show that the conditional distribution of $x - y$ is the same as the distribution of x.

7. Show that the mean of the negative binomial distribution is rq/p and the variance is rq/p^2 by using the moment generating function.

8

Basic Continuous Distributions

There is nothing in the whole world which is permanent. Everything flows onward.

OVID (43 B.C.–17 A.D.)

Uniform or Rectangular
Distribution

Cauchy Distribution

Exponential and Erlang
Distributions

Sums of Random Variables

Normal Distribution

Central Limit Theorem

Reproduction Properties of the
Normal Distribution

Bivariate Normal Distribution

Gamma Distribution

Beta Distribution

Chi-Square, *F*, and *t* Distributions

UNIFORM OR RECTANGULAR
DISTRIBUTION

The most simple distribution for a continuous random variable y is the rectangular, or uniform, distribution. Its density function is constant within a finite interval, and zero outside this interval:

$$g(y) = \frac{1}{b - a}, \qquad a \leqslant y \leqslant b.$$

Thus the density function consists of a rectangle on the admissible range. See Figure 8.1. We shall also say that y is uniformly distributed over this range. By a linear transformation of the variable, the range of the distribution can always be transferred to any given interval. For example, $x = (y - b)/(b - a)$ has the density function

$$f(x) = g(y) \frac{dy}{dx} = \frac{1}{b - a} (b - a) = 1 \qquad \text{for } 0 \leqslant x \leqslant 1.$$

This density is uniform over the unit interval $(0,1)$. Thus the cumulative distribution function of a uniform random variable on the interval $(0,1)$ is $F(x) = 0$ for $x < 1$, $F(x) = x$ for $0 \leqslant x \leqslant 1$, and $F(x) = 1$ for $x > 1$.

An important use of the uniform distribution is that it can be used to generate random observations having any desired distribution. For example, suppose we want to generate a discrete random variable y with the geometric distribution $g(y) = \frac{1}{2}, \frac{1}{4}, \frac{1}{8}, \cdots$ for $y = 0, 1, 2, \cdots$ respectively. First we pick an observation x from a uniform distribution over $0 \leqslant x \leqslant 1$. Such uniform random observations are obtained by various electronic devices and appear in tabulated form as "tables of random numbers." Next we divide up the interval $0 \leqslant x \leqslant 1$ according to the desired probability mass function $g(y)$. For example, we want $g(0) = \frac{1}{2}$, so one-half the interval must be assigned to $y = 0$. We want $g(1) = \frac{1}{4}$, so the next one-fourth of the interval is assigned to $y = 1$. We want $g(2) = \frac{1}{8}$, so we assign the next $\frac{1}{8}$ of the interval to $y = 2$, and so on. Thus, given an observation x from the tables, we arrive at the corresponding value of y according to the scheme

$$y = \begin{cases} 0 & \text{if } 0 \leqslant x < \frac{1}{2}, \\ 1 & \text{if } \frac{1}{2} \leqslant x < (\frac{1}{2}) + (\frac{1}{4}), \\ 2 & \text{if } (\frac{1}{2}) + (\frac{1}{4}) \leqslant x \leqslant (\frac{1}{2}) + (\frac{1}{4}) + (\frac{1}{8}) \\ \vdots & \vdots \end{cases}$$

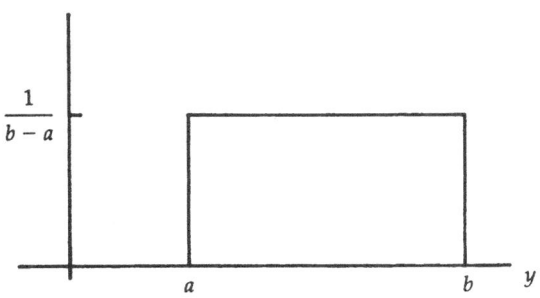

FIGURE 8.1 Uniform probability density.

Suppose we draw from the tables the value $x = 0.61406$. Because this value of x falls between 0.5 and 0.75, the corresponding value of y is $y = 1$. From the above scheme, it follows that

$$P(y = 0) = P(0 \leqslant x < {}^{1}/_{2}) = {}^{1}/_{2},$$
$$P(y = 1) = P({}^{1}/_{2} \leqslant x < {}^{3}/_{4}) = {}^{1}/_{4},$$
$$P(y = 2) = P({}^{3}/_{4} \leqslant x < {}^{7}/_{8}) = {}^{1}/_{8},$$
$$\vdots \qquad \vdots$$

Therefore the value of y so determined represents an observation from the required geometric distribution.

The same procedure works in the continuous case as in the discrete case. We thus have:

THEOREM 8.1. Let $G(y)$ be a given cumulative distribution function defined on the interval $a \leqslant y \leqslant b$ (where a may be $-\infty$ and/or b may be infinite). Suppose that $G(y)$ is continuous and strictly increasing so that there is a well-defined inverse $y = G^{-1}(x)$. If x is a random variable with a uniform distribution on the interval $0 \leqslant x \leqslant 1$, then the corresponding $y = G^{-1}(x)$ is a random variable with the distribution $G(y)$.

PROOF. The event $[y \leqslant t] = [G^{-1}(x) \leqslant t]$ is the same as the event $[G(y) \leqslant G(t)] = [x \leqslant G(t)]$. Because x is uniform, the probability of the event $[x \leqslant G(t)]$ is

$$\int_{0}^{G(t)} dx = G(t).$$

Therefore the probability of the event $[y \leqslant t]$ is also $G(t)$, which means that G is the cumulative distribution of y.   Q.E.D.

PROBLEM. If x is uniform, find y as a function of x such that y has the probability density $g(y) = 1/y^2$ for $y \geqslant 1$ and $g(y) = 0$ elsewhere.

SOLUTION. The cumulative distribution function is

$$G(y) = \int_{-\infty}^{y} g(y) \, dy = \begin{cases} 0 & \text{for } y \leqslant 1, \\ 1 - 1/y & \text{for } y > 1. \end{cases}$$

If $x = G(y) = 1 - (1/y)$ for $y > 1$, then the inverse is $y = G^{-1}(x) = 1/(1 - x)$ for $0 < x < 1$. Thus $y = 1/(1 - x)$ has the required probability density $g(y)$.

A change of variables of fundamental importance is the *probability transformation*. It is defined by the equation $y = F(x)$, where $F(x)$ is the distribution function of x. This transformation exists for all distributions. In the case when the probability density function exists, then y has the interesting property of being uniformly distributed. In order to prove this result, we see from the definition that the range of y is 0 to 1; and in case the density $f(x)$ exists we have

$$\frac{dy}{dx} = f(x) \qquad \text{and} \qquad \frac{dx}{dy} = \frac{1}{f(x)}.$$

Therefore

$$g(y) = f(x) \frac{dx}{dy} = \frac{f(x)}{f(x)} = 1 \qquad \text{for } 0 \leqslant y \leqslant 1$$

which shows that y has a uniform distribution. We state this result as:

THEOREM 8.2. Suppose x is a random variable with the density function $f(x)$. Then the random variable y defined as

$$y(x) = \int_{-\infty}^{x} f(x') \, dx'$$

is uniformly distributed on $0 \leqslant y \leqslant 1$.

For example, if $f(x) = 1/x^2$ for $x \geqslant 1$ and $f(x) = 0$ elsewhere, then y defined as

$$y = \int_1^x \frac{1}{x'^2} dx' = 1 - \frac{1}{x} \qquad \text{for } x \geq 1$$

has a uniform distribution for $0 \leq y \leq 1$.

The probability transformation forms a theoretical link between any pair of distributions having probability densities. Theoretically, any distribution for which the probability density exists can be transformed into any other by applying the probability transformation to each and thus bringing the variables into one-to-one correspondence. The value of this lies in the fact that an initial variable having a distribution that is not suitable for certain types of standard statistical analyses can be replaced by another derived from the first by a definite transformation and having a distribution that satisfies the conditions under which the proposed techniques apply. Even when the distribution functions cannot be expressed as explicit closed functions of the variable values, they can be obtained in tabular form by numerical integration. The transformation is then arrived at in the form of a table exhibiting the corresponding values of the two variables for selected values of the equated distribution functions.

PROBLEM. Find y as a function of x such that y has probability density $g(y) = 3(1 - \sqrt{y})$ for $0 \leq y \leq 1$, given that x has probability density $f(x) = 6x(1 - x)$ for $0 \leq x \leq 1$.

SOLUTION.

$$F(x) = \int_0^x f(x) \, dx = 3x^2 - 2x^3,$$

$$G(y) = \int_0^y g(y) \, dy = 3y - 2y^{3/2}.$$

Setting $G(y) = F(x)$ we get

$$3y - 2y^{3/2} = 3x^2 - 2x^3.$$

Since this must be satisfied identically, the solution is unique and is found by inspection to be $y = x^2$.

EXERCISES

1. Find and draw the cumulative form of the uniform distribution.

2. Show that if the random variable x is uniformly distributed on $0 \leq x \leq 1$, then its kth moment is $E(x^k) = 1/(k + 1)$ and its moment generating function is $M(\theta;x) = (e^\theta - 1)/\theta$.

3. Show that the mean of a uniform variable on (a,b) is $(a + b)/2$. Also show that the variance is $(b - a)^2/12$. Find and make use of the moment generating function.

4. Suppose that x is uniformly distributed on $(0,1)$. Show that:
   (a) $y = 1/x$ has density $g(y) = 1/y^2$ for $y \geqslant 1$ and $g(y) = 0$ elsewhere.
   (b) $y = x^2$ has density $g(y) = 1/(2\sqrt{y})$ for $0 < y \leqslant 1$ and $= 0$ elsewhere.
   (c) $y = \sqrt{x}$ has density $g(y) = 2y$ for $0 \leqslant y \leqslant 1$ and $= 0$ elsewhere.

5. If x is uniformly distributed over $0 \leqslant x \leqslant 1$, show that $1/x$ and $1/(1 - x)$ each have the same distribution.

6. Fill in the steps for the following alternate proof of Theorem 8.2: Let $F(x)$ be the cumulative distribution function of X. If X is a random observation from this distribution, then Y defined as $Y = F(X)$ is a random variable that must fall in the interval $0 \leqslant y \leqslant 1$ with probability one. We must show that for any point y in $0 \leqslant y \leqslant 1$ we have $P(Y \leqslant y) = y$. But the event $(Y \leqslant y)$ is exactly the same as the event $(X \leqslant x)$ because $Y = F(X)$ and $y = F(x)$. Draw a picture to illustrate this relationship. Hence $P(Y \leqslant y) = P(X \leqslant x)$. But $P(X \leqslant x) = F(x)$ and by definition $F(x) = y$. Therefore $P(Y \leqslant y) = y$, which says that the distribution function of Y is uniform.

7. Given the random variable X with density function $f(x) = (x - 1)/2$ for $1 < x < 3$ and $f(x) = 0$ elsewhere. What transformation will change X to a random variable Y with a uniform distribution over $0 < y < 1$? What interval for y corresponds to $1.1 < x < 2.9$?

8. Let X have a uniform distribution in the interval $(0,1)$. Expressing the value of X in decimal notation, what is the probability that the digit in the seventh position is a 6?
   ANSWER. $1/10$.

9. Let Z be a continuous random variable with cumulative distribution function $H(z)$ and density function $h(z)$. Show that the random variable X defined by the equation $X = H(Z)$ has the uniform density $f(x) = 1$ for $0 \leqslant x \leqslant 1$.

CAUCHY DISTRIBUTION

In Figure 8.2, let Θ be a random variable with uniform density function $g(\theta) = 1/\pi$ in the interval $-\pi/2 \leqslant \theta \leqslant \pi/2$. The values of Θ can be obtained by means of a spinner. The radius OP is extended to the point Q on the y-axis. Each value of θ yields a unique value of y.

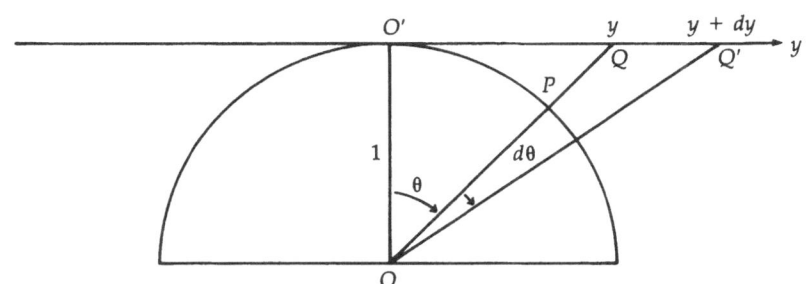

FIGURE 8.2 Derivation of the Cauchy distribution.

The set of all y obtained in this fashion yields a random variable Y whose distribution we wish to obtain. If Y lies in the range $(y, y+dy)$, then Θ lies in the range $(\theta, \theta+d\theta)$, and conversely, where $y = \tan \theta$. Thus

$$h(y)\, dy = g(\theta)\, d\theta = \frac{1}{\pi}\, d\theta.$$

Because $\theta = \tan^{-1} y$, we have $d\theta = dx/(1 + x^2)$. Thus

$$h(y) = \frac{1}{\pi} \frac{1}{1 + y^2}, \qquad -\infty < y < \infty.$$

This is the required density function, and the random variable Y is said to have a standard Cauchy distribution. The most general Cauchy probability density function $f(x; \mu, a)$ is obtained from $h(y)$ by the linear transformation $y = (x - \mu)/a$. Thus

$$f(x; \mu, a) = h(y)\frac{dy}{dx} = \frac{1}{\pi} \frac{1}{1 + [(x - \mu)/a]^2} \frac{1}{a},$$

which is

$$f(x; \mu, a) = \frac{a}{\pi} \frac{1}{a^2 + (x - \mu)^2}, \qquad -\infty < x < \infty.$$

This density function is characterized by the parameters μ (which can take any value) and a (which must be positive). See Figure 8.3.

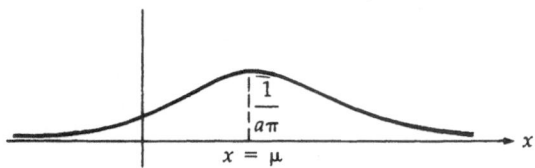

FIGURE 8.3 The Cauchy density function.

EXERCISES

1. For what values of x does the Cauchy density function reach a maximum value? Are there any points of inflection?

   ANSWER. $x = \mu$, $x = \mu \pm a/\sqrt{3}$.

2. Show that for the Cauchy distribution

$$\lim_{c \to \infty} \int_{\mu-c}^{\mu+c} xf(x;\mu,a)\, dx = \mu,$$

but that the mean, as given by its general definition

$$E(x) = \int_{-\infty}^{\infty} xf(x;\mu,a)\, dx,$$

does not exist. Thus the Cauchy has a mean only in a restricted sense as given by the first integral.

3. Show that the cumulative distribution function of the Cauchy is

$$F(x;\mu,a) = \frac{1}{2} + \frac{1}{\pi} \tan^{-1} \frac{x - \mu}{a}.$$

Relate this result to the cumulative distribution function of the uniform random variable Θ described in the text.

4. Show that the Cauchy distribution reproduces itself by convolution, so that we have the addition theorem

$$f(x;\mu_1,a_1) * f(x;\mu_2,a_2) = f(x;\mu_1+\mu_2,a_1+a_2).$$

Hence, deduce the following interesting property of the distribution: If $x_1, x_2, \cdots, x_n$ are independent, and all have the same Cauchy distribution, then the arithmetic mean $\bar{x} = (x_1 + x_2 + \cdots + x_n)/n$ has the same distribution as every x_i.

5. Show that the Cauchy distribution is autoreciprocal; that is, if X has a Cauchy distribution then $1/X$ has the same Cauchy distribution.

EXPONENTIAL AND
ERLANG DISTRIBUTIONS

The variety of possible continuous distributions is potentially inexhaustible. In this chapter we consider some of the better-known continuous distributions. As with discrete distributions, such distributions arise operationally (i.e., from considering some physical or operational process) as well as out of sampling problems. In both cases, the continuous distribution may emerge as a convenient approximation to a discrete distribution. Computational advantage may be gained by such substitution, but usually some clarity of reference to the original discrete process is sacrificed. One of the most satisfying continuous distributions, in that it has direct operational models, is the exponential distribution, which is at the same time a continuous version of the geometric distribution. We consider a sequence of Bernoulli trials, with probability p for success on an individual trial.

The probability that the first success occurs on trial k is given by the Pascal-geometric mass function $q^{k-1}p$ where $q = 1 - p$ and $k = 1$, 2, $\cdots$. Thus the probability that the first success occurs after the kth trial is

$$\sum_{j=k+1}^{\infty} q^{j-1}p = q^k p(1 + q + q^2 + \cdots) = \frac{q^k p}{1 - q} = q^k.$$

Since $q = 1 - p$, this probability is $(1 - p)^k$. Now we will make use of the *Poisson process* as follows. Suppose that each trial takes time Δt to perform, so that k trials take $k \, \Delta t$. We shall let $\Delta t \to 0$ and $k \to \infty$ so that the total time consumed remains constant. That is, we hold $k \, \Delta t = t$. As in the derivation of the Poisson distribution, we must let $p \to 0$ in such a way that the mean number of successes per unit time remains constant. To do this, we set $p = \lambda \, \Delta t$, where λ is the mean rate of occurrence. Then

$$(1 - p)^k \to \left(1 - \frac{\lambda t}{k}\right)^k \to e^{-\lambda t} \qquad \text{as } k \to \infty. \tag{8.1}$$

Thus the probability that first success occurs after time t is $e^{-\lambda t}$. In other words, if τ is a random variable denoting time until first success, then

$$P(\tau > t) = e^{-\lambda t}. \tag{8.2}$$

Thus the cumulative distribution function is

$$F(t) = P(\tau \leqslant t) = 1 - e^{-\lambda t}. \tag{8.3a}$$

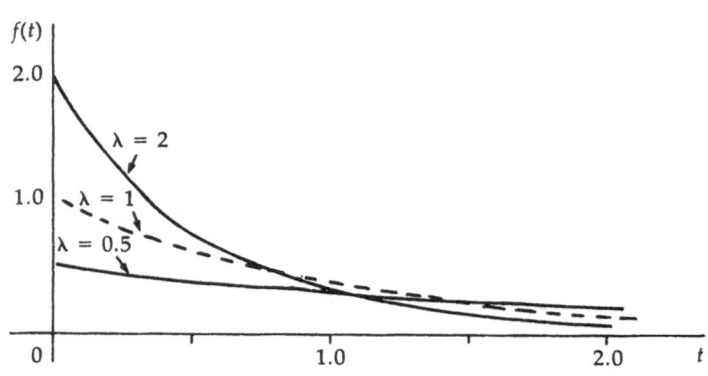

FIGURE 8.4 The exponential density functions for $\lambda = 0.5$, $\lambda = 1$, and $\lambda = 2$.

We can immediately derive the density function

$$f(t) = \frac{d}{dt} P(\tau \leq t) = \lambda e^{-\lambda t}, \qquad t \geq 0, \tag{8.3b}$$

which is the probability density that first success occurs at time t. Either the cdf (Equation 8.3a) or the density (Equation 8.3b) is known as the *exponential distribution* with parameter λ. See Figure 8.4.

We want to present two additional derivations (called the second and third) of the exponential density. Of course, all the derivations are mathematical equivalents. Let us now give the second derivation. According to Theorem 7.3, the probability of no success in time t for a Poisson process with parameter λ is given by the Poisson term $p(0;\lambda t)$. The probability of success in the next time increment is $\lambda\, dt$. Because of the independence of events in nonoverlapping time intervals, it follows that the probability of the first success in the interval $(t, t+dt)$ is

$$f(t)\, dt = p(0;\lambda t)\lambda\, dt = e^{-\lambda t}\, \lambda\, dt,$$

which yields the required exponential density $f(t)$ as given in Equation (8.3b).

The third derivation makes use of the Poisson approximation to the binomial. As before, divide the time interval $(0,t)$ into k increments of length Δt, so $k\,\Delta t = t$. The probability of success in any increment is $\lambda\,\Delta t$. Because of the mutual independence of events occurring in different increments, the probability of no successes in the k increments is given by the binomial term $b(0;k,\lambda\,\Delta t)$. The probability of success in the $(k + 1)$st increment is $\lambda\,\Delta t$. Thus the probability of the first success

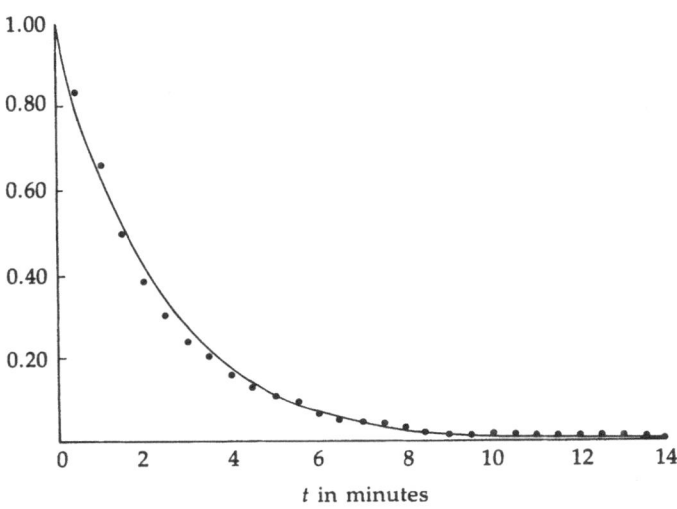

FIGURE 8.5 Distribution of lengths of 7387 local telephone calls. The small circles give the observed frequency, and the solid curve gives the exponential fit $P(T \geq t) = \exp(-t/2.26)$.

in the increment $(t,t+\Delta t)$ is $b(0;k,\lambda\,\Delta t)\lambda\,\Delta t$. In the limit as $k \to \infty$, $\lambda\,\Delta t \to \lambda\,dt$, and $k\lambda\,\Delta t = \lambda t$, this quantity approaches $p(0;\lambda t)\,\lambda\,dt$, which is the same expression we directly obtained in the second derivation.

In a study made by E. C. Molina at the Bell Telephone Laboratories ("Telephone Trunking Problems," Bell System Technical Journal, Vol. VI, 1927, p. 463) a frequency distribution was plotted for the lengths of telephone conversations in calls between subscribers in two adjacent New Jersey towns. The observed frequencies and Molina's "exponential fit" are shown in Figure 8.5.

Let us now consider some of the properties of the exponential distribution. The mode occurs at $t = 0$, which is often a surprise in practical applications. To get the mean and variance, let us note first that the moment generating function is

$$M(\theta;t) = \int_0^{\infty} e^{\theta t}\lambda e^{-\lambda t}\,dt = \frac{\lambda}{\lambda - \theta}. \tag{8.4}$$

Thus

$$E(t) = \frac{d}{d\theta}M(\theta;t)\bigg|_{\theta=0} = \frac{1}{\lambda} \tag{8.5}$$

and

$$\text{var}\,(t) = \frac{d^2}{d\theta^2}\,M(\theta;t)\,\bigg|_{\theta=0} - [E(t)]^2 = \frac{2}{\lambda^2} - \frac{1}{\lambda^2} = \frac{1}{\lambda^2}. \qquad (8.6)$$

There are two other important properties of an exponential variable t. Thinking of t as time until completion for some operation in progress, let us suppose that the operation has been going on for t_0 units of time, with completion not yet reached, and let us ask for the probability density that τ additional time is required until completion. This is the conditional probability density $\phi\{t = t_0 + \tau | t \geq t_0\}$, which is just

$$\frac{f(t_0 + \tau)}{\displaystyle\int_{t_0}^{\infty} f(y)\,dy} = \frac{\lambda e^{-\lambda(t_0+\tau)}}{e^{-\lambda t_0}} = \lambda e^{-\lambda\tau}. \qquad (8.7)$$

Thus this conditional probability is still the same exponential. Applied to the above case of telephone calls, this means that the probability that a call requires still τ units of time to complete is independent of how long the call is known to have been in progress. As a special case, if we let $\tau = 0$ in Equation (8.7) above, then we find that the value of the conditional probability density of completion at any moment is λ. This constant value characterizes the exponential distribution.

Except for the geometric distribution (including, of course, the geometric-Pascal and other such geometric distributions), the exponential distribution is the only one with this property of independence between additional waiting time and elapsed waiting time. We may describe this phenomenon as follows. If we have already spent some time in waiting, the distribution of further waiting time is the same as that of the waiting time that we initially faced. Thus the elapsed time spent in waiting has been in vain. A suggestive way of stating this property is that the exponential variable, as well as the geometric variable, has no memory.

If an empirical set of data is suspected to be exponential in distribution, a quick examination can be made by plotting the data on a semilog graph. If the variable is indeed exponentially distributed, the data will approximate a straight line when so plotted. Since $\log f(t) = \log(\lambda e^{-\lambda t}) = \log \lambda - \lambda t$, (i.e., a linear function of t), the slope of this line will, in numerical value, be $-\lambda$. We will discuss methods of evaluating unknown constants so as to give a "best" fit later, under the general topic of regression in Chapter 9.

In the exponential distribution, we found the probability density for the time at which success first occurs. Let us now generalize this to the time t at which the rth success occurs (assuming a Poisson process as before). Let us imagine a succession of r independent experiments in which a new experiment is begun as soon as each success is obtained. It follows that the total time t needed for r successes (where t is called the Erlang variable) is precisely the sum of times t_i of the r individual experiments. Thus the Erlang variable is $t = t_1 + t_2 + \cdots + t_r$, where each t_i is an independent identically distributed exponential variable. Since the mean of the common exponential variable is $1/\lambda$ and its variance is $1/\lambda^2$, it follows that the mean and variance of the Erlang variable is

$$E(t) = r/\lambda \qquad \text{and} \qquad \text{var}(t) = r/\lambda^2.$$

It also follows that the Erlang density function is the convolution of r identical exponential density functions $\lambda e^{-\lambda t_i}$, and that the Erlang mgf is the product of r exponential mgf's

$$M(\theta;t) = (1 - \theta/\lambda)^{-r}.$$

From Theorem 6.6, we recognize this $M(\theta;t)$ to be the moment generating function of the probability density

$$f(t;r,\lambda) = \frac{\lambda(\lambda t)^{r-1}e^{-\lambda t}}{(r - 1)!} \qquad \text{for } 0 \leqslant t < \infty. \tag{8.8}$$

This density function for integral r is the probability density of the *Erlang distribution*, with parameters r and λ. The Erlang distribution is a special case of the more general gamma distribution which is defined for all positive r, integral or nonintegral. The gamma distribution will be treated later in this chapter. The Erlang variable t is the time required for r successive completions of a Poisson process.

A second derivation of the Erlang density function is similar to derivation given for the Pascal probability mass function on page 259. In this derivation, we write

$$f(t;r,\lambda)\, dt = P(t < T \leqslant t + dt) = P(A)P(B|A),$$

where A is the event that there are exactly $r - 1$ successes in the interval $0 \leqslant T \leqslant t$ and B is the event that the rth success occurs in the interval $t < T \leqslant t + dt$. We now make use of the Poisson process.

Since events occurring on nonoverlapping time intervals in a Poisson process are independent, it follows that $P(B|A) = P(B)$. Now $P(A)$ is given by the Poisson distribution for $r - 1$ successes in time t, and $P(B)$ is given by $\lambda\, dt$. Thus we have the Erlang probability element

$$f(t;r,\lambda)\, dt = p(r-1;\lambda t)\lambda\, dt, \qquad 0 \leq t < \infty, r = 1, 2, \cdots.$$

Thus the Erlang probability density function is

$$f(t;r,\lambda) = \frac{\lambda(\lambda t)^{r-1}}{(r - 1)!}\, e^{-\lambda t}, \qquad 0 \leq t < \infty, r = 1, 2, \cdots.$$

In particular, for $r = 1$ we obtain the exponential density function

$$f(t;1,\lambda) = \lambda e^{-\lambda t}, \qquad 0 \leq t < \infty.$$

A third derivation of the Erlang density is obtained by using the Poisson approximation to the binomial. After the subdivision of time t into k subintervals of length Δt, the probability that the rth success occurs in the $(k + 1)$st interval is given by the Pascal probability mass function $b(r-1;k,\lambda\, \Delta t)\lambda\, \Delta t$. In the limit as $k \to \infty$, $\lambda\, \Delta t \to \lambda\, dt$, and $k\lambda\, \Delta t = \lambda t$, the binomial is approximated by the Poisson distribution. Thus the Pascal mass function tends to $p(r-1;\lambda t)\lambda\, dt$, which is the required probability element of the Erlang distribution.

The formula in (8.8) looks suggestively like a term from the Poisson distribution. Actually it is not, but there is nevertheless a close relation. The Erlang cdf $F(\tau;r,\lambda) = \int_0^\tau f(t;r,\lambda)\, dt$ gives the probability that the rth completion occurs before time τ (measured from a start at $\tau = 0$). If we now require not merely that r completions occur in time τ, but that *only* r occur (i.e., no more than r), then we can write

$$\tau = t + (\tau - t),$$

where the rth completion occurs at time t and *no* completion occurs during $\tau - t$. To have r and only r completions in time τ, we see that t can obviously be chosen anywhere from 0 to τ, so that we will have to integrate over all such t. The Erlang density $f(t;r,\lambda)$ gives the probability density of r completions in the time interval from 0 to t. From equation (8.2), we have P(no completions in the interval t to τ) $= e^{-\lambda(\tau-t)}$. Thus

$$P\{r \text{ and only } r \text{ completions in } \tau\} = \int_0^\tau f(t;r,\lambda)e^{-\lambda(\tau-t)}\, dt \qquad (8.9)$$

$$= \int_0^\tau \frac{\lambda(\lambda t)^{r-1}}{(r-1)!} e^{-\lambda\tau}\, dt$$

$$= \frac{(\lambda\tau)^r}{r!} e^{-\lambda\tau},$$

which is the corresponding term of the Poisson distribution.

We now summarize our results in the form of the following theorem.

THEOREM 8.3. In a Poisson process with parameter λ, the time intervals t between successive completions (or arrivals or counts) are independent exponential variables, all with mean $1/\lambda$. Conversely, if we have a process of completions (or arrivals) with the time intervals t between successive completions being independent exponential variables, all with mean $1/\lambda$, then the number of arrivals in a fixed time t is a Poisson variable with mean λt.

We can thus characterize the Poisson process by saying that the interval between successive successes is exponentially distributed and independent in each case. Moreover, no other distribution has this property, for note that in the Poisson distribution,

$$p(0;\lambda t) = e^{-\lambda t}.$$

That is, the probability of no occurrences in time t is just the probability that the interval until the next occurrence exceeds t is exponentially distributed. Note finally that λ is the mean number of arrivals per unit time for the Poisson distribution, and $1/\lambda$ is the mean time between arrivals, which further checks the correspondence. If the time interval is one, then the number of counts (the Poisson variable) has mean λ, which agrees with the definition of λ as the mean number of counts per unit time. On the other hand, if the number of counts is set equal to one, then the waiting time (the exponential variable) has mean $1/\lambda$, which says that $1/\lambda$ is the mean waiting time per count.

EXERCISES

1. If t is an exponential variable with mean 1, find the probability that:
   (a) $t > 1$.

(b) $t \leqslant 0.5$.

(c) in 10 random observations of t at least one value of t is greater than 5.

ANSWERS. (a) 0.368. (b) 0.394. (c) 0.067.

2. If $f(t) = \lambda e^{-\lambda t}$, $t \geqslant 0$, find $E(t^n)$ by direct integration. (HINT. Integrate by parts repeatedly reducing n.) Compare your result with the coefficient of $\theta^n/n!$ in the expansion of

$$M(\theta;t) = \frac{\lambda}{\lambda - \theta} = \frac{1}{1 - \theta/\lambda} = 1 + \frac{\theta}{\lambda} + \left(\frac{\theta}{\lambda}\right)^2 + \cdots.$$

3. A clock runs continuously for times that are exponentially distributed with a mean of 3 months. If upon stopping it is instantly restarted, find

(a) the probability of exactly 6 stops in a year.

(b) the probability of no stops in 2 years.

(c) the probability density that the 4th stop occurs at time t.

ANSWERS. (a) 0.114. (b) 0.0003.

4. If $f(t) = \lambda^r t^{r-1} e^{-\lambda t}/(r - 1)!$, then $t = t_1 + t_2 + \cdots + t_r$, where t_i for each $i = 1, \cdots, r$, has exponential density $f(t_i) = \lambda e^{-\lambda t_i}$. Use this result to find $E(t)$ and var (t).

ANSWER. $E(t) = r/\lambda$, var $(t) = r/\lambda^2$.

5. If $f(t)$ is defined as in Exercise 4:

(a) show on intuitive grounds that

$$\int_0^\tau f(t) \, dt = \sum_{x=r}^\infty p(x;\lambda\tau),$$

where $p(x;\lambda t)$ is the Poisson distribution.

(b) Verify (a) also by integration for $r = 1, 2, 3$.

6. If t_1 and t_2 are independent exponential variables with means $1/\lambda_1$ and $1/\lambda_2$ respectively, find the distribution of $t = t_1 - t_2$

(a) if $\lambda_1 \neq \lambda_2$.

(b) if $\lambda_1 = \lambda_2$.

7. Suppose that x and y are independent random variables, each exponentially distributed with the same parameter λ. Show that $x/(x + y)$ is uniformly distributed over the interval $[0,1]$.

8. (a) Given that the waiting time t for some chance event is exponentially distributed, and letting t_0 be an arbitrary fixed value of t, find the conditional distribution of waiting time under the restriction that $t > t_0$. Deduce that $P(t > t_0 + \tau | t > t_0) = P(t > \tau)$. Express this proposition in words.

ANSWER. $\phi(t|t > t_0) = \lambda e^{-\lambda(t-t_0)}$ for $t_0 < t < \infty$.

(b) The proposition proved in (a) for the exponential distribution may be stated analytically as follows:

$$\frac{f(x + y)}{\int_x^\infty f(z)\,dz} \equiv f(y) \qquad \text{for } y \geq 0.$$

Prove, conversely, that if this identity holds for all $y \geq 0$, and the variable is nonnegative, then the distribution is exponential. This fact that the conditional probability of exceeding a given arbitrary value by a stated margin (or more) depends only on the margin stipulated and not on the arbitrary starting point is thus a definitive property of the exponential distribution.

HINT. Simplify and differentiate with respect to x. To prove uniqueness, it is then sufficient to show that the differential equation obtained by setting $y = 0$ is satisfied only by an exponential function.

9. Suppose that the orders for a certain article of merchandise come from two main sources, the demands per unit time being independent Poisson variables with respective parameters λ_1, λ_2. Show that the elapsed time between orders is exponentially distributed with $\lambda = \lambda_1 + \lambda_2$.

10. Suppose that $x_1, \cdots, x_n$ are independent random variables, exponentially distributed with parameters $\lambda_1, \cdots, \lambda_n$. Show that min $(x_1, \cdots, x_n) = y$ is exponentially distributed with parameter $\lambda = \lambda_1 + \cdots + \lambda_n$.

HINT.

$$\begin{aligned}
P(y \leq t) &= 1 - P(y > t) \\
&= 1 - P[(x_1 > t) \cap \cdots \cap (x_n > t)] \\
&= 1 - P(x_1 > t) \cdots P(x_n > t).
\end{aligned}$$

11. Show that an exponentially distributed random variable x has the following property:

$$P(0 \leq x - u \leq t | 0 \leq x - u) = P(x \leq t)$$

for any positive constants t and u. In other words, if x is exponential, then the conditional distribution of the amount by which x exceeds u, given that it exceeds u, is the same as the distribution of x itself.

12. Suppose that x and y are independent random variables, each ex-

ponentially distributed with the same parameter λ. Let $u = \min (x,y)$, $v = \max (x,y)$.

(a) Show that the joint density of u and v is

$$f(u,v) = \begin{cases} 2\lambda^2 e^{-\lambda(u+v)} & \text{for } 0 \leqslant u \leqslant v, \\ 0 & \text{otherwise.} \end{cases}$$

HINT.

$$P[(u,v) \text{ in } A] = P[(x,y) \text{ in } A) \cap (x \leqslant y)]$$
$$+ P[(y,x) \text{ in } A) \cap (y < x)]$$
$$= 2 \int \int_A f(u,v) \, du \, dv.$$

(b) Show that u and $v - u$ are independent, exponential, with parameters 2λ and λ, respectively.

HINT. Let f be as in (a):

$$P[(u \leqslant r) \cap (v - u \leqslant s)] = \int_{\substack{0 \leqslant u \leqslant r \\ 0 \leqslant v - u \leqslant s}} \int 2\lambda^2 e^{-\lambda(u+v)} \, du \, dv.$$

Make a change of variables, $u = x$, $v - u = y$.

13. Cars arrive at a tunnel entrance according to the following model. Successive waiting times between car arrivals are independent exponential random variables, each with the same parameter λ (λ is in units of counts per minute).

(a) What is the expected number of cars to arrive per hour?

ANSWER. 60λ.

(b) What is the expected value of the maximum interarrival time among the first 10 cars? (Count the wait for the first car as one of the times.)

ANSWER. $(1/\lambda)(1 + {}^1\!/_2 + \cdots + {}^1\!/_{10})$ minutes.

(c) What is the expected value of the minimum interarrival time among the first 10 cars?

ANSWER. $1/(10 \, \lambda)$.

14. Suppose that $x_1, x_2, \cdots, x_n$ are mutually independent random variables, each exponential with the same parameter λ.

(a) Let $M_n = \max (x_1, x_2, \cdots, x_n)$. Show that the density of M_n is

$$f(x) = \begin{cases} n\lambda(1 - e^{-\lambda x})^{n-1} e^{-\lambda x} & \text{for } x \geqslant 0, \\ 0 & \text{otherwise.} \end{cases}$$

(b) Manipulate the integral for expected value

$$E(M_n) = \int_0^\infty n\lambda x(1 - e^{-\lambda x})^{n-1}e^{-\lambda x}\,dx$$

to show that

$$E(M_n) - E(M_{n-1}) = \frac{1}{n\lambda}.$$

HINT. Integrate by parts to obtain

$$E(M_n) = \int_0^\infty [1 - (1 - e^{-\lambda x})^n]\,dx.$$

Then write this integral as

$$\int_0^\infty [1 - (1 - e^{-\lambda x})^{n-1}(1 - e^{-\lambda x})]\,dx = E(M_{n-1}) + \frac{1}{n\lambda}.$$

(c) Conclude from (b) that

$$E(M_n) = \frac{[1 + (1/2) + \cdots + (1/n)]}{\lambda}.$$

15. Consider the following process. A point u_1 is picked at random in $[0,1]$. Then a second point u_2 is picked at random in $[0,u_1]$. Then a third point u_3 is picked at random in $[0,u_2]$, and so forth.

(a) Convince yourself that an appropriate model for the distribution of $u_1, \cdots, u_n, \cdots$ is that

$$u_1 = x_1,\ u_2 = x_1 x_2,\ \cdots,\ u_n = x_1 x_2 \cdots x_n,\ \cdots,$$

where $x_1, \cdots, x_n, \cdots$ are mutually independent, each uniformly distributed on $[0,1]$.

(b) Let z be the number of points $u_1, u_2, \cdots$ that fall in $[c,1]$ where $0 < c < 1$. Show that z is Poisson distributed with parameter $\mu = -\log c$.

HINT.

$$P(z = k) = P[(x_1 \cdots x_k \ge c) \cap (x_1 \cdots x_{k+1} < c)]$$

$$= P\left[\left(\sum_{i=1}^{k} (-\log x_i)\right.\right.$$

$$\left.\le (-\log c)\right) \cap \left(\sum_{i=1}^{k+1} (-\log x_i) > (-\log c)\right)\right].$$

Now use the fact that $-\log x_1, -\log x_2, \cdots$ are mutually independent, each exponentially distributed with parameter $\lambda = 1$, and consider the relationship between the Poisson and exponential distributions.

16. Suppose that x and y are independent, nonnegative, random variables where x has an arbitrary density function and y is exponential with parameter $\lambda = 1$. Show that the distribution of $y - x$ has the following property:

$$P(0 \leqslant y - x \leqslant t | y - x \geqslant 0) = \lambda \int_0^t e^{-\lambda y} \, dy.$$

HINT. The required conditional probability is

$$\frac{P(0 \leqslant y - x \leqslant t)}{P(0 \leqslant y - x)} = \frac{P[1 \leqslant (e^{-x}/e^{-y}) \leqslant e^t]}{P[1 \leqslant (e^{-x}/e^{-y})]},$$

where e^{-y} is uniform on $[0,1]$.

SUMS OF RANDOM VARIABLES

In deriving distributions of sums of random variables, it is not always necessary to use the method of moment generating functions. In particular, let t_1 and t_2 be random variables with a joint density function $h(t_1,t_2)$. Now let $t = t_1 + t_2$. Then t can be written as

$$t = (t - t_2) + t_2,$$

where $t_1 = t - t_2$. By letting t_2 range over all its possible values, with t_1 then ranging over all its possible corresponding values $t - t_2$, we get all possible contributions to the variable t. Hence

$$f(t) = \int_{-\infty}^{\infty} h(t-t_2,t_2) \, dt_2. \qquad (8.10)$$

The integral in (8.10) is a type of convolution integral, as can be seen geometrically in Figure 8.6. The density function $f(t)$ is obtained by integrating the joint density function $h(t_1, t_2)$ over all pairs of values adding up to t, i.e., along the line $t = t_1 + t_2$.

PROBLEM. If $f_1(t_1) = \lambda e^{-\lambda t_1}$, $t_1 \geqslant 0$, and $f_2(t_2) = [\lambda(\lambda t_2)^n \, e^{-\lambda t_2}]/n!$, $t_2 \geqslant 0$, find $f(t)$ by Equation (8.10), where $t = t_1 + t_2$, assuming t_1 and t_2 are independent.

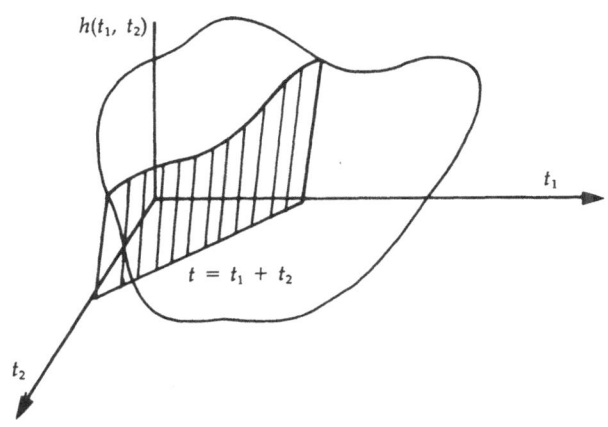

FIGURE 8.6 Integration of joint density function to find density function of sum.

SOLUTION. Since $t_1 \geqslant 0$ and $t_2 \geqslant 0$, (8.10) becomes

$$f(t) = \int_0^t f_1(t - t_2)f_2(t_2)\, dt_2$$

$$= \int_0^t \lambda e^{-\lambda(t - t_2)} \frac{\lambda(\lambda t_2)^n e^{-\lambda t_2}}{n!}$$

$$= \frac{\lambda^{n+2} e^{-\lambda t}}{n!} \int_0^t t_2^n\, dt_2$$

$$= \frac{\lambda(\lambda t)^{n+1} e^{-\lambda t}}{(n + 1)!}.$$

Thus by induction on n, we have established Equation (8.8) for the Erlang density in the case when $r = n + 2$ without using moment generating function methods. We give some illustrative examples.

PROBLEM. If the length of time required to load ships at a port of call is exponentially distributed with a mean of 48 hours:
  (a) Find the probability that a single ship requires between 24 and 48 hours to load.
  (b) Find that time of loading which will be exceeded in only 5 percent of all cases.
  (c) Find the probability that one ship takes longer than another to load.

SOLUTION.

(a) The probability is

$$\int_{24}^{48} \frac{e^{-t/48}}{48} \, dt = e^{-0.5} - e^{-1} = 0.238.$$

(b) If t is the time, then

$$\int_{t}^{\infty} \frac{e^{-t/48}}{48} = 0.05 = e^{-t/48}.$$

Thus $-t/48 = \ln 0.05 = -2.996$, so $t = 144$ hours.

(c) The probability is

$$\int_{0}^{\infty} f(t) \int_{0}^{t} f(y) \, dy \, dt = \int_{0}^{\infty} \frac{e^{-t/48}}{48} [1 - e^{-t/48}] \, dt$$

$$= 1 - (^1/_2) \int_{0}^{\infty} (^1/_{24}) e^{-t/24} \, dt$$

$$= 1 - (^1/_2) = ^1/_2.$$

Actually $^1/_2$ is the answer for any continuous distribution.

PROBLEM. Suppose that a random variable t is the sum of two independent exponential variables t_1 and t_2 having different parameters λ_1 and λ_2, respectively. What is the distribution of t?

SOLUTION. $t = t_1 + t_2 = (t - t_2) + t_2$; i.e., let $t_1 = t - t_2$.

$$f(t) = \int_{0}^{t} \lambda_1 e^{-\lambda_1(t - t_2)} \lambda_2 e^{-\lambda_2 t_2} \, dt_2 \tag{8.11}$$

$$= \lambda_1 \lambda_2 e^{-\lambda_1 t} \int_{0}^{t} e^{(\lambda_1 - \lambda_2) t_2} \, dt_2$$

$$= \frac{\lambda_1 \lambda_2}{\lambda_1 - \lambda_2} (e^{-\lambda_2 t} - e^{-\lambda_1 t}).$$

Thus the distribution is an algebraic sum of exponential terms, under the condition $\lambda_1 \neq \lambda_2$. In contrast when $\lambda_1 = \lambda_2$, the distribution is Erlang with parameters $r = 2$ and $\lambda = \lambda_1 = \lambda_2$.

PROBLEM. If t_1 and t_2 are independent exponentially distributed variables with parameters λ_1 and λ_2, respectively, find the probability that $t_1 > t_2$.

SOLUTION. The probability is

$$\int_0^\infty \lambda_1 e^{-\lambda_1 t_1}(1 - e^{-\lambda_2 t_1})\, dt_1 = 1 - \frac{\lambda_1}{\lambda_1 + \lambda_2} = \frac{\lambda_2}{\lambda_1 + \lambda_2}.$$

EXERCISES

1. Let x_1, x_2, $\cdots$, x_n be independent random variables each uniformly distributed over $(0,1)$. The sum $s = x_1 + x_2 + \cdots + x_n$ is confined to the interval $(0,n)$. Show that the density function of s is

$$h_n(s) = \frac{1}{(n-1)!}\left[s^{n-1} - \binom{n}{1}(s-1)^{n-1} + \binom{n}{2}(s-2)^{n-1} + \cdots \right],$$

where $0 < s < n$, and the summation is continued as long as the arguments s, $s - 1$, $s - 2$, $\cdots$, are positive. For example,

$$h_2(s) = \begin{cases} s & \text{for } 0 < s < 1, \\ s - 2(s-1) & \text{for } 1 < s < 2. \end{cases}$$

$$h_3(s) = \begin{cases} s^2/2 & \text{for } 0 < s < 1, \\ [s^2 - 3(s-1)^2]/2 & \text{for } 1 < s < 2, \\ [s^2 - 3(s-1)^2 + 3(s-2)^2]/2 & \text{for } 2 < s < 3. \end{cases}$$

Plot these curves. Show that the mean and variance of the sum are $n/2$ and $n/12$ respectively. [Later in this chapter we will study the Central Limit Theorem from which it follows that as $n \to \infty$, the density function of the standardized sum $(s - n/2)/\sqrt{n/12}$ approaches the standard normal density function.]

2. Suppose that x and y are independent, Erlang-distributed random variables, with parameters r_1, λ and r_2, λ, respectively. (The parameters r_1 and r_2 need not be the same, but the parameter λ is the same for both x and y.) Show that the sum $x + y$ is Erlang with parameters $r_1 + r_2, \lambda$.

NORMAL DISTRIBUTION

The normal probability density function is the most frequently used of all probability laws, both because the normal random variable does often occur in practical problems and, as we shall see, because it provides an accurate approximation to a large number of other probability laws.

We say that x is a normal random variable, or simply that x is normally distributed, with parameters μ and σ^2 if the density of x is given by

$$f(x) = \frac{1}{\sqrt{2\pi}\sigma} e^{-(x-\mu)^2/2\sigma^2} \qquad \text{for } -\infty < x < \infty. \tag{8.12}$$

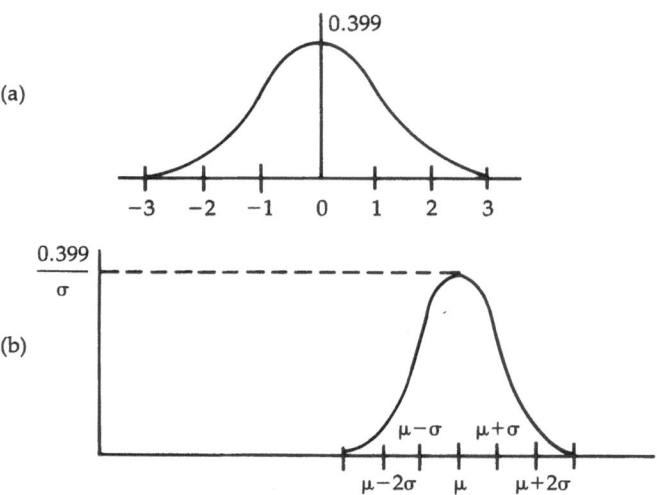

FIGURE 8.7 The normal density function: (a) with $\mu = 0$, $\sigma = 1$; and (b) arbitrary μ, σ^2.

Since the exponential function is nonnegative, we obviously have $f(x) \geq 0$ for all x. We will soon show that the area under $f(x)$ is one, as is required in order for $f(x)$ to be a probability density function. The normal density function is a bell-shaped curve that is symmetric about μ (see Figure 8.7). The values μ and σ^2 represent, in some sense, respectively the average value and the possible variation of x. (These concepts will be made precise shortly.) Due to the bell shape of the density function, a normally distributed random variable has the biggest probability of taking on a value close to μ (in a sense) and correspondingly less of taking on values further from μ (on either side).

Figure 8.8 shows the graphs of some typical density functions for various values of the parameters μ and σ^2. Notice that the density function is a symmetric, bell-shaped curve centered at μ (thus we would guess that μ is, in fact, the mean of the random variable; this is proved below.) The parameter σ controls the relative spread of the bell. Keeping μ constant and decreasing σ causes the density function to become more concentrated, thus giving higher probabilities of x being close to μ. Increasing σ causes the density function to spread out, thus giving lower probabilities of x being close to μ. If σ is held constant and μ is varied, the density function's shape is held constant with its midpoint occurring at the location of μ. As we will see, the parameter σ is the standard deviation of the random variable.

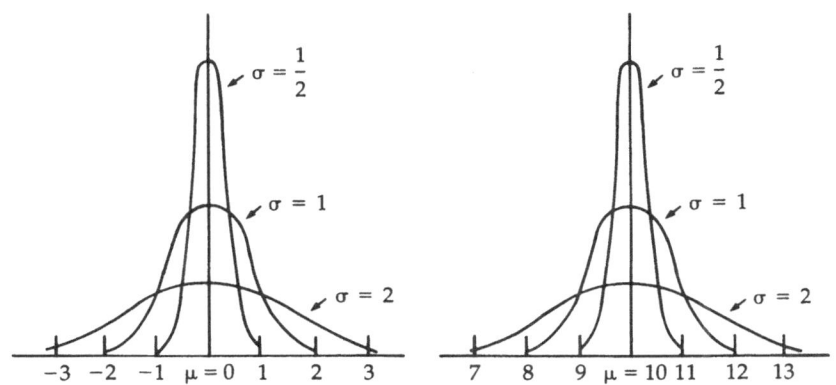

FIGURE 8.8 Normal density functions.

If we make the change of variable

$$z = \frac{x - \mu}{\sigma},$$

then $dx = \sigma \, dz$ and the normal density function $f(x)$ becomes

$$\phi(z) = \frac{1}{\sqrt{2\pi}} e^{-z^2/2} \qquad \text{for } -\infty < z < \infty. \tag{8.13}$$

This distribution is the *standard normal distribution*, since (as we shall soon verify) μ and σ^2 are indeed the mean and variance, respectively, of x.

To verify that $\phi(z)$ is indeed a probability density function, we need to show that

$$\int_{-\infty}^{\infty} \frac{1}{\sqrt{2\pi}} e^{-z^2/2} \, dz = 1.$$

We prove instead that the square of the integral is 1. Now,

$$\left[\int_{-\infty}^{\infty} \frac{e^{-z^2/2}}{\sqrt{2\pi}} \, dz \right]^2 = \frac{2}{\pi} \int_0^{\infty} e^{-y^2/2} \, dy \int_0^{\infty} e^{-x^2/2} \, dx$$

$$= \frac{2}{\pi} \int_0^{\infty} \int_0^{\infty} e^{-(x^2+y^2)/2} \, dx \, dy.$$

Changing to polar coordinates, let $x = r \cos \theta$, $y = r \sin \theta$, $x^2 + y^2 = r^2$. Then $dx \, dy$ goes into an element $r \, dr \, d\theta$ of area, and we have to evaluate just

$$\frac{2}{\pi} \int_0^{\pi/2} \int_0^\infty e^{-r^2/2} r \, dr \, d\theta = \frac{2}{\pi} \int_0^{\pi/2} d\theta = 1.$$

Thus $\phi(z)$ is in fact a proper probability density, and so, consequently, is $f(x)$.

Let us now find the moment generating function. The mgf of z is

$$M(\theta;z) = \int_{-\infty}^\infty \frac{e^{\theta z} e^{-z^2/2}}{\sqrt{2\pi}} \, dz.$$

Completing the square in the exponent, we have

$$\theta z - \frac{z^2}{2} = -\frac{1}{2}(z^2 - 2\theta z + \theta^2 - \theta^2) = -\frac{1}{2}(z - \theta)^2 + \frac{\theta^2}{2}.$$

Thus

$$M(\theta;z) = e^{\theta^2/2} \int_{-\infty}^\infty \frac{e^{-(1/2)(z-\theta)^2}}{\sqrt{2\pi}} \, dz.$$

But we have just proved that the last integral here is 1, so that

$$M(\theta;z) = e^{\theta^2/2}. \tag{8.14}$$

Now

$$E(z) = \frac{d}{d\theta} M(\theta;z) \bigg|_{\theta=0} = \theta e^{\theta^2/2} \bigg|_{\theta=0} = 0$$

and

$$\sigma_z^2 = \frac{d^2}{d\theta^2} M(\theta;z) \bigg|_{\theta=0} - 0^2 = 1.$$

Thus z is indeed a standard variable, and since $z = (x - \mu)/\sigma$, the mean and standard deviation of x are respectively

$$E(x) = \mu \quad \text{and} \quad \sigma_x = \sigma, \tag{8.15}$$

as was claimed above.

For brevity, instead of saying that x is a normal variable with mean μ and variance σ^2, we shall often just say x is normal (μ, σ^2), or simply $N(\mu, \sigma^2)$.

We note that $f(x)$ has an axis of symmetry at $x = \mu$. The mode is determined by

$$f'(x) = -\frac{x - \mu}{\sigma^2} f(x) = 0,$$

so that the mode, mean, and median coincide. Further, $f(x)$ has points of inflection at

$$f''(x) = 0 = \left(\frac{x - \mu}{\sigma^2}\right)^2 f(x) - \frac{f(x)}{\sigma^2} = 0,$$

which has the solution $x = \mu \pm \sigma$. These are just the points where the standard variable z is plus one or minus one.

Extensive tables of the normal distribution are available for the standard form of the variable. Often these are given in the form

$$A(z) = \int_0^z \frac{e^{-y^2/2}}{\sqrt{2\pi}} \, dy. \tag{8.16}$$

We see that $A(z)$ gives the area under the standard normal density from 0 to z for nonnegative values of z. Owing to the symmetry of the normal density, that is, owing to $\phi(z) = \phi(-z)$, the distribution function for z is $\Phi(z) = 0.5 + A(z)$ for $z \geq 0$, and $\Phi(z) = 0.5 - A(-z)$ for $z \leq 0$. Table 8.1 gives numerical values of the standard normal distribution functions $\Phi(z)$ for nonnegative z. For negative values of z, we use the equation $\Phi(z) = 1 - \Phi(-z)$. See Figure 8.9.

Since $z = (x - \mu)/\sigma$ is a standard normal random variable whenever x is normally distributed with parameters μ and σ^2, it follows that the distribution function of x can be expressed as

$$F(x) = \Phi\left(\frac{x - \mu}{\sigma}\right).$$

PROBLEM. If X is a normal random variable with parameters $\mu = 3$ and $\sigma^2 = 9$, find $P\{2 < X < 5\}$, $P\{X > 0\}$, and $P\{|X - 3| > 6\}$.

TABLE 8.1 Cumulative Distribution Function $\Phi(z)$ for Standard Normal Distribution for z = 0.00, 0.01, 0.02, $\cdots$, 3.48, 3.49 (That is, the entries $\Phi(z)$ are equal to the area under the standard normal density from $-\infty$ to z.)

z	.00	.01	.02	.03	.04	.05	.06	.07	.08	.09
.0	.5000	.5040	.5080	.5120	.5160	.5199	.5239	.5279	.5319	.5359
.1	.5398	.5438	.5478	.5517	.5557	.5596	.5636	.5675	.5714	.5753
.2	.5793	.5832	.5871	.5910	.5948	.5987	.6026	.6064	.6103	.6141
.3	.6179	.6217	.6255	.6293	.6331	.6368	.6406	.6443	.6480	.6517
.4	.6554	.6591	.6628	.6664	.6700	.6736	.6772	.6808	.6844	.6879
.5	.6915	.6950	.6985	.7019	.7054	.7088	.7123	.7157	.7190	.7224
.6	.7257	.7291	.7324	.7357	.7389	.7422	.7454	.7486	.7517	.7549
.7	.7580	.7611	.7642	.7673	.7704	.7734	.7764	.7794	.7823	.7852
.8	.7881	.7910	.7939	.7967	.7995	.8023	.8051	.8078	.8106	.8133
.9	.8159	.8186	.8212	.8238	.8264	.8289	.8315	.8340	.8365	.8389
1.0	.8413	.8438	.8461	.8485	.8508	.8531	.8554	.8557	.8599	.8621
1.1	.8643	.8665	.8686	.8708	.8729	.8749	.8770	.8790	.8810	.8830
1.2	.8849	.8869	.8888	.8907	.8925	.8944	.8962	.8980	.8997	.9015
1.3	.9032	.9049	.9066	.9082	.9099	.9115	.9131	.9147	.9162	.9177
1.4	.9192	.9207	.9222	.9236	.9251	.9265	.9279	.9292	.9306	.9319
1.5	.9332	.9345	.9357	.9370	.9382	.9394	.9406	.9418	.9429	.9441
1.6	.9452	.9463	.9474	.9484	.9495	.9505	.9515	.9525	.9535	.9545
1.7	.9554	.9564	.9573	.9582	.9591	.9599	.9608	.9616	.9625	.9633
1.8	.9641	.9649	.9656	.9664	.9671	.9678	.9686	.9693	.9699	.9706
1.9	.9713	.9719	.9726	.9732	.9738	.9744	.9750	.9756	.9761	.9767
2.0	.9772	.9778	.9783	.9788	.9793	.9798	.9803	.9808	.9812	.9817
2.1	.9821	.9826	.9830	.9834	.9838	.9842	.9846	.9850	.9854	.9857
2.2	.9861	.9864	.9868	.9871	.9875	.9878	.9881	.9884	.9887	.9890
2.3	.9893	.9896	.9898	.9901	.9904	.9906	.9909	.9911	.9913	.9916
2.4	.9918	.9920	.9922	.9925	.9927	.9929	.9931	.9932	.9934	.9936
2.5	.9938	.9940	.9941	.9943	.9945	.9946	.9948	.9949	.9951	.9952
2.6	.9953	.9955	.9956	.9957	.9959	.9960	.9961	.9962	.9963	.9964
2.7	.9965	.9966	.9967	.9968	.9969	.9970	.9971	.9972	.9973	.9974
2.8	.9974	.9975	.9976	.9977	.9977	.9978	.9979	.9979	.9980	.9981
2.9	.9981	.9982	.9982	.9983	.9984	.9984	.9985	.9985	.9986	.9986
3.0	.9987	.9987	.9987	.9988	.9988	.9989	.9989	.9989	.9990	.9990
3.1	.9990	.9991	.9991	.9991	.9992	.9992	.9992	.9992	.9993	.9993
3.2	.9993	.9993	.9994	.9994	.9994	.9994	.9994	.9995	.9995	.9995
3.3	.9995	.9995	.9995	.9996	.9996	.9996	.9996	.9996	.9996	.9997
3.4	.9997	.9997	.9997	.9997	.9997	.9997	.9997	.9997	.9997	.9998

SOLUTION.

$$P\{2 < X < 5\} = P\left\{\frac{2-3}{3} < \frac{X-3}{3} < \frac{5-3}{3}\right\} = P\left\{-\frac{1}{3} < Z < \frac{2}{3}\right\}$$

$$= \Phi\left(\frac{2}{3}\right) - \Phi\left(-\frac{1}{3}\right) = \Phi\left(\frac{2}{3}\right) - \left[1 - \Phi\left(\frac{1}{3}\right)\right] = 0.3779.$$

$$P\{X > 0\} = P\{Z > -1\} = 1 - \Phi(-1) = \Phi(1) = 0.8413.$$

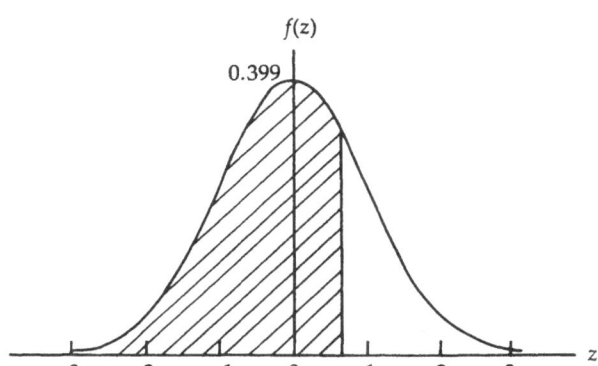

FIGURE 8.9 Standard normal curve with area $\Phi(z)$ shaded in.

$$P\{|X - 3| > 6\} = P\{X > 9\} + P\{X < -3\} = P\{Z > 2\} + P\{Z < -2\}$$
$$= 1 - \Phi(2) + \Phi(-2) = 2[1 - \Phi(2)] = 0.0456.$$

We have seen that the exponential distribution $\lambda e^{-\lambda t}$ is the limiting form of the Pascal-geometric pq^{n-1} (or the geometric pq^n equally as well) as trials are performed more and more nearly continuously, such that

$$p = \lambda \frac{t}{n}, \qquad n \rightarrow \infty.$$

It is natural to inquire if the binomial distribution has in a similar way some continuous approximation.

Now in the binomial distribution we want the probability of S_n successes in n Bernoulli trials. As $n \rightarrow \infty$, the mean number np of successes will also go to ∞, so that we cannot expect a reasonable distribution for S_n as $n \rightarrow \infty$. However, the value S_n/n might nevertheless remain under control, i.e., the fraction of successes in n trials. This turns out to be the case. In fact, another variable is also found to remain reasonable as $n \rightarrow \infty$, namely the variable

$$\frac{(S_n - np)}{\sqrt{npq}}.$$

This latter variable will be recognized as a standard variable.

Historically, the first result obtained was the following famous theorem:

THEOREM 8.4. *DeMoivre Laplace Limit Theorem.* Let S_n be the number of successes in n Bernoulli trials, where p is the probability for success at each trial. Let S_n^* be the standard variable corresponding to S_n; that is, let

$$S_n^* = \frac{S_n - np}{\sqrt{npq}}.$$

Then for any real numbers a, b, where $a < b$, the probability that S_n^* lies between a and b has the limiting form given by

$$P\{a \leq S_n^* \leq b\} \rightarrow \int_a^b \frac{1}{\sqrt{2\pi}} e^{-z^2/2}\, dz \qquad \text{as } n \rightarrow \infty. \tag{8.17}$$

In this theorem we recognize the function $(1/\sqrt{2\pi})e^{-z^2/2}$ as the density function $\phi(z)$ of the *standard normal distribution*, so Equation (8.17) can also be written as $P\{a \leq S_n^* \leq b\} \rightarrow \Phi(b) - \Phi(a)$ as $n \rightarrow \infty$.

When n is finite but large, the numerical approximation can be quite good. As a result, we can adapt (8.17) in the following form for computation:

$$P\{h \leq S_n \leq k\} = \Phi\left(\frac{k - np)}{\sqrt{npq}}\right) - \Phi\left(\frac{h - np}{\sqrt{npq}}\right). \tag{8.18}$$

Because $S_n \rightarrow \infty$ as $n \rightarrow \infty$, this form becomes of course meaningless in the limit. An even better numerical approximation is sometimes obtained by replacing h by $h - \frac{1}{2}$ and k by $k + \frac{1}{2}$ on the *right side* of Equation (8.18) as modified limits of integration. This so-called *continuity correction* helps adjust for the fact that a step function is being approximated by a continuous function. In Table 8.2, numerical approximation provided by Equation (8.18) is illustrated for $n = 40$. It will be observed that the relative error of the approximation is the smallest in the neighborhood of the mean np.

PROBLEM. Find the normal approximation to the central term of the binomial distribution $b(x;100,0.5)$.

SOLUTION. Here $x = 50$, and

$$\frac{x + 0.5 - np}{\sqrt{npq}} = \frac{50.5 - 50}{\sqrt{100(\frac{1}{2})(\frac{1}{2})}} = \frac{0.5}{5} = 0.1.$$

By symmetry, $P(49.5 \leq x \leq 50.5) = 2[\Phi(0.1) - 0.5]$ which, from Table 8.1, is $2(0.0398) = 0.0796$. If this value seems small, note

TABLE 8.2 Normal Approximation to Binomial, $n = 40$

Range of x	Binomial Probability	Corresponding $a \leqslant z \leqslant b$, $z = \dfrac{x - np}{\sqrt{npq}}$	$\displaystyle\int_a^b \dfrac{e^{-z^2/2}}{\sqrt{2\pi}}$
		$p = 0.25$	
$-0.5 \leqslant x \leqslant 4.5$	0.01604	$-3.8341 \leqslant z \leqslant - 2.0083$	0.02226
$4.5 \leqslant x \leqslant 9.5$	0.41950	$-2.0083 \leqslant z \leqslant - 0.1826$	0.40523
$9.5 \leqslant x \leqslant 14.5$	0.51002	$-0.1826 \leqslant z \leqslant + 1.6432$	0.52226
$14.5 \leqslant x \leqslant 19.5$	0.05386	$+1.6432 \leqslant z \leqslant + 3.4689$	0.04993
$19.5 \leqslant x \leqslant 24.5$	0.00057	$+3.4689 \leqslant z \leqslant + 5.2946$	0.00026
$24.5 \leqslant x \leqslant 29.5$	0.00001	$+5.2946 \leqslant z \leqslant + 8.1981$	0.00000
$29.5 \leqslant x \leqslant 34.5$	0.00000	$+8.1981 \leqslant z \leqslant +10.3001$	0.00000
$34.5 \leqslant x \leqslant 40.5$	0.00000	$+10.3001 \leqslant z \leqslant +12.8226$	0.00000
	1.00000		0.99994[a]
		$p = 0.5$	
$-0.5 \leqslant x \leqslant 4.5$	0.00000	$-6.4827 \leqslant z \leqslant -4.9015$	0.00000
$4.5 \leqslant x \leqslant 9.5$	0.00034	$-4.9015 \leqslant z \leqslant -3.3204$	0.00045
$9.5 \leqslant x \leqslant 14.5$	0.04000	$-3.3204 \leqslant z \leqslant -1.7393$	0.04057
$14.5 \leqslant x \leqslant 19.5$	0.39697	$-1.7393 \leqslant z \leqslant -0.1581$	0.39617
$19.5 \leqslant x \leqslant 24.5$	0.48576	$-0.1581 \leqslant z \leqslant +1.4230$	0.48544
$24.5 \leqslant x \leqslant 29.5$	0.07582	$+1.4230 \leqslant z \leqslant +3.0042$	0.07604
$29.5 \leqslant x \leqslant 34.5$	0.00111	$+3.0042 \leqslant z \leqslant +4.5853$	0.00133
$34.5 \leqslant x \leqslant 40.5$	0.00000	$+4.5853 \leqslant z \leqslant +6.4827$	0.00000
	1.00000		1.00000

[a] The missing 0.00006 probability is the left tail of the normal corrected to $z \leqslant -3.8341$.

however that there are 101 possible values for x, so that it is relatively large.

We will give a proof of the DeMoivre Laplace Limit Theorem in the next section.

The normal distribution was first discovered in 1733 by Abraham de Moivre (1667–1754) in connection with his derivation of the limiting form of the binomial distribution. He was a French Huguenot who settled in London after the revocation of the Edict of Nantes (1685) and earned a living by private tutoring. However, de Moivre's discovery seems to have passed unnoticed and it was not until long afterward that the normal distribution was rediscovered, in 1809 by Carl Friedrich Gauss (1777–1855) and in 1812 by Pierre Simon Laplace (1749–1827). Actually Laplace had already introduced the normal distribution as early as 1780, but his contribution is dated by his great book *Theorie Analytique des Probabilites* published in 1812. Because of his monumental book, Laplace may be regarded as the founder of the modern phase of probability. Although both de Moivre and Laplace have priority, the alternate name of the normal distribution is the Gaussian distribution. Gauss and Laplace were both led to the normal distribution in connection with their works on the theory of errors of physical observations.

EXERCISES

1. The annual rainfall (in inches) over a certain region is normally distributed with $\mu = 39.5$, $\sigma = 3.5$. Considering the annual amounts independent (an assumption that the actual observations seem to support), find the probability that in four consecutive years:
   (a) the first two amounts will be less than 36.7 and the next two greater than 42.3.
   (b) some pair or other will be less than 36.7 and the remaining pair greater than 42.3.

   ANSWERS. (a) 0.002. (b) 0.012.

2. Suppose that x and y are independent and normally distributed: x is $N(5,4)$, y is $N(-4,1)$.
   (a) What are the values of

   $$P(x > 5), \quad P(y < -4), \quad P(x + y > 1), \quad P\left(\frac{3x + 2y - 7}{12} \geq 0\right)?$$

   ANSWERS. $1/2$ in each case.
   (b) Use tables for approximate values of the following:

   $$P(x + y > 1 + \sqrt{5}) \quad \text{and} \quad P(3x + 2y - 7 > \sqrt{40}).$$

   ANSWERS. 0.1587 in each case.

3. A fair coin is tossed 50 times. Using the normal approximation, find the probability of getting:
   (a) 50 heads.
   (b) more than 39 and less than 60 heads.
   (c) within 6 of the expected number of heads.

   ANSWERS. (a) 0.8354. (b) 0.7540. (c) 0.8740.

4. A salesman finds that he makes a sale on a call with probability $1/10$. How many calls should he make in a year so that the probability that he makes less than 80 sales is no more than 0.05? (Assume Bernoulli trials.)

   ANSWER. Approximately, at least 952.

5. Suppose that the height, in inches, of a 25-year-old man is a normal random variable with parameters $\mu = 71$, $\sigma^2 = 6.25$.

(a) What percentage of 25-year-old men are over 6 feet 2 inches tall?

(b) What percentage of men in the 25-year-old 6-footer club are over 6 foot 5 inches?

ANSWERS. (a) $1 - \Phi(3/2.5)$. (b) $[1 - \Phi(6/2.5)]/[1 - \Phi(1/2.5)]$.

6. A student entering a university is assumed to choose a humanities major with probability $3/5$ and a science major with probability $2/5$. What is the approximate probability that in an entering freshman class of 500 students between 175 and 225 students will choose a science major? (Assume Bernoulli trials.)

ANSWER. Approximately 0.9774.

7. Find the probability density function of $y = e^x$ when x is normally distributed with parameters μ and σ^2. The random variable y is said to have a lognormal distribution (since log y has a normal distribution) with parameters μ and σ^2.

8. Assuming absenteeism on the part of a worker to be a Bernoulli trial each day, what is the probability that more than 10 persons are absent on the same day in a factory of 500 employees where attendance averages 99 percent?

ANSWERS. 0.0215.

9. A manufacturer of radio tubes knows that 5 percent of his product is defective. The tubes are sold in lots of 100.

(a) What number of defective tubes per lot is exceeded by only 5 percent of the lots?

(b) Between what limits will the percentage of defectives lie 95 percent of the time?

10. If it was desired to estimate the proportion of republicans in a certain town and if it was desired that the estimate be accurate to within 0.02 with a probability of 90 percent, how large a sample should be taken?

ANSWER. $n = 1333$.

11. If z is a standard normal random variable, show that, for every $a > 0$,

$$\lim_{x \to x} \frac{P\{z \geq x + a/x\}}{P\{z \geq x\}} = e^{-a}.$$

12. In firing at a target, the horizontal error is found to be normally distributed with $\sigma = 2$ feet.

(a) In 200 shots, how many would be expected to miss a target 10 feet wide?

(b) How many shots would you need to fire to be sure with probability 0.95 that 50 or more fell within 3 feet of the centerline?

ANSWERS. (a) About $1/2$. (b) 85.

13. Consider that the diameter of ball bearings of a certain manufacture is normally distributed with a mean and standard deviation of 0.0020 inches and 0.0002 inches, respectively.

(a) Between what limits will the diameter lie 95 percent of the time?

(b) Approximately what will be the 95 percent limits for the mean of a sample of 25 ball bearings whether x is normal or not?

ANSWERS. (a) $0.0016 \leqslant d \leqslant 0.0024$. (b) $0.0019 \leqslant d \leqslant 0.0021$.

14. A balanced coin is flipped 400 times. Determine the number x such that the probability that the number of heads is between $200 - x$ and $200 + x$ is approximately 0.85. (Assume Bernoulli trials.)

ANSWER. 14.

CENTRAL LIMIT THEOREM

We return now to the question of the normal distribution as a limiting form of the binomial, which is essentially the import of the DeMoivre Laplace limit theorem. That theorem is actually a special case of a much more powerful theorem which shows that the normal distribution is potentially of very wide applicability. This more powerful theorem is known as the *central limit theorem*, which we can state and prove in outline.

THEOREM 8.5. *Central Limit Theorem.* Let $u_1, u_2, \cdots, u_n, \cdots$ be independent standard random variables all with the same distribution function $F(u)$ and moment generating function $M(\theta; u)$. Let

$$S_n^* = \frac{u_1 + u_2 + \cdots + u_n}{\sqrt{n}}.$$

Then as $n \to \infty$, we have

$$P(S_n^* \leqslant z) \to \Phi(z), \tag{8.19}$$

where $\Phi(z)$ is the standard normal distribution function.

PROOF. Because $u_1, u_2, \cdots, u_n$ are independent with mean zero and variance one, it follows that $u_1 + u_2 + \cdots + u_n$ has mean zero and variance n. Hence S_n^* has mean zero and variance one; that is, S_n^* is a standard variable. By Theorems 6.7 and 6.8 in Chapter 6, we have

$$M(\theta;S_n^*) = M\left(\frac{\theta}{\sqrt{n}};u_1+u_2+\cdots+u_n\right) = \left[M\left(\frac{\theta}{\sqrt{n}};u\right)\right]^n.$$

We now make use of the series expansion of the moment generating function, and thus obtain

$$M(\theta;S_n^*) = \left[1 + \left(\frac{\theta}{\sqrt{n}}\right)E(u) + \left(\frac{\theta}{\sqrt{n}}\right)^2\frac{1}{2!}E(u^2) + \cdots\right.$$

$$\left. + \left(\frac{\theta}{\sqrt{n}}\right)^n\frac{1}{n!}E(u^n) + \cdots\right]^n.$$

By hypothesis, μ is a standard variable, that is, $\mu_u = E(u) = 0$, and $\sigma_u^2 = 1$. Hence

$$E(u^2) = \sigma_u^2 + \mu_u^2 = 1.$$

Thus

$$M(\theta;S_n^*) = \left[1 + \frac{\theta^2}{2n} + \left(\frac{\theta}{\sqrt{n}}\right)^3\frac{E(u^3)}{3!} + \cdots\right]^n.$$

Because all the terms within the brackets after the first two contain a factor $n^{-3/2}$, they are of a smaller order of magnitude than $1 + \theta^2/2n$. By a limiting argument whose details we will omit here, it follows that these smaller terms become negligible. Hence we have

$$\lim_{n\to\infty} M(\theta;S_n^*) = \lim_{n\to\infty}\left(1 + \frac{\theta^2}{2n}\right)^n = e^{\theta^2/2}.$$

By Equation (8.14) of the preceding section, we recognize exp $(\theta^2/2)$ to be the mgf of a standard normal variable z. Because of the uniqueness of moment generating functions, it follows that the limiting distribution function of S_n^* is the standard normal distribution function $\Phi(z)$ and thus the theorem is proved.   Q.E.D.

As was indicated, this theorem was first discovered in its application to Bernoulli trials and the binomial distribution. Let y_i correspond to the outcome of an individual Bernoulli trial; that is, $y_i = 1$ if success occurs (with probability p) and 0 if failure occurs (with probability $q = 1 - p$). Then, $E(y_i) = 1\cdot p + 0\cdot q = p$ and the variance of y_i is $(1 - p)^2\cdot p + (-p)^2(1 - p) = p(1 - p) = pq$. Consequently, in the statement of the central limit theorem, let

$$u_i = \frac{y_i - E(y_i)}{\sigma(y_i)} = \frac{y_i - p}{\sqrt{pq}}.$$

Then $E(u_i) = 0$ and $\sigma^2(u_i) = 1$, so the hypothesis that u_i is a standard variable is fulfilled. Further, let $S_n = y_1 + \cdots + y_n$ (i.e., S_n is the number of successes in n trials), and let

$$S_n^* = \frac{u_1 + \cdots + u_n}{\sqrt{n}} = \frac{S_n - np}{\sqrt{npq}}.$$

Then by the central limit theorem, we have

$$P\{a \leqslant S_n^* \leqslant b\} = P\left\{a \leqslant \frac{S_n - np}{\sqrt{npq}} \leqslant b\right\} \to \int_a^b \frac{1}{\sqrt{2\pi}} e^{-z^2/2} \, dz \quad (8.20)$$

as $n \to \infty$. This result is just the DeMoivre Laplace theorem. That is, we have shown that if S_n is the number of successes in n Bernoulli trials, then $(S_n - np)\sqrt{npq}$ is a standard normal variable in the limit as $n \to \infty$.

We can get still a further result about Bernoulli trials. In n trials with S_n successes, S_n/n is the *proportion* of successes, i.e., S_n/n is the relative frequency of success. Denoting this proportion by ρ_n, we note that

$$S_n^* = \frac{\rho_n - p}{\sqrt{pq/n}} \quad (8.21)$$

is a standard normal variable in the limit as $n \to \infty$.

As we have noted, the limiting form of the binomial distribution was discovered in 1733 by de Moivre. This result is a special case of the central limit theorem. In his great book of 1812, Laplace gave the first (incomplete) statement of the general form of the central limit theorem, and in addition made a large number of important applications of the normal distribution to various questions in the theory of probability. The fundamental importance of the result of de Moivre was completely revealed by Laplace. Under the influence of the work of Laplace it was for a long time more or less regarded as an axiom that probability distributions of practically all kinds would approach the normal distribution as an ideal limiting form. Even if this view was definitely exaggerated and has had to be considerably modified, it is undeniable that we often meet in applications distributions which are normal or approximately normal. Such is the case with many empirical distributions of physical, biological, and economic measurements and observations. The central limit theorem affords a theoretical explanation of these empirical facts. The explanation rests on the hypothesis that the random variable in question is the sum of a large number of mutually independent small effects. None of these effects may predominate. Then, by the central limit theorem, the random variable should be approximately normally distributed. For example, the total consumption of electric energy delivered by the utility company is the sum of the quantities consumed by all its customers. The total gain or loss on the risk business of an insurance company is the sum of the gains or losses of each single policy. In cases of this character, we should expect to find at least approximately normal distributions.

There is a famous remark to the effect that "everybody believes in the normal distribution: the experimenters because they think it is a mathematical theorem, and the mathematicians because they think it is an experimental fact." In a sense, both groups are correct, provided that their beliefs are not too absolute. Mathematical proof tells us that often we are justified in expecting a normal distribution *under certain qualifying conditions,* while statistical experience shows us that in fact many distributions are often *approximately normal.*

The name "central limit theorem" is used generally to designate any convergence theorem in which the normal distribution appears as the limit. More particularly it applies to sums of random variables under the so-called "normal" conditions. The version of the central limit theorem we have proved here makes use of quite strong normal conditions; in more advanced versions, the normal conditions are considerably weakened. Thus the narrow context of Theorem 8.5 can be generalized in several directions: the assumptions of a moment generating function, of a common distribution, and of strict independence can all be relaxed. However, if the normal conditions are radically altered, then the central limit theorem will no longer apply, and random phenomena abound in which the limit distribution is no longer normal. The Poisson limit theorem may be considered as one such example. In fact, there is a whole family of limit theorems dealing with so-called "stable" and "infinitely divisible" distributions, and the central limit theorem as well as the Poisson limit theorem are particular members of this family.

It should be stressed that the central limit theorem as stated in Theorem 8.5 does not give an estimate of the "error." In other words, it asserts convergence without indicating any speed of convergence. This renders the result useless in accurate numerical computations. However, under specified conditions it is possible to obtain bounds for the error. For example, in the DeMoivre-Laplace case it can be shown that the error does not exceed $C/\sqrt{n}$ where C is a numerical constant involving p but not a or b. In simple applications, the error is simply ignored, as we will do in our examples.

The fact that in the central limit theorem the normal distribution is reached in the limit as $n \to \infty$ implies that for suitably large *finite n*, the random variable S_n^* will be "approximately" normally distributed. Obviously, however, the degree of approximation will depend not only on n, but also on the distribution function $F(u)$. Let us list, therefore, several useful corollaries of the central limit theorem.

COROLLARY 8.5A. Let $y_1, y_2, \cdots, y_n$ be independent identically distributed random variables each with mean μ, variance σ^2, and mgf $M(\theta;y)$. Let $x = y_1 + y_2 + \cdots + y_n$. Then for suitably large n, the random variable x is approximately normal with mean $n\mu$ and variance $n\sigma^2$; that is,

$$P(z_1\sigma\sqrt{n} < x - n\mu < z_2\sigma\sqrt{n}) \approx \Phi(z_2) - \Phi(z_1), \qquad (8.22)$$

where $\Phi(z)$ is the standard normal distribution function. In case we have a symmetrical spread about the mean, i.e., $z_2 = -z_1 = z > 0$, then Equation (8.22) becomes

$$P(|x - n\mu| < z\sigma\sqrt{n}) \approx 2\Phi(z) - 1. \qquad (8.23)$$

PROOF. We see that $u_i = (y_i - \mu)/\sigma$ is a standard variable. Thus by the central limit theorem, $S_n^* = \Sigma u_i/\sqrt{n}$ is approximately a normal variable with mean 0 and $\sigma = 1$. But

$$\frac{\Sigma u_i}{\sqrt{n}} = \frac{\Sigma y_i - n\mu}{\sigma\sqrt{n}} = \frac{x - n\mu}{\sigma\sqrt{n}}.$$

Thus x is approximately a normal variable with mean $n\mu$, and variance $n\sigma^2$. Q.E.D.

The above corollary states the important fact that sums of arbitrary identically distributed independent variables are asymptotically normally distributed. The remarkable degree of approximation is illustrated in Table 8.3 for the case where each y_i is an exponential variable with a mean $\mu = 1/\lambda = 1$, and where $n = 25$. We know that if $x = y_1 + y_2 + \cdots + y_{25}$, then x is an Erlang variable with parameters $n = 25$ and $\lambda = 1$, so its density function is

$$f(x) = \frac{x^{24}e^{-x}}{24!} \qquad \text{for } x \geq 0.$$

The mean of x is $n/\lambda = 25$ and the variance of x is $n/\lambda^2 = 25$. Table 8.3 gives the probabilities that x lie in certain ranges as computed directly from this Erlang density function, in comparison with the prob-

TABLE 8.3 Normal Approximation to the Sum of 25 Identical Unit-Mean Exponential Variables

Range of x	Erlang probability	Normal approximation
$x \leqslant 10$	0.00004	0.00135
$10 \leqslant x \leqslant 15$	0.01037	0.02140
$15 \leqslant x \leqslant 20$	0.14088	0.13590
$20 \leqslant x \leqslant 25$	0.36725	0.34134
$25 \leqslant x \leqslant 30$	0.31991	0.34134
$30 \leqslant x \leqslant 35$	0.12803	0.13590
$35 \leqslant x \leqslant 40$	0.02886	0.02140
$40 \leqslant x \leqslant 45$	0.00419	0.00132
$45 \leqslant x \leqslant 50$	0.00043	0.00004
$x \geqslant 50$	0.00004	0.00000

abilities computed from the normal distribution having the same mean and variance.

COROLLARY 8.5B. In Bernoulli trials with probability p for success on each trial, let x be the number of successes in n trials. Then

a. $\dfrac{x - np}{\sqrt{npq}}$ is approximately a standard normal variable.

b. x is approximately normal with mean np, variance npq.

c. $\dfrac{x}{n}$ is approximately normal with mean p, variance $\dfrac{pq}{n}$.

EXAMPLE. For Bernoulli trials with $n = 100$, $p = \frac{1}{2}$, $P(35 < x < 65)$ is approximately equal to

$$P\left\{ \frac{35 - 50}{5} < z < \frac{65 - 50}{5} \right\} = \Phi(3) - \Phi(-3) = 2\Phi(3) - 1$$

$$= 0.9974, \quad \text{from Table 8.1}$$

In this case, one unit of deviation in z corresponds to five units in x, for $\sqrt{npq} = 5$. Also

$$P\left(\frac{x}{n} \leqslant 0.4 \right) = \Phi\left\{ \frac{(0.4 - 0.5)}{\sqrt{(0.5)(0.5)/100}} \right\}$$

$$= \Phi(-2) = 1 - \Phi(2)$$

$$= 0.0228, \quad \text{from Table 8.1.}$$

COROLLARY 8.5C. (Exercise) Show how a Poisson distribution can be approximated by a normal. Compute the case when $\mu = 100$.

PROBLEM. A quantity with true value μ is measured many times for accuracy. Each measurement is subject to a random error that is uniformly distributed between -1 and $+1$. If we take the arithmetical mean [average] of n measurements, what is the probability that the average differs from the true value by less than δ?

SOLUTION. Let there be n measurements. Denote each measurement by x_i and the associated error by v_i where $i = 1, 2, \cdots,$ n. The hypothesis says that the model is $x_i = \mu + v_i$ where v_i is a random variable which has the uniform distribution in $(-1, +1)$. Thus

$$E(v_i) = \int_{-1}^{+1} \frac{1}{2}\, dv = 0,$$

$$\sigma^2(v_i) = E(v_i^2) = \int_{-1}^{+1} \frac{1}{2} v^2\, dv = \frac{1}{3},$$

$$E(x_i) = \mu,$$

$$\sigma^2(x_i) = {}^1\!/_3.$$

The arithmetic average of the n measurements is defined to be $\bar{x} = (x_1 + x_2 + \cdots + x_n)/n$. If we define $y_i = x_i/n$, then $\bar{x}$ can be expressed as $\bar{x} = y_1 + y_2 + \cdots + y_n$ where each y_i has mean $E(y_i) = E(x_i)/n = \mu/n$ and variance $\sigma^2(y_i) = \sigma^2(x_i)/n^2 = 1/3n^2$. Thus by Corollary 8.5A, the arithmetic average $\bar{x}$ is approximately normal with mean μ and variance $1/3n$. We have

$$P(|\bar{x} - \mu| < z/\sqrt{3n}) \approx 2\Phi(z) - 1.$$

Hence $\delta = z/\sqrt{3n}$, so $z = \delta\sqrt{3n}$. Thus the required probability is $2\Phi(\delta\sqrt{3n}) - 1$. For instance, if $n = 25$ and $\delta = {}^1\!/_5$, then the result is equal to

$$2\Phi(\sqrt{3}) - 1 \approx 2\Phi(1.73) - 1 \approx 0.9164.$$

In words, if 25 measurements are taken, then we are 92 percent sure that their average is within one fifth of a unit from the true value.

PROBLEM. Let us use the same model as given in the preceding problem, but let us now turn around the question. Let us ask how many measurements should we take in order that the probability will exceed α (the "significance level") that the average will differ from the true value by at most δ?

SOLUTION. This means we must first find the value z_α such that $2\Phi(z_\alpha) - 1 = \alpha$, or $\Phi(z_\alpha) = (1 + \alpha)/2$, and then choose n to make $\delta\sqrt{3n} > z_\alpha$. For instance, if $\alpha = 0.95$, then Table 8.1 shows that $z_\alpha \approx 1.96$. Thus $n > z_\alpha^2/3\delta^2$. If $\delta = 1/5$, this gives $n > 32$, which means that eight more measurements than 25 (i.e., 25 + 8 = 33) should increase our degree of confidence from 92 percent to 95 percent. Whether this is worthwhile may depend on the cost of doing the additional work as well as other factors.

From the above two problems, we see that there are three variables involved in questions of this kind, namely: δ, α, and n. If two of them are fixed, we can solve for the third. In the first problem, n and δ were given and we solved for α. In the second problem, α and δ were given and we solved for n. In a third type of problem, n and α are fixed and we find δ. For example, if $n = 25$ is fixed because the measurements are found in recorded data and not repeatable, and our credulity demands a high degree of confidence α, say 99 percent, then we must compromise on the coefficient of accuracy δ.

EXERCISES

1. An encyclopedia salesman finds that on a given call he makes no sale with probability $1/2$; he sells the regular edition at \$50 with probability $3/8$; and he sells the custom edition at \$100 with probability $1/8$. How many calls should he make within a year to insure total sales of \$5000 with an approximate probability of 0.95? (Assume independent trials.)

2. Suppose that a balanced die is thrown 2000 times and that these trials are independent. What is the approximate probability that the total score is no less than 7000 and no more than 8000? Within what limits of the form $7000 - x$ and $7000 + x$ will the score be with an approximate probability of 0.5?

3. Suppose that a random sample of size 1000 is drawn from a population and that the numerical quantity being studied has an unknown population average μ and a known population variance of $\sigma^2 = 16$.
   (a) With what approximate probability will the sample average deviate by no more than $1/8$ from the population average?
   (b) Determine a positive ϵ for which the approximate probability that

the sample average differs from the population average by more than ϵ is 0.05.

ANSWER. (a) 0.6778.

4. (a) Suppose a random sample of size 20000 is drawn from a standard deck of cards with replacement. What is the approximate probability that the total number of aces that are drawn is no less than 1400 and no more than 1600? Is no less than 1400?

ANSWERS. 0.9483, 0.9999.

(b) Consider an experiment consisting of drawing a random sample of size five from a standard deck of cards without replacement. If 4000 independent trials of this experiment are made under identical conditions, what is the approximate probability that the total number of aces drawn is no less than 1400 and no more than 1600? Is no less than 1400? Explain the difference between the answers to (a) and (b).

5. When an unbiased die is tossed repeatedly, estimate the minimum number of tosses required for the probability to be at least 0.95 that the relative frequency of one spot showing on the tossed die will lie between $9/60$ and $11/60$.

ANSWER. 1860 (or 1921), with (or without) continuity correction.

6. Let z have the normal distribution with mean 0 and variance 1. For $\epsilon = 1/2, 1, 2, 3$, and 4, compute exact values of $P\{|z| > \epsilon\}$, and compare these values with those obtained from the upper bound $1/\epsilon^2$ provided by Chebyshev's inequality.

ANSWER. 0.617, 4; 0.317, 1; 0.046, 0.25; 0.0027, 0.111; 0.00006, 0.063.

7. Let S_n have the binomial distribution with parameters n and $p = q = 1/2$. On ordinary (x, y) coordinates sketch the graph of the standard normal distribution function $y = \Phi(x)$, and also sketch the graphs represented in parametric form by $x = (t - np)/\sqrt{npq}$ and $y = P\{S_n \leq t\}$ for a few values of n, e.g., $n = 4, 8, 12$.

8. Write an exact expression for the probability that the number x of heads which occur when a true coin is tossed 1600 times will satisfy $790 \leq x \leq 810$. Compute an approximate value of this probability.

ANSWER. 0.40.

9. The probability is 0.04 that a bolt chosen at random from the output of a certain machine will be defective. Let A be the event that a keg of 100 bolts made by this machine will contain at most 2 defective bolts. Evaluate exactly the probability of the event A, and also use the normal and Poisson approximations to the binomial distribution

to find approximate values of this probability. Which approximation is closer to the exact value?

ANSWER. 0.232, 0.222, 0.238. Poisson.

REPRODUCTION PROPERTIES OF THE NORMAL DISTRIBUTION

We recall from Equation (8.14) that the moment generating function of a standard normal variable z is $M(\theta;z) = \exp(\theta^2/2)$. Now z can be $(x - \mu)/\sigma$, where x is normal (μ,σ^2). Consequently $x = \mu + \sigma z$, so the mgf of x is

$$M(\theta;x) = \int_{-\infty}^{\infty} e^{\theta(\mu+\sigma z)}f(z)\,dz = e^{\theta\mu}M(\theta\sigma;z).$$

That is,

$$M(\theta;x) = e^{\theta\mu}\,e^{\theta^2\sigma^2/2}. \tag{8.24}$$

This is then the mgf of a normal variable x having mean μ, variance σ^2.

THEOREM 8.6. If x is normal (μ,σ^2), then $ax + b$ (with $a \neq 0$) is normal with mean $a\mu + b$ and variance $(a\sigma)^2$.

PROOF.

$$\begin{aligned}
M(\theta;ax+b) &= e^{\theta b}M(a\theta;x)\\
&= e^{\theta b}e^{a\theta\mu}e^{(a\theta)^2\sigma^2/2}\\
&= e^{\theta(a\mu+b)}e^{\theta^2(a\sigma)^2/2}.
\end{aligned}$$

Comparison with (8.24) above establishes the result.

THEOREM 8.7. If x_1 and x_2 are independent normal variables with means μ_1 and μ_2 and variances σ_1^2 and σ_2^2, respectively, then the variable $x = x_1 + x_2$ is normal with mean $\mu_1 + \mu_2$ and variance $\sigma_1^2 + \sigma_2^2$.

PROOF.

$$\begin{aligned}
M(\theta;x) &= M(\theta;x_1)\,M(\theta;x_2)\\
&= e^{\theta(\mu_1+\mu_2)}e^{\theta^2(\sigma_1^2+\sigma_2^2)/2}.
\end{aligned}$$

THEOREM 8.8. If $x_1, \cdots, x_n$ are independent normal variables, then any linear combination $x = a_1x_1 + \cdots + a_nx_n + b$ is normal with mean and variance given by

$$\mu_x = a_1\mu_1 + \cdots + a_n\mu_n + b,$$
$$\sigma_x^2 = (a_1\sigma_1)^2 + \cdots + (a_n\sigma_n)^2.$$

where the mean and variance of x_i are μ_i and σ_i, respectively.

If a sample of n random observations are made of a variable x, we recall that the sample mean

$$\bar{x} = \frac{x_1 + \cdots + x_n}{n}$$

always has the properties

$$E(\bar{x}) = E(x), \operatorname{var}(\bar{x}) = \frac{1}{n}\sigma_x^2, \tag{8.25}$$

because $E(\bar{x}) = E[(\Sigma x_i)/n] = nE(x)/n = E(x)$, and $\operatorname{var}(\bar{x}) = [\operatorname{var}(\Sigma x_i)]/n^2 = n\sigma_x^2/n^2 = \sigma_x^2/n$. In case x is a normal variable, it follows further by Theorem 8.8 that $\bar{x}$ is normal. We note then, for emphasis:

THEOREM 8.9. If x is normal (μ,σ^2), then the mean $\bar{x}$ of a random sample of n observations of x is normal $(\mu,\sigma^2/n)$.

EXAMPLE. If x is normal with mean 20, $\sigma = 5$, then $P\{5 \leqslant x \leqslant 35\} = 2\Phi(3) - 1$ which from Table 8.1 is 0.9974. For a sample of 25 observations of x, $P\{17 \leqslant \bar{x} \leqslant 23\} = 0.9974$.

If the original variable x were not normally distributed, $\bar{x}$ would not be necessarily normal. However, the central limit theorem tells us that as n increases in size, x is more and more nearly normally distributed. In fact, we can summarize as follows:

THEOREM 8.10. If x has any distribution with mean μ, variance σ^2 (and moment generating function), then $\bar{x}$ is approximately normally distributed with mean μ, variance σ^2/n, in that $(\bar{x} - \mu)/(\sigma/\sqrt{n})$ approaches a standard normal variable as $n \to \infty$.

Already, in Table 8.3, we have illustrated how $x_1 + \cdots + x_n$ would be nearly normally distributed for $n = 25$. Consequently, $\bar{x}$ would be nearly normally distributed in this case.

EXERCISES

1. Let x_1 and x_2 be independent normal variables with means 6 and 8 and variances 20 and 4, respectively. Write the density function for the random variable $y = 3 + x_1 - 2x_2$.

   ANSWER. Normal density function with mean -7 and variance 36.

2. The function Φ has the property that $\Phi(x) \to 1$ as $x \to \infty$ and $\Phi(-x) \to 0$ as $x \to \infty$. Using this, show that the weak law of large numbers is a consequence of the central limit theorem.

3. Let $\bar{z}_n = (z_1 + z_2 + \cdots + z_n)/n$ where $z_1, z_2, \cdots, z_n$ are mutually independent, identically distributed normal random variables with mean 0 and variance 1. Compute the values of $p\{|z_n| > \varepsilon\}$ for $\varepsilon = {}^1\!/_2$ and $n = 4, 9, 16, 36$ and also for $\varepsilon = {}^1\!/_{10}$ and $n = 100, 225, 400, 900$.

   ANSWER. For both cases: 0.3174, 0.1336, 0.0456, 0.0026.

4. In a large community, $^2\!/_3$ of the adult males own cars. If a random sample of size 300 is drawn with replacement from among adult males, find the number x such that the approximate probability that the number of car owners in the sample is between $200 - x$ and $200 + x$ is 0.90.

5. Let S_n be the number of successes in n independent Bernoulli trials with probability p of success for each trial. How large a sample size n is required to bring S_n/n within 0.02 of p with approximate probability 0.95? To bring S_n/n within 0.001 of p with approximate probability 0.95?

6. How large a sample size n is required in Bernoulli trials so that S_n/n is at least $p - 0.01$ with approximate probability 0.95?

7. The average value $\bar{x}$ of a random sample of n observations from a certain population is normally distributed with

$$\mu = 20 \qquad \sigma = 5/\sqrt{n}.$$

   How large a sample should we draw in order to have a probability of at least 0.90 that $\bar{x}$ will lie between 18 and 22?

   ANSWER. 17.

8. If x is normally distributed with parameters (μ, σ^2) then $\bar{x}$, the average of n independent observations of x, is also normally distributed with corresponding parameters $(\mu, \sigma^2/n)$. This result is not merely approximate, in this case, but exact. Given a normal population with $\sigma = 1$, plot the distance between the lower and upper 98 percent confidence limits for μ as a function of the sample size

n on which the sample mean $\bar{x}$ is based.

ANSWER. Graph of $4.65/\sqrt{n}$.

9. Suppose x has the $N(\mu,\sigma^2)$ distribution.
   (a) Show that $(x - \mu)/\sigma$ has the $N(0,1)$ distribution.
   (b) Show that $ax + b$ has the $N(a\mu+b,a^2\sigma^2)$ distribution.

10. If $x_1, \cdots, x_n$ are mutually independent, each distributed $N(0,1)$, and if $a_1, \cdots, a_n$ are constants, show that $a_1x_1 + \cdots + a_nx_n$ is $N(0,a_1^2 +\cdots+a_n^2)$.

11. Suppose that X and Y are independent, each distributed $N(0,1)$.
    (a) Show that

$$U = \frac{X + Y}{\sqrt{2}} \quad \text{and} \quad V = \frac{X - Y}{\sqrt{2}}$$

are also independent $N(0,1)$.

HINT.

$$P[(U \leq u) \cap (V \leq v)] = \frac{1}{2\pi} \iint\limits_{\substack{(x+y)/\sqrt{2}\leq u \\ (x-y)/\sqrt{2}\leq v}} \exp\left[\frac{-(x^2 + y^2)}{2}\right] dx\, dy.$$

Make a change of variables, $(x + y)/\sqrt{2} = r$, $(x - y)/\sqrt{2} = s$. The Jacobian is 1 and $x^2 + y^2 = r^2 + s^2$. Hence, the required probability is

$$\iint\limits_{\substack{r\leq u \\ s\leq v}} \left[\frac{1}{(2\pi)^{1/2}} \exp\left(\frac{-r^2}{2}\right)\right]\left[\frac{1}{(2\pi)^{1/2}} \exp\left(\frac{-s^2}{2}\right)\right] dr\, ds.$$

(b) Show that $2XY$ and $X^2 - Y^2$ each have the same distribution.

HINT. $X^2 - Y^2 = 2[(X - Y)/\sqrt{2}][(X + Y)/\sqrt{2}]$. Now use (a).

12. The breaking strength (in pounds) of string manufactured by an established process is normally distributed with $\mu = 15$, $\sigma = 3$. A new process is tried, and a random sample of 9 pieces of standard length have an average breaking strength of 18. Under the null hypothesis that these pieces are random selections from the original population, what is the probability that their average would differ from μ as much as this or more in either direction?

ANSWER. 0.0027.

13. If x and y are independent normal random variables, find the joint density of the random variables r and θ defined by $x = r \cos \theta$, $y = r \sin \theta$. Are they independent?

ANSWER. They are independent, with r having a Rayleigh distribution and θ a uniform distribution.

BIVARIATE NORMAL DISTRIBUTION

If two random variables x and y each are normally distributed, and if in addition they are independent, then their joint density function is the product of the two marginal normal density functions. In symbols, the joint density function is

$$f(x,y) = \frac{\exp\{-\tfrac{1}{2}[(x - \mu_x)/\sigma_x]^2\}}{\sqrt{2\pi}\,\sigma_x} \cdot \frac{\exp\{-\tfrac{1}{2}[(y - \mu_y)/\sigma_y]^2\}}{\sqrt{2\pi}\,\sigma_y}$$

$$= \frac{\exp\left(-\tfrac{1}{2}\{[(x - \mu_x)/\sigma_x]^2 + [(y - \mu_y)/\sigma_y]^2\}\right)}{2\pi\sigma_x\sigma_y}.$$

However, now we wish to consider the case when x and y are still normal but not independently distributed. In such a case we must modify the above joint density function so as to take into account the relationship between x and y. We observe that the exponent of $f(x,y)$ as it stands has a sum of two squares. The required modification is done by introducing a cross-product term in the exponent, and by introducing a scale factor. We therefore give the following definition.

DEFINITION. The bivariate normal density function is given by the following formula, where $-1 < \rho < 1$,

$$f(x,y) = \frac{\exp\left\{-\dfrac{1}{2(1 - \rho^2)}\left[\left(\dfrac{x - \mu_x}{\sigma_x}\right)^2 - 2\rho\left(\dfrac{x - \mu_y}{\sigma_x}\right)\left(\dfrac{y - \mu_y}{\sigma_y}\right) + \left(\dfrac{y - \mu_y}{\sigma_y}\right)^2\right]\right\}}{2\pi\sigma_x\sigma_y\sqrt{1 - \rho^2}}.$$

We note that when $\rho = 0$ the cross-product term vanishes and the scale factor $\sqrt{1 - \rho^2}$ reduces to one. We also note that when $\rho^2 = 1$ the scale factor becomes zero. It is for this reason that we restrict the possible values of ρ to the range $-1 < \rho < 1$.

It can be verified that bivariate normal density $f(x,y)$ as defined possesses the required properties of a joint density function. If we compute the moments of $f(x,y)$ by evaluating the necessary integrals, we find $E(x) = \mu_x$, $E(y) = \mu_y$, var $(x) = \sigma_x^2$, var $(y) = \sigma_y^2$, cov $(x,y) = \rho\sigma_x\sigma_y$. Thus the parameter ρ in the expression for $f(x,y)$ is in fact the correlation coefficient.

The marginal distribution of x is given by

$$f_1(x) = \int_{-\infty}^{\infty} f(x,y)\, dy,$$

where $f(x,y)$ is the bivariate normal density. In order to simplify the integration, let $u = (x - \mu_x)/\sigma_x$ and introduce the change of variable $v = (y - \mu_y)/\sigma_y$. Then $dy = \sigma_y\, dv$ and the above integral reduces to

$$f_1(x) = \frac{1}{2\pi\sigma_x\sqrt{1 - \rho^2}} \int_{-\infty}^{\infty} \exp\left\{ -\frac{1}{2(1 - \rho^2)} [u^2 - 2\rho uv + v^2] \right\} dv.$$

If we add and subtract $\rho^2 u^2$ to the exponent in order to complete the square in v, then we obtain

$$f_1(x) = \frac{1}{2\pi\sigma_x\sqrt{1 - \rho^2}} \int_{-\infty}^{\infty} \exp\left\{ -\frac{1}{2(1 - \rho^2)} [v^2 - 2\rho uv + \rho^2 u^2 - \rho^2 u^2 + u^2] \right\} dv$$

$$= \frac{e^{-u^2/2}}{2\pi\sigma_x\sqrt{1 - \rho^2}} \int_{-\infty}^{\infty} \exp\left\{ -\frac{1}{2(1 - \rho^2)} [v - \rho u]^2 \right\} dv.$$

If we make the change of variable $z = (v - \rho u)/\sqrt{1 - \rho^2}$, then $dv = \sqrt{1 - \rho^2}\, dz$. Thus $f_1(x)$ reduces to

$$f_1(x) = \frac{e^{-u^2/2}}{2\pi\sigma_x} \int_{-\infty}^{\infty} e^{-z^2/2}\, dz.$$

Because

$$\frac{1}{\sqrt{2\pi}} \int_{-\infty}^{\infty} e^{-z^2/2} = 1,$$

we obtain

$$f_1(x) = \frac{1}{\sqrt{2\pi}\sigma_x} e^{-u^2/2}.$$

Finally we replace the value of u by its expression in terms of x, and thus obtain

$$f_1(x) = \frac{1}{\sqrt{2\pi}\sigma_x} \exp\left[-\frac{1}{2}\left(\frac{x - \mu_x}{\sigma_x} \right)^2 \right].$$

This equation shows that the marginal distribution of x is normal with mean $E(x) = \mu_x$ and variance var $(x) = \sigma_x^2$. The corresponding result for y follows by the symmetric roles played by x and y in the bivariate normal distribution. This result, that the marginal distributions of a joint normal distribution are normal, was to be expected. One would certainly be unhappy with the definition of a joint normal distribution if the individual variables were not normally distributed.

If we set $\rho = 0$ in the bivariate normal distribution $f(x,y)$ we see that it reduces to the product of the two marginal distributions; that is, if $\rho = 0$, then

$$f(x,y) = f_1(x)f_2(y).$$

This shows that if two normal variables are uncorrelated, they are independently distributed. From the discussion of correlation given in Chapter 5, it should be clear that a lack of linear correlation does not ordinarily guarantee a lack of relationship of every kind between two variables. Thus normal distributions are special in this respect.

Let us next find the conditional density function $\phi_2(y|x)$. As before, let $u = (x - \mu_x)/\sigma_x$ and $v = (y - \mu_y)/\sigma_y$. Then we have

$$\phi_2(y|x) = \frac{f(x,y)}{f_1(x)} = \frac{\exp\left\{-\dfrac{1}{2(1-\rho^2)}[u^2 - 2\rho uv + v^2]\right\}}{2\pi\sigma_x\sigma_y\sqrt{1-\rho^2}}\left[\frac{e^{-u^2/2}}{\sqrt{2\pi}\sigma_x}\right]^{-1}$$

$$= \frac{1}{\sqrt{2\pi}\sigma_y\sqrt{1-\rho^2}}\exp\left\{-\frac{1}{2(1-\rho^2)}[v^2 - 2\rho uv - \rho^2 u^2]\right\}$$

$$= \frac{1}{\sqrt{2\pi}\sigma_y\sqrt{1-\rho^2}}\exp\left\{-\frac{1}{2}\left[\frac{v - \rho u}{\sqrt{1-\rho^2}}\right]^2\right\}.$$

Finally, we insert the values of u and v in terms of x and y, and thus obtain

$$\phi_2(y|x) = \frac{1}{\sqrt{2\pi}\sigma_y\sqrt{1-\rho^2}}\exp\left\{-\frac{1}{2}\left[\frac{y - \mu_y - \rho(\sigma_y/\sigma_x)(x - \mu_x)}{\sigma_y\sqrt{1-\rho^2}}\right]^2\right\}.$$

Since x has a fixed value and y is the random variable, this equation shows that the conditional random variable y has a normal distribution with mean $\mu_y + \rho(\sigma_y/\sigma_x)(x - \mu_x)$ and standard deviation $\sigma_y\sqrt{1-\rho^2}$. By symmetry, the corresponding result holds for x and y interchanged.

Thus, the conditional distributions of a joint normal distribution are also normal.

Since by definition the curve of regression is the locus of the means of the conditional distribution, it follows that the curve of regression of y on x, for x and y jointly normally distributed, is the straight line whose equation is

$$E(y|x) = \mu_y + \rho \frac{\sigma_y}{\sigma_x} (x - \mu_x).$$

Likewise, the curve of regression of x on y, for x and y jointly normally distributed, is the straight line

$$E(x|y) = \mu_x + \rho \frac{\sigma_x}{\sigma_y} (y - \mu_y).$$

This property of a joint normal distribution, namely, that the curves of regression are straight lines, helps justify the frequent use of linear regression, because variables that are normally distributed, or approximately normally distributed, are frequently encountered in applications.

Let us think in terms of probability density in the plane, and draw contour lines for various levels of probability density. Denote a fixed value of density by z. For this fixed z, the quantity in brackets in the exponent of the bivariate normal density function $f(x,y)$ must also have a fixed value, say c. Thus we obtain the contour curve

$$\left(\frac{x - \mu_x}{\sigma_x} \right)^2 - 2\rho \left(\frac{x - \mu_x}{\sigma_x} \right) \left(\frac{y - \mu_y}{\sigma_y} \right) + \left(\frac{y - \mu_y}{\sigma_y} \right)^2 = c,$$

where c corresponds to the selected value of z. Since this is a quadratic function in x and y, the contour curves must be conic sections. Furthermore, since the type of conic section depends only on the quadratic terms, the discriminant for testing conic sections may be applied directly to give

$$B^2 - AC = \left(\frac{\rho}{\sigma_x \sigma_y} \right)^2 - \frac{1}{\sigma_x^2} \frac{1}{\sigma_y^2} = \frac{(\rho^2 - 1)}{\sigma_x^2 \sigma_y^2} < 0.$$

This result shows that the intersecting curves are ellipses, because by definition, $\rho^2 < 1$. Allowing c to assume different values will merely change the sizes of the ellipses; consequently, these ellipses have the same centers and the same orientation of principal axes. It will be found, upon rotating axes properly to eliminate the xy term, that the two principal axes of these ellipses are not parallel to either of the two lines of regression as might be supposed. Instead the regression line $E(y|x)$ lies on the locus formed by bisecting the vertical chords of an ellipse (i.e., those chords parallel to the y-axis.) Similarly, the regression line $E(x|y)$ lies on the locus formed by bisecting the horizontal chords of an ellipse.

Let us write the bivariate normal density as

$$f(x,y) = f_1(x)\phi_2(y|x) = f_1(x) \frac{\exp\left\{ -\frac{1}{2}\left[\frac{y - \mu_y - \rho\,(\sigma_y/\sigma_x)(x - \mu_x)}{\sigma_y\sqrt{1 - \rho^2}} \right]^2 \right\}}{\sqrt{2\pi}\sigma_y\sqrt{1 - \rho^2}}.$$

We now want to consider this density along a vertical line $x = $ constant, i.e., a line parallel to the y-axis. See Figure 8.10. Except that the total vertical density $\int_{-\infty}^{\infty} f(x,y)\,dy$ is not one [but is equal to $f_1(x)$], the above equation shows that the vertical density $f(x = $ constant, $y)$ has the shape of the normal curve (i.e., the familiar bell shape.) Thus if we slice $f(x,y)$ along vertical lines, we preserve the bell-like shape of the normal curve. Likewise, if we slice $f(x,y)$ along horizontal lines, we also preserve the bell-like shape. These results are due to the fact that the two conditional distributions are each normal.

Also it is interesting to observe that the conditional distribution $\phi_2(y|x)$ has the same standard deviation $\sigma_y\sqrt{1 - \rho^2}$ whatever the value of $x = $ constant. The regression line $E(y|x)$ falls on the locus of the maximum vertical densities, and the two curves defined to be at one standard deviation $\sigma_y\sqrt{1 - \rho^2}$ on each side of the regression line are straight lines parallel to the regression line. A similar result holds for the other regression line. The grand maximum value of $f(x,y)$ occurs at the mean (μ_x, μ_y) and both lines of regression pass through this point. The axes of the concentric ellipses also pass through this point.

The normal density function is a very useful mathematical model for distributions of a single continuous variable. As would be expected the joint normal density function for two or more variables also proves to be very useful in applications.

GAMMA DISTRIBUTION

The exponential distribution is a special case of a class of continuous distributions known as gamma distributions. The *gamma density function* has the form

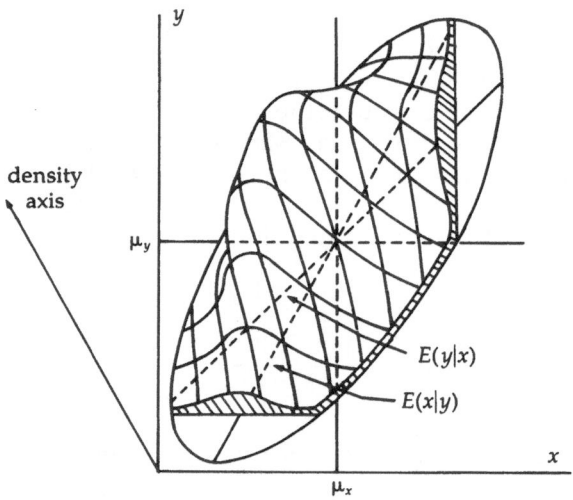

FIGURE 8.10 Bivariate normal density.

$$\gamma(x;r,\lambda) = \frac{\lambda^r x^{r-1} e^{-\lambda x}}{\Gamma(r)}, \qquad 0 \leqslant x < \infty.$$

The parameters r and λ are required to be positive and $\Gamma(r)$ denotes the Euler *gamma function* (see Chapter 6, page 197) defined for $r > 0$ by

$$\Gamma(r) = \int_0^\infty t^{r-1} e^{-t} \, dt.$$

It follows by integration by parts that

$$\Gamma(r) = (r - 1)\Gamma(r - 1);$$

and that, for positive integers r,

$$\Gamma(r) = (r - 1)!$$

The exponential distribution is the special case of the gamma distribution when $r = 1$; that is,

$$\gamma(x;1,\lambda) = \lambda e^{-\lambda x}, \qquad 0 \leqslant x < \infty.$$

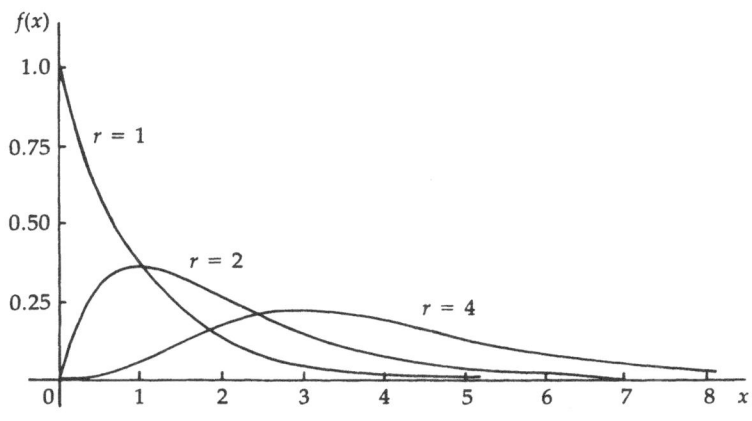

FIGURE 8.11 Various gamma density functions for $\lambda = 1$ and several values of r.

The Erlang distribution is the special case when r is a positive integer. Because the probability element for the gamma distribution can be written as

$$\gamma(x;r,\lambda)\,dx = \frac{(\lambda x)^{r-1}e^{-(\lambda x)}}{\Gamma(r)}\,d(\lambda x) = \gamma(\lambda x;r,1)\,d(\lambda x)$$

we see that $1/\lambda$ is a scale parameter for the $\gamma(x;r,\lambda)$ family.

Various gamma density functions are indicated schematically in Figure 8.11. We recall from Theorem 6.6 that the moment generating function of $\gamma(x;r,\lambda)$ is $M(\theta;x) = (1 - \theta/\lambda)^{-r}$. Either by direct integration or by moment generating function methods, it is easy to show that

$$E(x) = r/\lambda, \quad \text{var}(x) = r/\lambda^2.$$

Extensive tables of

$$\int_0^x y^{r-1}e^{-y}\,dy$$

have been tabulated as tables of the incomplete gamma function. As we have seen in the section on the Erlang distribution (page 279), when r is an integer greater than 1, the gamma variable x is interpretable as the sum of r random values of an exponential variable y having a mean of $1/\lambda$. When r is not an integer, this interpretation fails. However, there is an interesting relation between the normal and the gamma distribution.

THEOREM 8.11. The square of a standard normal variable z is a gamma variable with $r = \frac{1}{2}$, $\lambda = \frac{1}{2}$.

PROOF. The density function of the standard normal variable is

$$\phi(z) = \frac{e^{-z^2/2}}{\sqrt{2\pi}}.$$

The transformation is $x = z^2$. Because the inverse transformation $z = \pm x^{1/2}$ is not single valued, we must write the change-of-variable equation as

$$f(x) = \phi(-z)\left|\frac{dz}{dx}\right| + \phi(z)\left|\frac{dz}{dx}\right|.$$

Since $\phi(-z) = \phi(z)$, and $dz/dx = \pm(\frac{1}{2})x^{-1/2}$ this equation is

$$f(x) = 2\phi(z)(\tfrac{1}{2})x^{-1/2} \qquad \text{for } x > 0$$

which is

$$f(x) = x^{-1/2}\phi(x^{1/2}) = \frac{x^{-1/2}}{\sqrt{2\pi}}e^{-x/2} \qquad \text{for } x > 0.$$

Since $\sqrt{\pi} = \Gamma(\frac{1}{2})$, we finally have

$$f(x) = \frac{(\frac{1}{2})^{1/2}x^{-1/2}e^{-x/2}}{\Gamma(\frac{1}{2})},$$

which shows that x is a gamma variable with parameters $\frac{1}{2}$, $\frac{1}{2}$.   Q.E.D.

THEOREM 8.12. The sum of independent gamma variables with common λ is a gamma variable with the same λ.

PROOF. If x_1 and x_2 are independent gamma variables with parameters r_1,λ and r_2,λ respectively, then the moment generating function of their sum $y = x_1 + x_2$ is the product of their individual moment generating functions:

$$M(\theta;y) = (1 - \theta/\lambda)^{-r_1}(1 + \theta/\lambda)^{-r_2},$$

which is

$$M(\theta;y) = (1 - \theta/\lambda)^{-(r_1+r_2)}.$$

We recognize this result as the moment generating function of a gamma distribution. Thus the sum y is a gamma variable with parameters $r_1 + r_2$ and λ.   Q.E.D.

Consequently sums of squares of standard normal variables are gamma variables, a fact that is of importance in theoretical statistics. A special name is given to such sums; they are called chi-square variables. A *chi-square distribution* with n degrees of freedom is defined as a gamma distribution with parameters $r = n/2$, $\lambda = 1/2$.

THEOREM 8.13. The sum of squares of n independent standard normal variables has a chi-square distribution with n degrees of freedom.

PROOF. The distribution of the square z^2 of a standard normal variable z, as we have seen in Theorem 8.11, is a gamma variable with $r = 1/2$, $\lambda = 1/2$. Thus its moment generating function is

$$M(\theta;z^2) = (1 - \theta/\lambda)^{-r} = (1 - 2\theta)^{-1/2}$$

Thus the moment generating function of the sum $y = z_1^2 + z_2^2 + \cdots + z_n^2$ where the $z_1, z_2, \cdots, z_n$ are independent standard normal variables is

$$M(\theta;y) = (1 - 2\theta)^{-n/2}$$

Thus the distribution of the sum y is a gamma distribution with $r = n/2$, $\lambda = 1/2$. By definition, this distribution is the chi-square distribution with n degrees of freedom, with density function

$$f(y) = \frac{1}{\Gamma(n/2)2^{n/2}} y^{(n/2)-1} e^{-y/2} \qquad \text{for } 0 \leqslant y < \infty.$$

The mean and variance of a chi-square distribution are

$$E(y) = \frac{r}{\lambda} = \left(\frac{n}{2}\right)(2) = n \qquad \text{and} \qquad \text{var }(y) = \frac{r}{\lambda^2} = \left(\frac{n}{2}\right)2^2 = 2n.$$

EXERCISES

1. Projectiles are filed at the origin of an (x,y) coordinate system. The mathematical model is that the point that is hit, (X,Y), consists of a pair of independent, $N(0, 1)$ random variables. Suppose that two projectiles are fired independently and let D be the distance between the two points that are hit. Show that D^2 is exponentially distributed with $\lambda = 1/4$.

HINT. X_1, Y_1, X_2, Y_2 are independent, $N(0,1)$. $D^2 = (X_1 - X_2)^2 + (Y_1 - Y_2)^2$ is the sum of two gamma variables.

2. The random variable x is distributed as follows:

$$f(x) = \frac{1}{\sigma\sqrt{2\pi}} e^{-(x-\mu)^2/2\sigma^2}, \qquad \sigma > 0; -\infty < x < \infty.$$

(a) Show that the total area is a unity.
(b) Determine the points of inflection.
(c) Locate the median and mode by inspection.
(d) Sketch the curve.

ANSWER. (a) Reduce to gamma function by substituting $u = (x - \mu)/\sigma\sqrt{2}$. Thus obtain area $= \Gamma(^1/_2)/\sqrt{\pi} = 1$. (b) Inflection points at $x = \mu + \sigma$ and $x = \mu - \sigma$. (c) Median $=$ mode $= \mu$. (d) Bell-shaped, symmetrical about $x = \mu$.

3. Using the definition of the gamma function, prove that $\Gamma(r) = (r - 1)\Gamma(r - 1)$. (Integrate by parts.)

4. Verify the values given in the text for the mean and variance of a gamma variable by direct integration (which is easy if one takes advantage of the relation given in Exercise 3).

5. (a) The distribution of x is known to be gamma with $\lambda = 1$ but unknown r. A random observation of x turns out to have the value 7.5. For what value of r would this value of x correspond to a fractile of 0.99? Of 0.01? These two values of r are called "98 percent confidence limits."

ANSWER. $r = 2.5$; $r = 15.0$.

(b) Under the same conditions as (a), suppose that instead of one observation, three are drawn at random, and their sum S is computed. If $S = 19.5$ what are the 98 percent confidence limits for r in the distribution of S? What are the corresponding limits for r in the distribution of x?

ANSWER. 10.5 to 31.0, 3.5 to 10.3.

6. (a) Suppose that x is a nonnegative random variable with density function $f(x)$. Show that $y = e^{-x}$ has the density $g(y) = f(-\ln y)/y$ for $0 < y < 1$.
(b) If x is gamma distributed with parameters $(r,1)$, show that $y = e^{-x}$ has the density $g(y) = (-\ln y)^{r-1}/\Gamma(r)$ for $0 < y < 1$.
(c) Show that if y is uniform on $[0,1]$, then $x = -\ln y$ is exponential.

HINT. In (b), put $r = 1$.

7. Let x_1, x_2, $\cdots$, x_n be independent variables that have chi-square distributions with r_1, r_2, $\cdots$, r_n degrees of freedom respectively. Show

that their sum $y = x_1 + x_2 + \cdots + x_n$ has a chi-square distribution with $r = r_1 + r_2 + \cdots + r_n$ degrees of freedom.

ANSWER. $M(\theta;y) = (1 - 2\theta)^{-r_1/2} \cdots (1 - 2\theta)^{-r_n/2} = (1 - 2\theta)^{-r/2}$.

8. The cumulative distribution function of a gamma variable with parameters $r,1$ (where r is an integer) is

$$P(X \leqslant \lambda) = \frac{1}{(r - 1)!} \int_0^\lambda x^{(r-1)} e^{-x} \, dx.$$

By repeated integration by parts, show that $P(X \leqslant \lambda)$ gives the probability of r or more successes in one time unit of a Poisson process with success rate λ; that is,

$$P(X \leqslant \lambda) = P(k \geqslant r) = \sum_{k=r}^{\infty} \frac{\lambda^k e^{-\lambda}}{k!}.$$

Note that the event $X \leqslant \lambda$ (i.e., the waiting time for r successes is less than or equal to λ) is the same as the event $k \geqslant r$ (i.e., r or more successes occur in time interval λ), and thus the above result merely states that the probabilities of these two events are the same.

BETA DISTRIBUTION

The class of beta distributions have densities of the form

$$f(x;r,s) = \frac{x^{r-1}(1 - x)^{s-1}}{\beta(r,s)} \qquad \text{for } 0 < x < 1. \tag{8.26}$$

The parameters r and s are required to be positive, and

$$\beta(r,s) = \Gamma(r)\Gamma(s)/\Gamma(r + s)$$

is the beta function (see Chapter 6, page 202). Straightforward integration by parts will show that

$$E(x) = \frac{r}{r + s}. \tag{8.27}$$

Moreover, the mode will be given by $f'(x) = 0 = (r - 1)x^{r-2}(1 - x)^{s-1} - x^{r-1}(s - 1)(1 - x)^{s-2} = (r - 1)(1 - x) - (s - 1)x$, which has the solution

$$x = \frac{r - 1}{r + s - 2}.$$

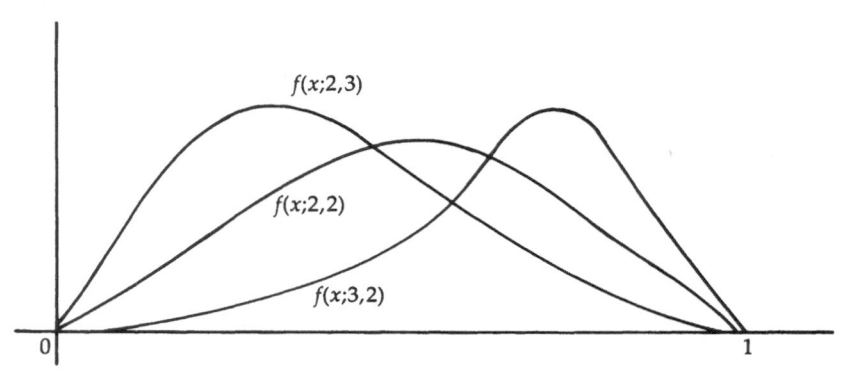

FIGURE 8.12 Various beta density functions.

Consequently, the variety of beta density functions can be indicated schematically as in Figure 8.12.

It is difficult to give a simple operational model for a beta variable. From Exercise 24 on page 244, we have the following theorem:

THEOREM 8.14. The beta distribution function $F(p;r,s)$ gives the probability of r or more successes in $r + s - 1$ Bernoulli trials where the probability of success at each trial is p; that is,

$$F(p;r,s) = \frac{1}{\beta(r,s)} \int_0^p y^{r-1}(1 - y)^{s-1} \, dy = \sum_{k=r}^{r+s-1} b(k;r+s-1,p),$$

where $b(k;r+s-1, p)$ denotes the binomial probability mass function for parameters $n = r + s - 1, p$.

For an operational connection between the binomial and beta variables, see Exercises 3 and 7 following this section.

The moment generating function for the beta distribution does not have a simple form, but the moments are

$$E(x^k) = \frac{\Gamma(r + s)}{\Gamma(r)\Gamma(s)} \int_0^1 x^{k+r-1}(1 - x)^{s-1} \, dx$$

$$= \frac{\Gamma(r + s)}{\Gamma(r)\Gamma(s)} \frac{\Gamma(k + r)\Gamma(s)}{\Gamma(k + r + s)} = \frac{\Gamma(r + s)\Gamma(k + r)}{\Gamma(r)\Gamma(k + r + s)}$$

Thus the mean, as we have seen, is $E(x) = r/(r + s)$. The variance is $rs/(r + s)^2(r + s + 1)$.

The beta distribution is often useful in fitting empirical data when the variable is limited to a finite range of values. While x in Figure 8.12 above is restricted to lie between 0 and 1, it would be easy to modify this so that x ranged from 0 to, say, a, the density function being then of the form

$$f(x) = kx^{m-1}(a - x)^{n-1}. \tag{8.28}$$

EXERCISES

1. Verify that if x is a beta variable with parameters r,s, then $E(x)$ and var (x) are as given in the text.

2. Show that the beta density is a uniform density in the case where $r = 1$, $s = 1$.

3. Suppose that n numbers X_1, X_2, $\cdots$, X_n are picked at random from the unit interval $[0,1]$. Let $Y_1 \le Y_2 \le \cdots \le Y_n$ be the values selected arranged in increasing order. Show that Y_r is beta distributed with parameters $r, n-r+1$.

   HINT. If k is the number of X_i's that are in $[0,p]$, then k is binomial with parameters n,p. Hence $P(Y_r \le p) = P(k \ge r)$. But in Exercise 24 on page 244, the binomial tail probability $P(k \ge r)$ is equal to the beta distribution function with parameters $r, n-r+1$.

4. Suppose that $n + 1$ points are picked at random from $[0,1]$. Let $Y_1 \le Y_2 \le \cdots \le Y_{n+1}$ be the value of these points arranged in order. Define

$$X = \begin{cases} Y_{r+1} & \text{with probability } \dfrac{r}{n+1}, \\ Y_r & \text{with probability } \dfrac{n-r+1}{n+1}. \end{cases}$$

   That is, after the $n + 1$ points are selected, either the $(r + 1)$st or rth smallest of these is selected with probabilities $r/(n + 1)$, $(n - r + 1)/(n + 1)$, respectively. Show that X has the beta density with parameters $r, n - r + 1$.

   HINT. If h is the density of Y_r, and g the density of Y_{r+1}, then

$$[r/(n + 1)]g + [(n - r + 1)/(n + 1)]h$$

   is the density of X.

5. Suppose that X is beta distributed with parameters r and s. Show that $1 - X$ is beta distributed with parameters s and r.

6. Suppose X is beta distributed with parameters r,s. Show that

(a) $E(X^2) = \dfrac{(r + 1)r}{(r + s + 1)(r + s)}$.

(b) $E[X^2(1 - X)^2] = \dfrac{(r + 1)r(s + 1)s}{(r + s + 3)(r + s + 2)(r + s + 1)(r + s)}$.

7. The cumulative distribution function of a beta variable X with integer parameters r and $n-r+1$ is

$$P(X \leqslant p) = \frac{n!}{(r - 1)!(n - r)!} \int_0^p x^{r-1}(1 - x)^{n-r}\, dx.$$

By repeated integration by parts, show that $P(X \leqslant p)$ gives the probability of r or more successes in n Bernoulli trials where the probability of success at each trial is p; that is, show that

$$P(X \leqslant p) = P(k \geqslant r) = \sum_{k=r}^{n} \binom{n}{k} p^k (1 - p)^{n-k}.$$

INTERPRETATION. If n numbers are picked at random from the unit interval and if they are arranged in increasing order, then the given beta variable X is the rth number in the ordered sequence. (See Exercise 3, where Y_r there is the same as X here.) The event $X \leqslant p$ (i.e., the rth number is less than or equal to p) is the same as the event $k \geqslant r$ (i.e., r or more of the ordered sequence occur in the interval p), and thus the above result merely states that the probabilities of these two events are the same.

8. In the text (page 280) we showed that for the Erlang density function (i.e., the gamma density γ with r an integer) satisfies

$$\gamma(t;r,\lambda)\, dt = p(r - 1;\lambda t)\, d(\lambda t),$$

where $p(r - 1;\lambda t)$ denotes the Poisson probability mass function. In a similar manner, show that the beta density function f satisfies

$$f(p;r,n-r+1)\, dp = b(r-1;n-1,p)\, d(np),$$

where $b(r-1;n-1,p)$ denotes the binomial probability mass function.

HINT. In Exercise 3, we saw that if n numbers are picked at random on the unit interval and arranged in order, then the rth one in the sequence is a beta variable with probability element $f(p;r,n-r+1)$ dp. This probability element gives the probability that $r - 1$ numbers fall in the left interval $[0,p]$, one number (the rth) falls in $[p,p+dp]$, and the remaining $n - r + 1$ numbers fall in the right interval $[p+dp,1]$. Since there are n choices to fall in the interval $[p,p+dp]$ the associated probability is ndp. The remaining $n - 1$ numbers are divided into two groups, one group of $r - 1$ numbers with individual probability p, the other group of $n - r + 1$ numbers with individual probability $1 - p - dp$, which is approximately $q = 1 - p$. Thus the *associated probability* is binomial, i.e., $b(r - 1; n - 1, p)$. Multiplying these two associated probabilities together, we get the required result.

9. Let $Y_1 \leqslant Y_2 \leqslant \cdots \leqslant Y_n$ be an ordered sequence of n points selected at random from $[0,1]$, so Y_r is a beta variable. Show that $E(Y_r) = r/(n + 1)$. Hence the sequence divides the unit interval into $n + 1$ random segments each with the same expected length $1/(n + 1)$. Compare this result with the corresponding result that if W_r is an Erlang variable with parameter r and λ, then $E(W_r) = r/\lambda$; hence the sequence $W_1, W_2, \cdots$ divides the time axis into random segments each with the same expected length $1/\lambda$.

**CHI-SQUARE, F, AND t
DISTRIBUTIONS**

In this section we derive the density functions of three of the most important distributions in theoretical statistics: chi-square, F, and t.

THEOREM 8.15. If X and Y are independent gamma variables with parameters r, λ and s, λ respectively, then $U = X + Y$ is a gamma variable with parameters $r+s, \lambda$ and $V = X/(X + Y)$ is a beta variable with parameters r, s. Moreover U and V are independent.

PROOF. If $\lambda = 1$, then the joint density of X and Y is

$$f(x,y) = [\Gamma(r)\Gamma(s)]^{-1} e^{-(x+y)} x^{r-1} y^{s-1} \qquad x > 0, \ y > 0.$$

The transformation

$$u = x + y \qquad v = x/(x + y)$$

is one-to-one between the admissible region $x > 0$, $y > 0$ and the admissible region $u > 0$, $0 < v < 1$. The inverse transform is

$$x = uv \qquad y = u - uv.$$

Therefore the Jacobian of the inverse transformation is

$$\frac{\partial(x,y)}{\partial(u,v)} = \begin{vmatrix} v & 1 - v \\ u & -u \end{vmatrix} = -u.$$

The joint density of the transformed variables is

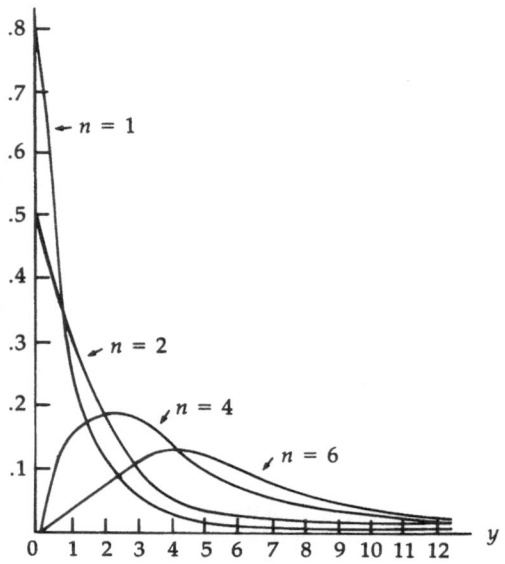

FIGURE 8.13 Chi-square density functions.

$$g(u,v) = [\Gamma(r)\Gamma(s)]^{-1}e^{-u}(uv)^{r-1}(u - uv)^{s-1}u$$

for $u > 0$, $0 < v < 1$. Simplifying, we obtain

$$g(u,v) = \gamma(u;r+s,1)f(v;r,s),$$

where $\gamma(u;r+s,1)$ is the gamma density function with parameters $r + s,1$; and $f(v;r,s)$ is the beta density function with parameters r,s. Thus the result is proved for $\lambda = 1$. If $\lambda \neq 1$, define $X' = \lambda X$ and $Y' = \lambda Y$. Because λ acts as a scale factor, it follows that X' and Y' are independent gamma variables with parameters $r,1$ and $s,1$ respectively. Our previous result tells us that $X' + Y'$ has a gamma distribution with parameters $r+s,1$. Since $X' + Y' = \lambda(X + Y)$, it follows that $X + Y$ has a gamma distribution with parameters $r+s,\lambda$. Finally, since $X'/(X' + Y') = X/(X + Y)$, it follows from the previous result that $X/(X + Y)$ has a beta distribution with parameters r,s.

By iterating the argument of Theorem 8.15 we obtain the following general result.

THEOREM 8.16. If $X_1, \cdots, X_n$ are n independent gamma variables with pa-

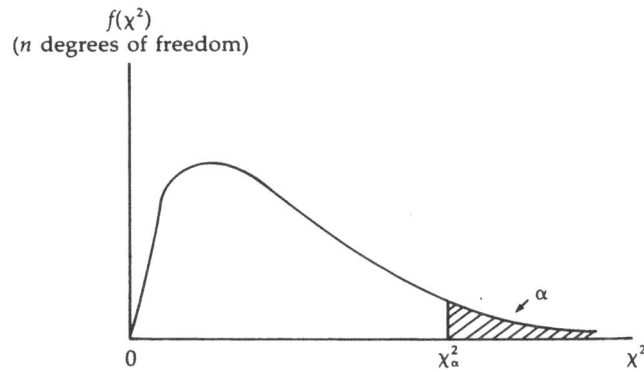

FIGURE 8.14 A chi-square density with shaded area showing tabulated values of chi-square.

rameters (r_1, λ), $\cdots$, (r_n, λ) respectively, then $X_1 + \cdots + X_n$ is a gamma variable with parameters $r_1 + \cdots + r_n, \lambda$.

Let us now introduce the three distributions that are fundamental to statistical theory. These distributions should be remembered in terms of their definitions and qualitative properties.

Throughout the rest of this section we assume that W_1, W_2, W_3, $\cdots$ are independent normal variables, each with zero mean and the same variance σ^2. We begin by investigating the distribution of the sum of their squares. The fundamental result is given in the following theorem. We recall that a chi-square variable with n degrees of freedom is defined as a gamma variable with parameters $r = n/2$ and $\lambda = 1/2$. See Figure 8.13 and Figure 8.14.

THEOREM 8.17. The random variable $Y = (W_1^2 + \cdots + W_n^2)/\sigma^2$ has a chi-square distribution with n degrees of freedom.

PROOF. Because $Y = Z_1^2 + \cdots + Z_n^2$, where $Z_i = W_i/\sigma$ are independent standard normal variables, the result follows directly from Theorem 8.13.

Let X and Y be independent chi-square variables with m and n degrees of freedom respectively. Then the random variable $F = (X/m)/(Y/n)$ is said to have an *F-distribution* with m and n degrees of freedom. See Figure 8.15.

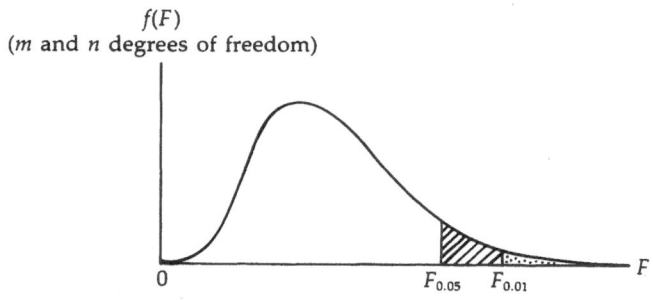

$f(F)$
(m and n degrees of freedom)

0 $F_{0.05}$ $F_{0.01}$ F

FIGURE 8.15 An F distribution with shaded areas showing tabulated values of F.

From this definition we see that the random variable defined as

$$\frac{(W_{n+1}^2 + \cdots + W_{n+m}^2)/m\sigma^2}{(W_1^2 + \cdots + W_n^2)/n\sigma^2} = \frac{(W_{n+1}^2 + \cdots + W_{n+m}^2)/m}{(W_1^2 + \cdots + W_n^2)/n},$$

has an F distribution. The important point here is that the variance σ^2 cancels out. This cancellation means that we do not have to know σ^2, and thus the F distribution becomes important in statistical applications of sampling from a normal distribution with unknown σ^2. In order to make the definition of the F distribution useful for computations we need its density function, which we now proceed to derive.

Note that if $V = X/(X + Y)$, then

$$F = \frac{n}{m}\frac{X}{Y} = \frac{n}{m}\frac{V}{1 - V}.$$

Since X is gamma with parameters $m/2$, $1/2$, and Y is gamma with parameters $n/2$ and $1/2$, and since X and Y are independent, it follows from Theorem 8.15 that V has a beta distribution with parameters $m/2$ and $n/2$. To obtain the density of F we can apply the change of variable formula to the density of V. The transformation is $F = (n/m)V/(1 - V)$. After some calculation we arrive at the density function of the F *distribution*

$$f(F;m,n) = \frac{(m/n)^{m/2}F^{(m-2)/2}[1 + (m/n)F]^{-(m+n)/2}}{\beta(m/2,n/2)}, \qquad F > 0,$$

where, of course, β denotes the beta function.

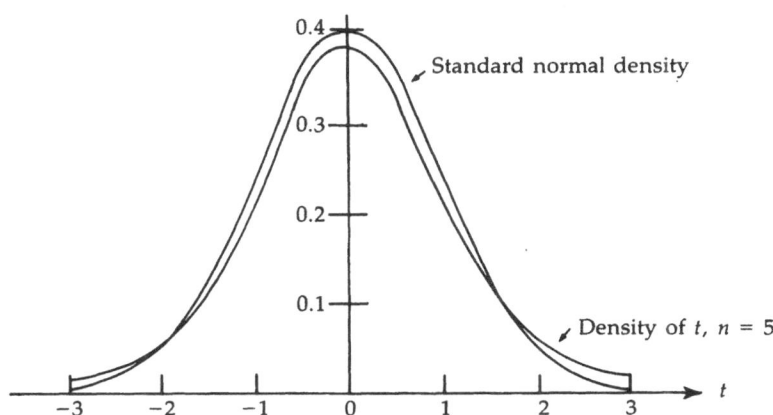

FIGURE 8.16 Comparison of standard normal density with t density for $n =$ 5.

We now introduce the t distribution with n degrees of freedom. By definition, it is the distribution of

$$T = \frac{Z}{(Y/n)^{1/2}}$$

where Z and Y are independent and are respectively standard normal and chi-square with n degrees of freedom. Thus the random variable

$$T = \frac{(W_{n+1}/\sigma)}{[(W_1^2 + \cdots + W_n^2)/n\sigma^2]^{1/2}} = \frac{W_{n+1}}{[(W_1^2 + \cdots + W_n^2)/n]^{1/2}}$$

has a t distribution. Again the σ cancels out. This cancellation of σ makes the t distribution valuable in statistical theory.

In order to derive the density of T we proceed as follows. Since $-Z$ has the same distribution as Z we may conclude that T and $-T$ are identically distributed. Thus we have

$$P(0 < T < t) = P(0 < -T < t),$$

so

$$P(0 < T < t) = P(-t < T < 0) = \tfrac{1}{2} P(0 < T^2 < t^2).$$

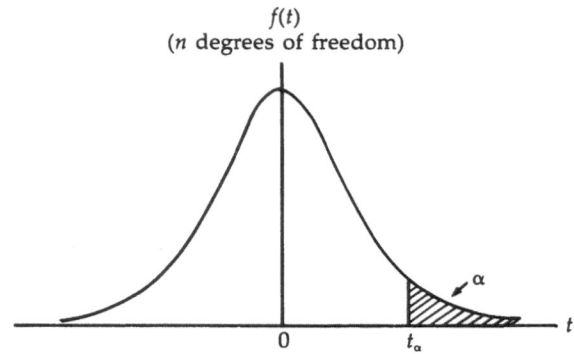

FIGURE 8.17 A t-density function with shaded area showing the probability that $t \geq t_\alpha$.

Differentiating, we obtain

$$f(t) = f(-t) = t\, g(t^2) \qquad \text{for } t > 0.$$

Since $g(t^2)$ is by definition an F distribution with 1 and n degrees of freedom, we can use the previously derived F density function to yield

$$f(t) = \frac{\Gamma[(n+1)/2][1 + (t^2/n)]^{-(n+1)/2}}{\sqrt{\pi n}\, \Gamma(n/2)} \qquad \text{for } -\infty < t < \infty.$$

This is the density function of the t *distribution* with n degrees of freedom. See Figure 8.16 and Figure 8.17.

EXERCISES

1. Perform the calculations omitted in the text in order to arrive at the density function of the F distribution.

    HINT.

$$V = [(m/n)F]/[1 + (m/n)F], \quad dV/dF = (m/n)(1 - V)^2.$$

2. Show that the t distribution with one degree of freedom is a Cauchy distribution.

3. For the t distribution with n degrees of freedom, show
    (a) that the kth moment $E(x^k)$ exists only for $k < n$;

(b) that if $k < n$ and k is odd, then $E(x^k) = 0$;

(c) that if $k < n$ and k is even, then

$$E(x^k) = n^{k/2} \frac{\Gamma[(k + 1)/2]\Gamma[(n - k)/2]}{\Gamma(^1/_2)\Gamma(n/2)}.$$

HINT. Use the beta function.

4. Show that the form of the density function of F given in this section is identical to that given previously in Chapter 3, page 68.

5. Show that if F is an F variable with m and n degrees of freedom, then its reciprocal $F' = 1/F$ is an F variable with n and m degrees of freedom. For a fixed probability, say 0.05, the value of F_1 such that $P(F < F_1) = 0.05$ is called the left critical point, and the value F_2 such that $P(F > F_2) = 0.05$ is called the right critical point. Why is it that only right critical points are tabulated? How does one find a left critical point from such tables?

9

An Advanced Treatment of Regression and Correlation

Mathematics is the door and key to the sciences.
ROGER BACON (1220–1292)

Regression Functions
General Properties of Regression Functions
Methods of Estimation
Least Squares
The Method of Moments
Inappropriate Hypotheses
Simple Correlation
Linear Hypotheses
Multiple and Partial Correlation
Properties of Sample Estimates

REGRESSION FUNCTIONS

In ordinary mathematical analysis, we are accustomed to think of rigorous functional relationships existing between two or more variables; and a major object of scientific research is the discovery of general laws connecting fundamental variables. Where random variables are concerned, however, the concept of functional dependence must be replaced by the broader one of statistical dependence. In such cases the relationships between variables are reflected in their joint distribution.

The mathematical representation of a variable y in terms of other variables $x_1, x_2, \cdots, x_n$ becomes a statistical problem when the causal system that produces variations in y is either imperfectly understood or incompletely specified; for y will behave observationally as a random variable so long as any of the pertinent variables remain unidentified or inaccessible to measurement. While scientific investigation of y should aim at establishing the relevant independent variables and determining the functional dependence of y upon them, this goal can be furthered and interim results of appreciable value obtained by statistical techniques that come under the headings of *regression* and *correlation*.

Regression is the statistical counterpart of the functional expression of one variable in terms of others. The strength of the statistical relationship, as represented by concentration about a single-valued function, is measured by the coefficient of correlation. Because strict functional dependence is rare among random variables, the terms *dependent variable* and *independent variables* are replaced by the less restrictive ones *regressand* and *regressors* respectively, or by *predictand* and *predictors* respectively if future time is involved.

The term *regression* was introduced into statistical literature by Sir Francis Galton[1] in 1886, who observed that sons of very tall or very short fathers tend to deviate in stature from the general average in the same direction as the fathers but to a lesser degree. For this reason, he said there was a "regression" toward the population mean. Through wide use, the term acquired an extended meaning and now refers to the average value assumed by one variable when others in the system considered are held fixed.

With a given set of regressors (or predictors as the case may be), the best possible specification of the regressand (predictand) y will be a statement of its conditional distribution, for this exhausts the entire fund of information. The smaller the variance of the conditional distribution, the more definite the specification of the regressand can

[1] Francis Galton: "Regression toward Mediocrity in Hereditary Stature," *Jour. Anthrop. Inst.*, vol. XV., 1886, p. 246. "Family Likeness in Stature," *Proc. Roy. Soc.*, vol. XL., 1886, p. 42.

be; and as the variance approaches zero, the range of likely values contracts more and more toward a unique value. Whatever the expected range of variation, one usually faces the practical necessity of selecting a single number as the estimated or predicted value of y. The best that can be done is to choose some measure of central tendency appropriate to the conditional distribution. Here, the mean value is usually the most satisfactory measure, because it is well adapted to empirical estimation and often has further important advantages over other parameters. The conditional mean value of the regressand (predictand) y as a function of the regressors (predictors) $x_1, x_2, \cdots, x_n$, is called the *regression* of y on $x_1, x_2, \cdots, x_n$. Just as the symbol $\phi(y|x_1, x_2, \cdots, x_n)$ denotes the conditional density function of y, we shall represent the regression function by the analogous symbol $\mu(y|x_1, x_2, \cdots, x_n)$, and when the context is clear, we shall abbreviate this as $\mu(x)$.

In this section we shall illustrate the regression function with a few examples, introduce the idea of a joint distribution of regressors and regressand, and point out the inherent one-sidedness of the regression function. In the examples, a single regressor x is hypothesized in order to bring out the essential principles without unnecessary complications.

EXAMPLE 9–1.

$$\phi(y|x) = \frac{1}{\sqrt{2\pi}} \exp\left[\frac{-(y - x^2)^2}{2}\right] \qquad \text{for } -\infty < y < \infty.$$

Here we recognize this conditional distribution as normal with unit variance and mean x^2. Hence, by inspection,

$$\mu(y|x) \equiv \int_{-\infty}^{\infty} y\phi(y|x)\, dy = x^2.$$

EXAMPLE 9–2.

$$\phi(y|x) = \frac{x^{n+1} y^n e^{-xy}}{\Gamma(n + 1)} \qquad \text{for } 0 \leqslant y < \infty.$$

This is a gamma distribution with parameters $r = n + 1$ and $\lambda = x$. Because the mean is r/λ, it follows that the regression function is

$$\mu(y|x) \equiv \int_{0}^{\infty} y\phi(y|x)\, dy = \frac{n + 1}{x}.$$

EXAMPLE 9–3.

$$\phi(y|x) = \frac{12}{x^4} y^2(x - y) \qquad \text{for } 0 \leqslant y \leqslant x.$$

This is a beta distribution with range $(0, x)$. The regression function is

$$\mu(y|x) \equiv \int_0^x y\phi(y|x) \, dy = (^3/_5)x.$$

The main reason for resorting to statistical methods of estimation is the impossibility (or at least, prohibitive difficulty) of coping with the problem deterministically. It sometimes happens that although the value of the regressand cannot be prescribed in advance, the identifiable regressors are subject to such control that their values can be assigned at will. More often, the regressors themselves are produced by the interplay of forces that are not under human control, a characteristic of astronomical, geophysical, biological, and sociological phenomena. In that case all of the variables taken together, regressors and regressand, may be regarded as belonging to a multivariate population. Even when the regressors are in fact controllable, no loss in generality is incurred by attributing a probability distribution to them, for the possibility of control would permit selection according to any desired proportions. In a prediction problem in which time t is used explicitly as a predictor, the variable t may be regarded as rectangularly distributed over the period considered.

For future reference, let us suppose that the regression functions of Examples 9–1 to 9–3 were derived from bivariate populations. In the statement of each of these examples the conditional distribution was given. For the sake of completeness let us now give for each example the respective marginal distribution of x and the joint distribution of x and y:

In Example 9–1:

$$f_1(x) = \frac{1}{\sqrt{2\pi}} e^{-x^2/2} \qquad\qquad \text{for } -\infty < x < \infty,$$

$$f(x,y) = \frac{1}{2\pi} \exp\left[\frac{-(x^4 + x^2 - 2x^2y + y^2)}{2}\right] \qquad \text{for } -\infty < x < \infty,$$

$$-\infty < y < \infty.$$

In Example 9–2:

$$f_1(x) = {}^1/_2\, x^2 e^{-x} \qquad\qquad \text{for } 0 < x < \infty,$$

$$f(x,y) = \frac{1}{2\Gamma(n + 1)} x^{n+3} y^n e^{-xy-x} \qquad \text{for } 0 < x < \infty,\ 0 < y < \infty.$$

In Example 9–3:

$$f_1(x) = 30x^2(1-x)^2 \qquad\qquad \text{for } 0 \leq x \leq 1,$$

$$f(x,y) = 360\frac{y^2}{x^2}(x-y)(1-x)^2 \qquad \text{for } 0 \leq x \leq 1, 0 \leq y \leq x.$$

The regression function is *not* a reciprocal relationship applying mutually to the regressors and regressand. In other words, the regression function for y in terms of $x_1, x_2, \cdots, x_n$ cannot be inverted to yield, say, the regression function for x_1 in terms of $y, x_2, \cdots, x_n$. The reason is that the mean value of one conditional distribution has no general functional connection with the mean value of another. Although this point is fairly obvious where multivariate distributions are concerned, it is often missed in considering bivariate distributions. Often people are puzzled when they discover that the function by which y is estimated from x does not serve equally well to estimate x from y. The principal cause of confusion is an erroneous replacement of $\mu(y|x)$ by y itself. For example, a linear regression function such as $\mu(y|x) = \alpha + \beta x$ is sometimes incorrectly written as $y = \alpha + \beta x$, from which a person could be misled to infer that $x = (y - \alpha)/\beta$. The fact is that $\mu(y|x)$ and $\mu(x|y)$ need not agree even in a general mathematical form, let alone in a one-to-one correspondence. As a matter of interest, let us compare the regression functions $\mu(y|x)$ and $\mu(x|y)$ in the examples thus far presented.

In Example 9–1, we cannot express $\mu(x|y)$ in closed form, for the definite integrals involved do not reduce to elementary functions.

In Example 9–2, the marginal distribution of y is given by

$$f_2(y) = \frac{(n+3)(n+2)(n+1)y^n}{2(1+y)^{n+4}} \qquad \text{for } 0 \leq y < \infty.$$

Hence the conditional distribution of x for a fixed value of y is

$$\theta(x|y) = \frac{(1+y)^{n+4}}{\Gamma(n+4)} x^{n+3} e^{-x(1+y)} \qquad \text{for } 0 \leq x < \infty$$

and the regression function is

$$\mu(x|y) = \int_0^\infty x\theta(x|y)\, dx = \frac{n+4}{1+y}.$$

We can compare this regression function with the previous one

$$\mu(y|x) = \frac{n+1}{x}.$$

In Example 9–3 the marginal distribution of y takes the form

$$f_2(y) = -180y^2[(2 + 4y) \ln y + (1 - y)(5 + y)] \qquad \text{for } 0 \leqslant y \leqslant 1.$$

(For the benefit of those who are puzzled by the negative sign, an analysis is given in the next paragraph.) The conditional distribution of x is

$$\theta(x|y) = \frac{-2(x - y)(1 - x)^2}{x^2[(2 + 4y) \ln y + (1 - y)(5 + y)]} \qquad \text{for } y \leqslant x \leqslant 1.$$

In contrast to the previous regression function $\mu(y|x) = \frac{3}{5} x$, the regression function is now

$$\mu(x|y) = \int_y^1 x\theta(x|y)\, dx = \frac{-[6y \ln y + (1 - y)(2 + 5y - y^2)]}{3[(2 + 4y) \ln y + (1 - y)(5 + y)]}.$$

The negative sign attached to the expression for $f_2(y)$ in the latter example is typical of logarithmic density functions defined in the range $0 \leqslant y \leqslant 1$. By inspection we see that $f_2(1) = 0$ and by L'Hôpital's rule, we find that $f_2(y) \to 0$ as $y \to 0$. To show that the density is positive elsewhere in the range of definition, let us consider the function

$$\lambda(y) = (2 + 4y) \ln y + (1 - y)(5 + y)$$

and write $f_2(y) = -180y^2\lambda(y)$. Differentiating $\lambda(y)$ twice we obtain

$$\lambda'(y) = 4 \ln y + \frac{(2 + 4y)}{y} - 2y - 4,$$

$$\lambda''(y) = \frac{-2(y - 1)^2}{y^2};$$

which shows that the second derivative is negative except at $y = 1$, where it vanishes. Hence the first derivative $\lambda'(y)$ is always decreasing; and since it is zero at $y = 1$, it must be positive throughout the range $0 < y < 1$. Therefore $\lambda(y)$ must increase steadily; but since $\lambda = 0$ at $y = 1$, it follows that $\lambda(y)$ must be negative. Thus $f_2(y)$ is positive in the range $0 < y < 1$.

GENERAL PROPERTIES
OF REGRESSION FUNCTIONS

For a fixed set of values of the regressors, the regression function has, of course, a unique value; but as the regressors vary over their permissible ranges, the values of the regression function change correspondingly. For this reason we may speak of statistical properties of the regression function, such as the mean and variance, in reference to the distribution of possible values generated by the regressors.

For brevity let us denote $\mu(y|x_1,x_2, \cdots,x_n)$ by $\mu(x)$, the joint density function $f(x_1,x_2,\cdots,x_n)$ by $f(x)$, the conditional density function $\phi(y|x_1,x_2,\cdots, x_n)$ by $\phi(y|x)$ and the joint differential element $dx_1\, dx_2 \cdots dx_n\, dy$ by $dx\, dy$. In this abbreviated notation, the joint probability element of y and $x_1, x_2, \cdots, x_n$ becomes $f(x)\, \phi(y|x)\, dx\, dy$.

The regression function $\mu(x)$ may be interpreted as that part of y that is attributable to or explained by the regressors, and the residual $y-\mu(x)$ may be interpreted as the unexplained part. Denoting this unexplained part by u, we have identically

$$y \equiv \mu(x) + [y - \mu(x)] \equiv \mu(x) + u. \tag{9.1}$$

Since by definition $\mu(x)$ yields the conditional mean value of y, it follows that the difference $u = y - \mu(x)$ has a mean value of zero for any set of values of the regressors; in symbols:

$$\mu(x) \equiv \int y\phi(y|x)\, dy, \qquad \int u\phi(y|x)\, dy = 0. \tag{9.2}$$

Hence, averaging over the whole joint distribution we obtain

$$E(y) = E[\mu(x)], \qquad E(u) = 0. \tag{9.3}$$

In order to establish (9.3), we write

$$E(y) = \int\int yf(x)\phi(y|x)\, dx\, dy = \int[\int y\phi(y|x)\, dy]f(x)\, dx$$
$$= \int\mu(x)f(x)\, dx = E[\mu(x)]$$

and

$$E(u) \equiv E[y - \mu(x)] = E(y) - E[\mu(x)] = 0;$$

or alternatively, by (9.2),

$$E(u) = \int\int uf(x)\phi(y|x)\, dx\, dy = \int[\int u\phi(y|x)\, dy]f(x)\, dx = 0$$

Moreover we can obtain the product relations

$$E[\mu(x)u] = 0, \qquad E[\mu^2(x)] = E[\mu(x)y]. \tag{9.4}$$

In order to establish (9.4), we write

$$E[\mu(x)u] = \iint \mu(x)uf(x)\phi(y|x)\,dx\,dy$$
$$= \int[\int u\phi(y|x)\,dy]\mu(x)f(x)\,dx = 0;$$

and the second result follows by taking expectations of both sides of the identity,

$$\mu(x)u \equiv \mu(x)[y - \mu(x)] \equiv \mu(x)y - \mu^2(x)$$

Equations (9.3) and (9.4) express fundamental properties of regression functions. Taken together they imply that *the covariance between the regression function and the residual is zero* and that *the variance of the regression function equals the covariance between the regression function and the regressand.*

The variance of y can be partitioned into the variance of the explained part $\mu(x)$ plus the variance of the unexplained part u:

$$\sigma^2(y) = \sigma^2[\mu(x)] + \sigma^2(u). \tag{9.5}$$

This equation follows from the facts that $y \equiv \mu(x) + u$ and the covariance of $\mu(x)$ and u is zero. On account of (9.5), the variance of the regression function is called the *explained variance*, the variance of the residual the *unexplained variance*, and the variance of the regressand the *total variance*. Hence Equation (9.5) may be expressed verbally as

total variance = explained variance + unexplained variance.   (9.6)

The explained variance is also called the *variance reduction*, and the ratio $\sigma^2[\mu(x)]/\sigma^2(y)$ of explained variance to total variance is called the *relative reduction*, or when expressed in percent, the *percent reduction*. From (9.5) it is clear that the reduction ratio has a range of (0, 1). The lower limit 0 is obtained when the regression function degenerates to a constant (this constant being the marginal mean of y) and hence this regression function does not explain any part of the variation of y about its mean. The upper limit 1 is obtained when the unexplained variance reduces to zero, implying for all practical purposes that y is an exact function of the regressors. The unexplained variance $\sigma^2(u)$ is also called either the *residual variance* or the *mean-square error*. Its positive square root $\sigma(u)$ is variously referred to as the *residual standard deviation* or the *root-mean-square error*.

As an important generalization of (9.4), we may derive in similar fashion the following far-reaching result, wherein $\psi(x)$ is an arbitrary continuous function of the regressors $x_1, x_2, \cdots, x_n$:

$$E[\psi(x)\mu(x)] = E[\psi(x)y].\qquad(9.7)$$

This equation follows from

$$E[\psi(x)y] = \iint \psi(x)yf(x)\phi(y|x)\,dx\,dy$$
$$= \int \psi(x)[\int y\phi(y|x)\,dy]f(x)\,dx$$
$$= \int \psi(x)\mu(x)f(x)\,dx = E[\psi(x)\mu(x)].$$

From this equation we may develop methods of estimating the regression function from observational data.

METHODS OF ESTIMATION

Thus far we have treated the regression function from the standpoint of joint or conditional distributions. Since these distributions are rarely known in practice, reliance must be placed on empirical estimation. Here a working hypothesis of some sort is both a logical and a practical necessity, for (1) no amount of observational data can determine with certainty even the marginal distribution of a single random variable, let alone the joint distribution of several; and (2) the appropriate numerical procedure depends upon the theoretical model. Once the mathematical structure of a function is hypothesized, except for some undetermined parameters, these constants usually can be estimated from numerical data by suitable techniques. Sometimes it is possible to hypothesize the mathematical form of the entire joint distribution, and in that case the conditional distribution and regression function can be deduced theoretically as soon as estimates of the parameters are obtained. Much more often, however, the best one can do is to hypothesize the mathematical form of the regression function without attempting to specify the joint distribution.

In the mathematical analysis of ordinary functions, differentiation plays an important role as a means of determining unknown constants. But in the study of random variables, techniques based on integration or summation are usually more satisfactory than those based on differentiation or differencing. One reason why the former are often better is that differentiation tends to emphasize irregularities, whereas integration or summation tends to smooth them out. On the other hand, all methods of analysis have some limitations as to the mathematical functions they can handle, and when integration breaks down, differentiation or differencing sometimes offers the only feasible approach.

Three standard techniques utilizing integration or summation are *maximum likelihood, least squares,* and the *method of moments.* When the form of the distribution itself can be hypothesized, maximum likelihood is ordinarily the best method of estimating the parameters, but it does not apply unless the distribution is hypothesized. Therefore, we shall omit consideration of maximum likelihood as applied to regression. In the practical situation wherein the hypothesis is confined to the regression function, the other two methods are in order. Commonly, these yield identical results. The method of least squares is a more nearly direct development from the concept of regression, but occasionally this method fails either because the integrals diverge or the equations become excessively involved. In that event, a satisfactory solution is sometimes obtainable by the method of moments.

In addition to the foregoing, there are also iterative-improvement procedures involving successive approximations. Such procedures are well adapted to additive nonlinear functions and can be applied to a product of positive functions by use of logarithms. Although many stages are sometimes required for convergence, they offer insight as to practical ways of handling certain types of nonlinear regression. The claim is sometimes made that an iterative solution does not require a working hypothesis, inasmuch as the curves emerge from the process itself. This argument is misleading. Since the assumption of superposition is implicit in the process, a competent practitioner must give considerable thought to the construction of suitable variables, which, in order to insure additive effects, are often chosen as composites of the initial regressors. As the computations progress, there is further latitude for the exercise of personal judgment.

Estimation and interpretation are sometimes facilitated by changing variables. The accuracy of the estimate, however, cannot be improved without introducing independent, pertinent information. A system of regressors $\xi_1, \xi_2, \cdots, \xi_n$ derived through a mathematical transformation of an initial system $x_1, x_2, \cdots, x_n$ cannot possibly improve upon the accuracy of estimation obtainable from the initial system. The reason is that the x's determine the ξ's and the accuracy of estimation depends solely upon the conditional distribution of the regressand for stated values of the regressors. Provided the transformation of the x's into the ξ's has a unique inverse, no accuracy is lost by changing variables, for then it is also true that the ξ's determine the x's. However, a transformation that does not have a unique inverse can lose accuracy by failing to distinguish between different points on a given locus in the x-space. These statements hold true for the exact regression function. On the other hand, if merely an approximation to this function is involved, it is altogether possible that with a prescribed number of terms one set of variables will yield a better approximation than another.

With regard to the regressand itself, a continuous monotonic transformation, say $\zeta(y)$, would not affect the ultimate accuracy of estimation—provided the conditional distribution of the transformed variable ζ were known—because the corresponding distribution of y could be derived from it. This, situation, of course, is seldom the case. Normally we have to be satisfied with regression functions. But here we find that the regression function of y usually cannot be recovered from that of ζ, for as a rule $E[\zeta(y)] \neq \zeta[E(y)]$, and so the inverse transformation applied to the regression function of ζ will not yield the regression function of y.

LEAST SQUARES

A definitive property of the mean (or conditional mean) of any random variable y is that of all possible constants c, the mean-square deviation $E[y - c)^2]$ is minimum when c equals the mean (or in the case of a conditional distribution, the conditional mean.) Now by definition, the regression function $\mu(x)$ yields the conditional mean of y for any selection of values of the regressors $x_1, x_2, \cdots, x_n$. Therefore, if $m(x)$ is such a function of these regressors as to minimize the conditional mean-square deviation for all sets of values of the regressors, then $m(x)$ must coincide with $\mu(x)$. This principle is the basis of *least squares*.

Let the mathematical form of the regression function be $\mu(x,\alpha) \equiv \mu(y|x_1,x_2,\cdots,x_n;\alpha_1,\alpha_2,\cdots,\alpha_0)$ where $\alpha_1,\alpha_2,\cdots,\alpha_p$ are parameters. If the correct mathematical form could be hypothesized, we could set up the general function $m(x,a) \equiv m(y|x_1,x_2,\cdots,x_n;a_1,a_2,\cdots,a_p)$; and, by minimizing the mean square deviation $D \equiv E\{[y - m(x,a)]^2\}$ with respect to the a's, obtain the α's and hence determine $\mu(x,\alpha)$.

In terms of our abbreviated notation, we have

$$D = \iint [y - m(x,a)]^2 f(x)\phi(y|x)\,dx\,dy. \tag{9.8}$$

The partial derivative with respect to a_i is

$$\frac{\partial D}{\partial a_i} = 2 \iint [y - m(x,a)]\left[\frac{-\partial m(x,a)}{\partial a_i}\right] f(x)\phi(y|x)\,dx\,dy.$$

Setting this derivative equal to zero we obtain

$$E\left[m(x,a) \frac{\partial m(x,a)}{\partial a_i} \right] = E\left[y \frac{\partial m(x,a)}{\partial a_i} \right] \tag{9.9}$$

This result applies equally well to discrete random variables. Equation (9.9) is the fundamental equation of least squares. Similar equations hold for each of the a's, and these simultaneous equations are called collectively the *normal equations*.

The point of all this is that the normal equations involve expected values and expected values can be estimated empirically by averaging appropriate functions of the data. Examples of least squares as applied to observational data will be given later on. For the present let us see how the method works out for two problems we have previously solved. Since we already know the true regression functions, we can save ourselves some integration by using Equation (9.7) to evaluate the right side of Equation (9.9).

EXAMPLE 9-4 (BASED ON EXAMPLE 9-1). The distributions are

$$f_1(x) = \frac{1}{\sqrt{2\pi}} e^{-x^2/2} \qquad \text{for } -\infty < x < \infty,$$

$$\phi(y|x) = \frac{1}{\sqrt{2\pi}} e^{-(y-x^2)^2/2} \qquad \text{for } -\infty < y < \infty,$$

and the regression function is

$$\mu(y|x) = x^2.$$

Hypothesizing a general quadratic $m(x,a) = a_0 + a_1 x + a_2 x^2$, we ought to obtain the result that $a_0 = 0$, $a_1 = 0$, $a_2 = 1$. Let us now see if we in fact get this result from the solution of the normal equations. The partial derivatives are

$$\frac{\partial m(x,a)}{\partial a_0} = 1, \qquad \frac{\partial m(x,a)}{\partial a_1} = x; \qquad \text{and} \qquad \frac{\partial m(x,a)}{\partial a_2} = x^2;$$

and the normal equations are

$$E[m(x,a)] = E(y),$$
$$E[xm(x,a)] = E(xy),$$
$$E[x^2 m(x,a)] = E(x^2 y);$$

or, in expanded form,

$$a_0 + a_1 E(x) + a_2 E(x^2) = E(y),$$
$$a_0 E(x) + a_1 E(x^2) + a_2 E(x^3) = E(xy),$$
$$a_0 E(x^2) + a_1 E(x^3) + a_2 E(x^4) = E(x^2 y).$$

For the distribution in question, we have

$$E(x^2) = 1, \quad E(x^4) = 3, \quad \text{but} \quad E(x^n) = 0 \quad \text{if } n \text{ is odd;}$$

and by employing (9.7) we get

$$E(x^k y) = E(x^k \mu) = E(x^k x^2) = E(x^{k+2}).$$

Thus the normal equations reduce to

$$a_0 + a_2 = 1 \qquad a_1 = 0 \qquad a_0 + 3a_2 = 3.$$

The second equation gives $a_1 = 0$ as it stands. Subtract the first equation from the third, obtaining $2a_2 = 2$, whence $a_2 = 1$, $a_0 = 0$. Therefore

$$m(x,a) = x^2 = \mu(y|x)$$

showing that we have obtained the correct regression function by least squares.

EXAMPLE 9-5 (BASED ON EXAMPLE 9-2). The distributions are

$$f_1(x) = \frac{1}{2} x^2 e^{-x} \qquad \text{for } 0 \leqslant x < \infty,$$
$$\phi(y|x) = \frac{x^{n+1} y^n e^{-xy}}{\Gamma(n+1)} \qquad \text{for } 0 \leqslant y < \infty;$$

and the regression function is

$$\mu(y|x) = \frac{n+1}{x}.$$

Hypothesizing the function $m(x,a) = \dfrac{a_1}{x} + a_2 + a_3 x$, we should obtain $a_1 = n + 1$, $a_2 = 0$, $a_3 = 0$. The partial derivatives are

$$\frac{\partial m(x,a)}{\partial a_1} = \frac{1}{x}, \quad \frac{\partial m(x,a)}{\partial a_2} = 1, \quad \text{and} \quad \frac{\partial m(x,a)}{\partial a_3} = x;$$

and the normal equations are

$$E[x^{-1}m(x,a)] = E(x^{-1}y),$$
$$E[m(x,a)] = E(y),$$
$$E[xm(x,a)] = E(xy).$$

In expanded form, the normal equations are

$$a_1E(x^{-2}) + a_2E(x^{-1}) + a_3 = E(x^{-1}y),$$
$$a_1E(x^{-1}) + a_2 + a_3E(x) = E(y),$$
$$a_1 + a_2E(x) + a_3E(x^2) = E(xy).$$

For the distribution considered, we have

$$E(x^k) = \tfrac{1}{2}\,\Gamma(k + 3),$$
$$E(x^ky) = E(x^k\mu) = E\{x^k(n + 1)x^{-1}\}$$
$$= (n + 1)E(x^{k-1}) = \frac{n + 1}{2}\,\Gamma(k + 2).$$

After simplification, the normal equations become

$$a_1 + a_2 + 2a_3 = n + 1,$$
$$a_1 + 2a_2 + 6a_3 = n + 1,$$
$$a_1 + 3a_2 + 12a_3 = n + 1.$$

Subtracting the first equation from the others and simplifying, we get

$$a_2 + 4a_3 = 0 \qquad a_2 + 5a_3 = 0.$$

By another subtraction we obtain $a_3 = 0$. Thus $a_2 = 0$ and $a_1 = n + 1$. Therefore

$$m(x,a) = \frac{a_1}{x} = \frac{n + 1}{x} = \mu(y|x),$$

and the correct regression function has been obtained. However, if the density function of x had been xe^{-x} instead of the one given, the integral for $E(x^{-2})$ would have diverged and the method of least squares would have broken down; but a solution could still be obtained by the method of moments.

THE METHOD OF MOMENTS

Among the general properties of regression functions, we previously derived the following equation:

$$E[\psi(x)\mu(x)] = E[\psi(x)y], \tag{9.7}$$

where $\psi(x)$ is any continuous function of x for which the expected values of the products in question are defined. The true regression function $\mu(x)$ is the only function for which Equation (9.7) holds under arbitrary choices of $\psi(x)$. Hence if $m(x,a)$ is some general form such that $\mu(x,\alpha)$ is a special case, the parameters $\alpha_1, \alpha_2, \cdots, \alpha_p$ can be determined by multiplying $m(x,a)$ and y each successively by arbitrary functions $\psi_1(x), \psi_2(x), \cdots, \psi_p(x)$; taking expected values, equating corresponding results; and solving the simultaneous equations for $a_1, a_2, \cdots, a_p$. One solution of the simultaneous equations,

$$E[\psi_i(x)m(x,a)] = E[\psi_i(x)y]   \quad \text{for } i = 1, 2, \cdots, p, \tag{9.10}$$

will certainly be $a_1 = \alpha_1, a_2 = \alpha_2, \cdots, a_p = \alpha_p$. Thus we obtain $m(x,a)$ $= \mu(x,\alpha)$. Under ordinary circumstances, the equations admit of only one solution, and the regression function is then uniquely determined. This is a generalization of the ordinary *method of moments*.

Although in principle any manageable functions $\psi_i(x)$ will do, empirical estimates are more reliable when the arbitrary functions stay within moderate ranges of variation. In the ordinary method of moments, the arbitrary functions are chosen as powers of x. Usually these powers are $1, x, x^2, \cdots, x^{p-1}$. The simultaneous equations for the method of moments are

$$E[x^k m(x,a)] = E(x^k y), \tag{9.11}$$

where as a rule $k = 0, 1, \cdots, p - 1$.

As an illustration of the method of moments, let us modify Example 9–2 by setting $f_1(x) = xe^{-x}$ while keeping the conditional distribution of y as it was. Then, although the regression function is the same as before, we cannot obtain a solution by least squares, because one of the integrals diverges.

EXAMPLE 9–6 (MODIFICATION OF EXAMPLE 9–2). The distributions are

$$f_1(x) = xe^{-x}   \quad \text{for } 0 \leqslant x < \infty,$$

$$\phi(y|x) = \frac{x^{n+1}y^n e^{-xy}}{\Gamma(n + 1)}   \quad \text{for } 0 \leqslant y < \infty;$$

and the regression function is

$$\mu(y|x) = \frac{n + 1}{x}.$$

As in Example 9–5, we hypothesize $m(x,a) = a_1/x + a_2 + a_3x$. The simultaneous equations for determining the parameters by the method of moments are

$$E[x^k m(x,a)] = E(x^k y) \qquad \text{for } k = 0, 1, 2$$

In expanded form, the simultaneous equations are

$$a_1 E(x^{-1}) + a_2 + a_3 E(x) = E(y),$$
$$a_1 + a_2 E(x) + a_3 E(x^2) = E(xy),$$
$$a_1 E(x) + a_2 E(x^2) + a_3 E(x^3) = E(x^2 y).$$

For the distribution considered,

$$E(x^k) = \Gamma(k + 2),$$
$$E(x^k y) = E(x^k \mu) = E\{x^k (n + 1)x^{-1}\}$$
$$= (n + 1)E(x^{k-1}) = (n + 1)\Gamma(k + 1).$$

After simplification, the simultaneous equations become

$$a_1 + a_2 + 2a_3 = n + 1,$$
$$a_1 + 2a_2 + 6a_3 = n + 1,$$
$$a_1 + 3a_2 + 12a_3 = n + 1.$$

Recognizing this set as the same set of equations solved in Example 9–5, we write down the solution at once: $a_1 = n + 1$, $a_2 = 0$, $a_3 = 0$. Hence we obtain the correct solution:

$$m(x,a) = \frac{a_1}{x} = \frac{n + 1}{x} = \mu(y|x)$$

Another permissible hypothesis for this problem is $m(x_1 a) = a_0 x^{a_1}$. Here the equations for the method of moments are

$$a_0 E(x^{a_1}) = E(y) \qquad \text{and} \qquad a_0 E(x^{a_1 + 1}) = E(xy),$$

which are

$$a_0 \Gamma(a_1 + 2) = n + 1 \qquad \text{and} \qquad a_0 \Gamma(a_1 + 3) = n + 1.$$

Dividing the second equation by the first we get

$$\frac{\Gamma(a_1 + 3)}{\Gamma(a_1 + 2)} = 1, \qquad \text{so } a_1 + 2 = 1, a_1 = -1.$$

Thus the first equation yields

$$a_0\Gamma(1) = n + 1, \qquad \text{whence } a_0 = n + 1.$$

Therefore, once again we find

$$m(x,a) = a_0 x^{a_1} = \frac{n + 1}{x} = \mu(y|x).$$

INAPPROPRIATE HYPOTHESES

In least squares and the method of moments alike, a unique solution for the a's is usually obtainable even if the hypothesized function does not include the true regression function as a special case. Consequently, the fact that a definite solution has been found does not guarantee that the regression function has been determined. Often the difference is not serious, because most mathematical functions lend themselves to approximation by a series of arbitrary functions. Familiar examples are the Taylor expansion, polynomials, and Fourier series.

According to Equation (9.7), the normal equations of least squares (9.9) can be expressed as

$$E\left[m(x,a)\frac{\partial m(x,a)}{\partial a_i}\right] = E\left[\mu(x,a)\frac{\partial m(x,a)}{\partial a_i}\right]. \tag{9.12}$$

Thus the result of determining the constants of the regression function for y by least squares is precisely the same as though we had set out to minimize the quantity

$$D^* = E[\mu(x,\alpha) - m(x,a)]^2 \tag{9.13}$$

with respect to the a's. In other words, the constants determined by least squares are identically those that yield the best representation, in the sense of minimum mean-square error, of the true regression function as an expression of the form hypothesized. Correspondingly, for the method of moments, the simultaneous equations (9.10) are equivalent to

$$E[\psi_i(x)m(x,a)] = E[\psi_i(x)\mu(x,\alpha)]. \tag{9.14}$$

Hence the outcome is an approximation of $\mu(x,\alpha)$ in terms of $m(x,a)$ by the method of moments.

If the hypothesis is so poor that it has almost nothing in common with the regression function, then that fact will become apparent to the investigator by total failure when applied to new data. Furthermore, by seeing to it that the number of constants estimated is small compared to the number of independent observations, one is able to appraise the hypothesis fairly well from the original data alone; for there are ways of compensating for the extent to which the function is arbitrarily forced into shape. The major difficulties of estimation arise with small samples. With large amounts of data, an empirical determination of constants is sufficiently close to the values which would be obtained from the true mathematical expectations that the functional representation can be considered fairly reliable as far as it goes.

Unless there are strong reasons to the contrary, it is wise to formulate $m(x,a)$ in such a way that it can degenerate to an arbitrary constant. If this is done, even a poorly chosen function will not behave wildly. The least-squares solution will then yield negligibly small coefficients for all variables, and the regression will tend toward the mean value of y in its marginal distribution. Generally speaking, the method of moments has the same advantage.

As an instance of a poor hypothesis, suppose that in Example 9–4 we had taken $m(x,a) = a_0 + a_1x$. Here $m(x,a)$ neither includes the true regression function as a special case nor behaves at all like it. Nevertheless this $m(x,a)$ is capable of reducing to a constant. Both methods of solution yield $a_0 = 1$, $a_1 = 0$, whence $m(x,a) \equiv 1$. This is precisely the mean value of y; for by (9.7), $E(y) = E[\mu(x,\alpha)] = E(x^2) = 1$. Incidentally, this problem is interesting in that we can find the mean value of y, even though we cannot express the marginal distribution in closed form, but can only express it as an integral. Here is also an instance of zero covariance between two random variables that are not independent. The formula for the covariance is $\sigma_{xy} = E(xy) - E(x)E(y)$. In order to evaluate this formula we use Equation (9.7), which gives $E(xy) = E(x\mu) = E(xx^2) = E(x^3) = 0$. Also, we have $E(x) = 0$; hence $\sigma_{xy} = 0$.

SIMPLE CORRELATION

In general usage, correlation denotes a mutual relationship between variables or events. In statistical terminology, correlation is understood as a form of statistical dependence.

Unless otherwise indicated, the technical term "correlation" refers to *simple, linear correlation* between two variables x, y. This is measured by the *coefficient of correlation* $\rho(x,y)$ defined as follows

$$\rho(x,y) = \frac{E[(x - \xi)(y - \zeta)]}{\sigma(x)\sigma(y)} = \frac{\sigma_{xy}}{\sigma(x)\sigma(y)}, \tag{9.15}$$

where ξ and ζ denote the population means of x and y respectively, where $\sigma(x)$ and $\sigma(y)$ denote the respective standard deviations, and where σ_{xy} is the covariance of x and y. If the original variables x,y are replaced by corresponding standard scores $z_1 = (x - \xi)/\sigma(x)$ and $z_2 = (y - \zeta)/\sigma(y)$, the coefficient of correlation reduces to the expected value of the product $z_1 z_2$, which in this case is also equal to the covariance σ_{12} of the standard scores. Hence

$$\rho(x,y) = E(z_1 z_2) = \sigma_{12} = \rho(z_1, z_2). \tag{9.16}$$

More generally, if x' and y' are any (nontrivial) linear functions of x and y respectively, for instance $x' = c_0 + c_1 x$, $y' = k_0 + k_1 y$ where neither c_1 nor k_1 is zero, then (as may be verified by substitution) the coefficient of correlation between x', y' will be numerically equal to that between x,y but the algebraic sign will be the same or different depending upon whether or not c_1 and k_1 agree in sign. Hence

$$\rho^2(x,y) = \rho^2(x',y'), \tag{9.17}$$

where x' is a linear function of x, and y' is a linear function of y.

From the Schwarz inequality, it follows that the admissible range of ρ is -1 to 1. This fact can be established also by a simple statistical argument. For, letting z_1, z_2 denote any two standardized variables, consider their sum and difference. The variance of the sum is

$$\sigma^2(z_1 + z_1) = \sigma_1^2 + \sigma_2^2 + 2\sigma_{12} = 1 + 1 + 2\rho = 2(1 + \rho)$$

and the variance of the difference is

$$\sigma^2(z_1 - z_2) = 2(1 - \rho).$$

Since both variances are greater than or equal to zero, we conclude that

$$-1 \leqslant \rho \leqslant 1. \tag{9.18}$$

The coefficient of simple linear correlation ρ is sometimes called the *Pearsonian* coefficient of correlation, after Karl Pearson, or else the *product-moment* coefficient of correlation (because the covariance is a product-moment) to distinguish it from other types, such as biserial or tetrachoric coefficients of correlation, which are also recognized. The word "simple" is used in contradistinction to "multiple" or "partial." The word "linear" refers to a linear relationship between two variables, or more precisely, a linear approximation to the regression function of one with respect to the other.

When the regression function of y with respect to x is approximated by a linear function, say $m(x,a) = a_0 + a_1 x$, with intercept a_0 and slope a_1, the normal equations of least squares become

$$E[m(x,a)] = E(y) \qquad E[xm(x,a)] = E(xy), \qquad (9.19)$$

that is,

$$a_0 + a_1 E(x) = E(y) \qquad a_0 E(x) + a_1 E(x^2) = E(xy). \qquad (9.20)$$

Using the first equation of (9.20), we express the intercept a_0 in terms of the slope a_1 as $a_0 = E(y) - a_1 E(x)$. Substituting in the second equation of (9.20), we find the slope to be

$$a_1 = \sigma_{xy}/\sigma^2(x) \equiv \rho(x,y)\sigma(y)/\sigma(x). \qquad (9.21)$$

Thus the slope is jointly proportional to ρ and the quotient of the standard deviations, and its sign agrees with that of ρ. In case the two variables x,y have equal standard deviations, as they would, for instance, in standard scores, the correlation coefficient may be interpreted geometrically as the slope of the straight line of best fit for y in terms of x. We note in passing that ρ also has a similar interpretation with reference to the linear approximation of the regression of x with respect to y, but unless $\rho = 1$, the two lines do not coincide, because the axes are in reversed roles.

For further development, let us denote $m(x,a)$ by m and rewrite the normal equations (9.19) in the following more suggestive form:

$$E(y - m) = 0 \qquad E[x(y - m)] = 0. \qquad (9.19a)$$

Multiplying the first of these equations by a_0 and the second by a_1 and adding results we obtain $E[a_0(y - m) + a_1 x(y - m)] = 0$, which is $E[m(y - m)] = 0$. This equation gives

$$E(m^2) = E(my). \qquad (9.22)$$

Thus even though $m(x,a)$ may be a poor approximation to the true regression function $\mu(x,\alpha)$ it nevertheless has certain important properties in common with the latter, two of which are that (1) the mean error is zero, i.e., $E(y - m) = 0$; and (2) the estimated values as given by $m(x,a)$ are linearly uncorrelated with the errors $y - m$, i.e., $E[m(y - m)] = 0$. Furthermore, as a consequence of Equations (9.19a) and (9.22), the decomposition of variance holds:

$$\sigma^2(y) = \sigma^2(m) + \sigma^2(y - m). \tag{9.23}$$

Returning to (9.22), subtract $[E(m)]^2$ from $E(m^2)$ and subtract the equal quantity $E(m)E(y)$ from $E(my)$, obtaining

$$\sigma^2(m) = \sigma_{my}, \tag{9.24}$$

where σ_{my} denotes the covariance of m and y. We thus find that the correlation between m and y is intrinsically nonnegative since their covariance equals the variance of m; in fact $\rho(m,y)$ reduces to the quotient of the standard deviations; that is,

$$\rho(m,y) = \sigma_{my}/\sigma(m)\sigma(y) = \sigma(m)/\sigma(y). \tag{9.25}$$

Moreover, $\rho^2(m,y) = \rho^2(x,y) = \rho^2$ because m is merely a linear function of x. Therefore, squaring the extreme members of (9.25), simplifying, and using (9.23), we arrive at the following conclusions:

$$\rho^2 = \frac{\sigma^2(m)}{\sigma^2(y)}; \quad \sigma^2(m) = \rho^2\sigma^2(y); \quad \sigma^2(m - y) = (1 - \rho^2)\sigma^2(y). \tag{9.26}$$

Since $\sigma^2(m - y)$ cannot be negative, the second part of (9.26) shows once again that ρ^2 is at most unity, whence $-1 \leqslant \rho \leqslant 1$. Furthermore, since $\sigma^2(m - y)$ reduces to zero when and only when y is equal to m (with probability one), it follows that $\rho^2 = 1$ and $\rho = \pm 1$ when and only when y is strictly a linear function of x (except, perhaps, at points of zero density). Finally, noticing that $\sigma^2(m - y)$ is less than $\sigma^2(y)$ except when $\rho = 0$, in which case $a_1 = 0$ and $a_0 = E(y)$, we conclude that the best-fitting straight line becomes degenerate when $\rho = 0$ and then contributes nothing to the estimation of y beyond the mean $E(y)$. On the other hand, it does not necessarily follow that y is independent of x when $\rho = 0$; for statistical dependence might exist in a nonlinear form (as we saw in Example 9–1), and even linear regression might be found when the effects of other regressors are taken into account. The latter possibility comes under the heading of *partial correlation*.

From N paired observations (x_1,y_1), (x_2,y_2), $\cdots$, (x_N,y_N) we may obtain a sample estimate r of the true correlation ρ. The accepted estimate is given by the following formula

$$r = \frac{\Sigma(x_i - \bar{x})(y_i - \bar{y})}{[\Sigma(x_i - \bar{x})^2 \Sigma(y_i - \bar{y})^2]^{1/2}},$$
(9.27)

where $\bar{x} = \mathrm{E}x_i/N$, $\bar{y} = \Sigma y_i/N$ and each summation is taken from $i = 1$ to $i = N$. For computational purposes it is usually convenient to re-write the formula for r as

$$r = \frac{C}{(AB)^{1/2}},$$
(9.28)

where

$$A = N\Sigma x_i^2 - (\Sigma x_i)^2, \quad B = N\Sigma y_i^2 - (\Sigma y_i)^2, \quad C = N\Sigma x_i y_i - \Sigma x_i \Sigma y_i. \quad (9.29)$$

EXAMPLE 9-7. As an illustrative computation of r let us consider a ficti-tious sample of 10 pairs of values.

x	3	5	4	2	5	1	4	2	1	3
y	4	0	2	2	1	3	3	1	4	0

Thus, $N = 10$, $\Sigma x_i = 30$, $\Sigma x_i^2 = 110$, $\Sigma y_i = 20$, $\Sigma y_i^2 = 60$, $\Sigma x_i y_i = 50$, and

$$A = (10)(110) - (30)^2 \quad = 200;$$
$$B = (10)(60) - (20)^2 \quad = 200;$$
$$C = (10)(50) - (30)(20) = -100;$$

so

$$r = \frac{C}{(AB)^{1/2}} = \frac{-100}{200} = -0.5.$$

LINEAR HYPOTHESES

Any hypothesis according to which the function considered is represented in terms of definite arbitrary functions in such a way that the unknown parameters enter the expression linearly is said to be a *linear hypothesis*. For instance if X_1, X_2, $\cdots$, X_p are arbitrary functions of the regressors x_1, x_2, $\cdots$, x_n and contain no unknown constants, the hypothesis that the regression function is of the form

$$m = a_0 + a_1X_1 + a_2X_2 + \cdots + a_pX_p, \tag{9.30}$$

where the a's may be unknown, is a linear hypothesis. In particular, the functions

$$m = a_0 + a_1x_1,$$
$$m = a_t + a_1x_1^2 + a_2x_1x_2,$$
$$m = a_0 + a_1e^{-2x_1} + a_2\sin x_2 + a_3x_1x_4e^{-x_3}$$

all come under this heading, whereas the function $m = e^{-a_1x_1}$ does not, because a_1 does not enter the expression linearly. The additive a_0 may be dropped if logical considerations warrant its omission, but we shall include it in the following development.

The method of least squares is best suited to linear hypotheses. Setting $D = E[(y - m)^2]$ and differentiating we obtain as *normal equations*,

$$E(y - m) = 0 \quad \text{and} \quad E[X_i(y - m)] = 0 \qquad \text{for } i = 1, 2, \cdots, p, \tag{9.31}$$

showing that the mean error is zero and that each X_i in turn is linearly uncorrelated with the error. Once the a_i are determined as definite numbers by solving these equations, we may insert them inside the expected value sign obtaining

$$E[a_0(y - m)] = 0 \quad \text{and} \quad E[a_iX_i(y - m)] = 0. \tag{9.32}$$

By summing over the individual equations, we find

$$E[m(y - m)] = 0. \tag{9.33a}$$

Therefore the covariance between m and $y - m$ is zero; that is, the estimated values of y are linearly uncorrelated with the errors of estimate. From (9.33a), we obtain

$$E(m^2) = E(my). \tag{9.33b}$$

If we substitute (9.30) into (9.33a), we find

$$E(m^2) = a_0E(y) + a_1E(X_1y) + \cdots + a_pE(X_py). \tag{9.33c}$$

Thus it follows also that

$$\sigma^2(y) = \sigma^2(m) + \sigma^2(y - m) \tag{9.34}$$

and

$$\sigma^2(m) = \sigma_{my}. \tag{9.35}$$

Thus the principal equations derived in the foregoing section for the simple linear hypothesis $m = a_0 + a_1 x$ hold for the general linear hypothesis.

Sample estimates say $\hat{a}_0, \hat{a}_1, \cdots, \hat{a}_p$ of the constants $a_0, a_1, \cdots, a_p$ (leading to the sample approximation $\hat{m}$ of the function m) may be obtained by replacing the expected values involved in the normal equations by corresponding sample averages. In a sample of N simultaneous values of y, $X_1, X_2, \cdots, X_p$ let t denote the serial number of an individual observation, so that a particular set of simultaneous values will be y_t, $X_{1t}, X_{2t}, \cdots, X_{pt}$. The corresponding value of $\hat{m}$ will be

$$\hat{m}_t = \hat{a}_0 + \hat{a}_1 X_{1t} + \hat{a}_2 X_{2t} + \cdots + \hat{a}_p X_{pt}. \tag{9.36}$$

The sample approximations to the *normal equations* (9.31) then become

$$\frac{1}{N} \sum_{t=1}^{N} (y_t - \hat{m}_t) = 0 \quad \text{and} \quad \frac{1}{N} \sum_{t=1}^{N} [X_{it}(y_t - \hat{m}_t)] = 0 \tag{9.37}$$

$$\text{for } i = 1, 2, \cdots, p,$$

showing (1) that the sample mean of the residual $\hat{y} - \hat{m}$ is zero in this sample and (2) that the sample estimates of the linear correlation between the residual $\hat{y} - \hat{m}$ and each X_i in turn is also zero. Clearly the factor $1/N$ is immaterial, and so we can multiply the normal equations (9.37) by N. If we then substitute (9.36) into the result, we obtain, in expanded form, the *normal equations*

$$N\hat{a}_0 + (\Sigma X_{1t})\hat{a}_1 + (\Sigma X_{2t})\hat{a}_2 + \cdots + (\Sigma X_{pt})\hat{a}_p = \Sigma y_t, \tag{9.38}$$

$$(\Sigma X_{1t})\hat{a}_0 + (\Sigma X_{1t}^2)\hat{a}_1 + (\Sigma X_{1t}X_{2t})\hat{a}_2 + \cdots + (\Sigma X_{1t}X_{pt})\hat{a}_p = \Sigma X_{1t}y_t,$$

$$(\Sigma X_{2t})\hat{a}_0 + (\Sigma X_{1t}X_{2t})\hat{a}_1 + (\Sigma X_{2t}^2)\hat{a}_2 + \cdots + (\Sigma X_{2t}X_{pt})\hat{a}_p = \Sigma X_{2t}y_t,$$

$$\vdots$$

$$(\Sigma X_{pt})\hat{a}_0 + (\Sigma X_{1t}X_{pt})\hat{a}_1 + (\Sigma X_{2t}X_{pt})\hat{a}_2 + \cdots + (\Sigma X_{pt}^2)\hat{a}_p = \Sigma X_{pt}y_t,$$

where each summation is over $t = 1, 2, \cdots, N$. As the reader may easily verify, exactly the same equations can be obtained by minimizing the sum of squares $\Sigma(y_t - \hat{m}_t)^2$ with respect to $a_0, a_1, \cdots, a_p$. (The summation is over $t = 1, 2, \cdots, N$.)

In what follows, we shall denote the sample mean value of $\hat{m}$ by $\bar{m}$ instead of using the more logical but awkward symbol $\bar{\hat{m}}$, and as usual, $\bar{y}$ will stand for the sample mean of y. The sample counterpart of the equation $E(m) = E(y)$ is, as we have seen, $\bar{m} = \bar{y}$ and by manipulating the sample sums in the same way as we did the expected values, we readily obtain further analogous results. Some of these are:

$$\Sigma(y_t - \bar{y})^2 = \Sigma(\hat{m}_t - \bar{m})^2 + \Sigma(y_t - \hat{m}_t)^2, \tag{9.39}$$

$$\Sigma\hat{m}_t(y_t - \hat{m}_t) = 0 \quad \text{and} \tag{9.40}$$
$$\Sigma\hat{m}_t^2 = \Sigma\hat{m}_ty_t = \hat{a}_0\Sigma y_t + \hat{a}_1\Sigma X_{1t}y_t + \cdots + \hat{a}_p\Sigma X_{pt}y_t,$$

and

$$\Sigma(\hat{m}_t - \bar{m})^2 = \Sigma(\hat{m}_t - \bar{m})(y_t - \bar{y}), \tag{9.41}$$

where each summation is over $t = 1, 2, \cdots, N$. Furthermore, letting R denote the sample value of the linear correlation between $\hat{m}$ and y (the symbol R is conventional in this connection) we have

$$R \equiv r(\hat{m}, y) = \frac{\sqrt{\Sigma(\hat{m}_t - \bar{m})^2}}{\sqrt{\Sigma(y_t - \bar{y})^2}}, \tag{9.42}$$

showing that R is nonnegative. We also have the following:

Explained sum of squares: $\Sigma(\hat{m}_t - \bar{m})^2 = R^2\Sigma(y_t - \bar{y})^2.$ (9.43)
Unexplained sum of squares: $\Sigma(y_t - \hat{m}_t)^2 = (1 - R^2)\Sigma(y_t - \bar{y})^2.$

While (9.43) is interesting theoretically in showing the relation between R^2 and the explained or unexplained sum of squares, in practice it is more convenient to compute the explained sum of squares as follows. As the first step we have

$$\Sigma(\hat{m}_t - \bar{m})^2 \equiv \Sigma\hat{m}_t^2 - \bar{m}\Sigma\hat{m}_t. \tag{9.44}$$

We then substitute into (9.44) the two equations

$$\bar{m}\Sigma\hat{m}_t = \bar{y}\Sigma y_t \quad \text{and} \quad \Sigma\hat{m}_t^2 = \Sigma\hat{m}_ty_t = \hat{a}_0\Sigma y_t + \hat{a}_1\Sigma X_{1t}y_t + \cdots + \hat{a}_p\Sigma X_{pt}y_t.$$

All of these quantities are known, as we have solved the normal equations. Therefore, the computational formula for the explained sum of squares is

$$\Sigma(\hat{m}_t - \bar{m})^2 = [\hat{a}_0\Sigma y_t + \hat{a}_1\Sigma X_{1t}y_t + \cdots + \hat{a}_p\Sigma X_{pt}y_t] - \bar{y}\Sigma y_t. \quad (9.45)$$

Having computed the explained sum of squares from Equation (9.45) we get the unexplained sum of squares by substitution in (9.39). Finally we compute R by use of (9.42).

EXAMPLE 9–8. For the case of two regressors, the foregoing points are illustrated by the following fictitious sample of 10 corresponding observations.

HYPOTHESIS. $\hat{m}_t = \hat{a}_0 + \hat{a}_1 X_{1t} + \hat{a}_2 X_{2t}$.

Data

t	1	2	3	4	5	6	7	8	9	10	Sum	Sum of Squares
X_{1t}	5	8	9	2	1	9	8	2	1	5	50	350
X_{2t}	8	10	9	4	7	15	12	6	1	8	80	780
y_t	7	12	15	4	1	7	8	4	5	7	70	638

CROSS-PRODUCTS. $\Sigma X_{1t}X_{2t} = 500$, $\Sigma X_{1t}y_t = 450$, $\Sigma X_{2t}y_t = 620$.
NORMAL EQUATIONS.

$$10\hat{a}_0 + 50\hat{a}_1 + 80\hat{a}_2 = 70,$$
$$50\hat{a}_0 + 350\hat{a}_1 + 500\hat{a}_2 = 450,$$
$$80\hat{a}_0 + 500\hat{a}_1 + 780\hat{a}_2 = 620.$$

The symmetrical form of these equations is characteristic of linear hypotheses. The solution is

$$\hat{a}_0 = 5, \quad \hat{a}_1 = 2, \quad \hat{a}_2 = -1,$$

so the empirical regression is $\hat{m}_t = 5 + 2X_{1t} - X_{2t}$. The various sums of squares are as follows:

Total sum of squares:

$$\Sigma(y_t - \bar{y})^2 = \Sigma y_t^2 - \bar{y}\Sigma y_t = 638 - 7(70) = 148.$$

Explained sum of squares:

$$\Sigma(\hat{m}_t - \bar{m})^2 = \hat{a}_0\Sigma y_t + \hat{a}_1\Sigma X_{1t}y_t + \hat{a}_2\Sigma X_{2t}y_t - \bar{y}\Sigma y_t$$
$$= 5(70) + 2(450) - 1(620) - 7(70) = 140.$$

Unexplained sum of squares:

$$\Sigma(y_t - \hat{m}_t)^2 = \Sigma(y_t - \bar{y})^2 - \Sigma(\hat{m}_t - \bar{m})^2 = 148 - 140 = 8$$

The reduction ratio is

$$\frac{\Sigma(\hat{m}_t - \bar{m})^2}{\Sigma(y_t - \bar{y})^2} = \frac{140}{148} = 0.946 = R^2,$$

which gives $R = \sqrt{0.946} = 0.97$.

MULTIPLE AND PARTIAL CORRELATION

We can regard the regression function $\mu(x)$ as representing that part of the regressand y that is explainable in terms of the regressors x_1, x_2, $\cdots$, x_n. Thus the strength of the combined influence of the x_i upon y may be measured by the coefficient of linear correlation $\rho(\mu,y)$ between the regression function and the regressand. Inasmuch as this correlation involves all of the x_i simultaneously, it is called the *coefficient of multiple correlation* between y and x_1, x_2, $\cdots$, x_n. Historically, the concept of multiple correlation was developed from the viewpoint of a linear hypothesis, but the foregoing definition is a direct extension to general regression. When we explored the general properties of regression functions, we discovered that the covariance between y and μ is equal to the variance of μ, so that the coefficient of multiple correlation can never be negative; that is

$$0 \leqslant \rho(\mu,y) \leqslant 1. \tag{9.46}$$

Specifically (as the reader might anticipate),

$$\rho(\mu,y) = \frac{\sigma(\mu)}{\sigma(y)}, \qquad \rho^2(\mu,y) = \frac{\sigma^2(\mu)}{\sigma^2(y)}, \tag{9.47}$$

so that the square of the coefficient of multiple correlation yields the relative reduction of variance. Moreover, from this fact and the partition of variance we may write

$$\sigma^2(\mu) = \rho^2(\mu,y)\sigma^2(y), \qquad \sigma^2(u) = [1 - \rho^2(\mu,y)]\sigma^2(y). \tag{9.48}$$

The first equation in (9.48) provides an analytic interpretation of multiple correlation in terms of the explained variance $\sigma^2(\mu)$; the second equation, in terms of the unexplained variance $\sigma^2(u)$. Using equations (9.34) and (9.35) of the previous section we arrive at directly analogous results for a linear hypothesis, namely

$$\rho(m,y) = \sigma(m)/\sigma(y), \tag{9.49}$$
$$\sigma^2(m) = \rho^2(m,y)\sigma^2(y),$$
$$\sigma^2(y - m) = [1 - \rho^2(m,y)]\sigma^2(y),$$

and we have already seen that corresponding equations hold for the sample multiple correlation R.

Since in practice the true regression function is seldom known, it is ordinarily necessary to hypothesize its mathematical form and determine the constants by least squares. Commonly, the approximating function is chosen to be linear in the unknown constants in order to facilitate the process of estimation. Hence the term multiple correlation is ordinarily interpreted in the sense of $\rho(m,y)$ where $m = a_0 + a_1X_1 + \cdots + a_pX_p$. Just as r is taken as a sample estimate of simple correlation, R is often used as a sample estimate of multiple correlation. When many constants are computed from the data, however, the question of degrees of freedom becomes important; and it would be desirable to correct for the number of constants fitted to the data. In the case when (1) the linear hypothesis actually suits the true regression function and (2) the errors are independent of the regressors (instead of merely uncorrelated with them), it may be shown that an unbiased estimate of the *residual variance* is given by

$$s^2(u) = \frac{\Sigma(y_t - \hat{m}_t)^2}{N - (p + 1)}; \tag{9.50}$$

and as usual, an unbiased estimate of $\sigma^2(y)$ is given by $s^2(y) = \Sigma(y_t - \bar{y})^2/(N - 1)$. Thus, provided $s^2(u) \leq s^2(y)$, we may define a corrected value $\hat{R}$ of R by the equation

$$s^2(u) = (1 - \hat{R}^2)s^2(y); \tag{9.51}$$

and, in case $s^2(u) > s^2(y)$, we take $\hat{R}$ as zero. When $s^2(u) \leq s^2(y)$, the connection between $\hat{R}$ and R, as the reader may verify, is given by

$$\hat{R}^2 = \frac{(N - 1)R^2 - p}{N - (p + 1)}. \tag{9.52}$$

Thus Equation (9.51) will break down when $R^2 < p/(N - 1)$.

EXAMPLE 9-9. Using the data of Example 9–8, we have $N - (p + 1) = 7$ and

$$s^2(u) = \frac{8}{7},$$

$$s^2(y) = \frac{148}{9},$$

$$1 - \hat{R}^2 = \frac{s^2(u)}{s^2(y)} = \frac{18}{259} = 0.069,$$

$$\hat{R}^2 = 0.931 \quad \text{and} \quad \hat{R} = 0.96.$$

Partial correlation is defined as the simple linear correlation between the residuals of two variables with respect to common regressors. Let y_1, y_2 denote two regressands and μ_1, μ_2 their respective regression functions with regard to the same set of regressors x_1, x_2, $\cdots$, x_n. Also let u_1, u_2 stand for the corresponding residuals $(y_1 - \mu_1), (y_2 - \mu_2)$. The partial correlation between y_1, y_2 is then defined as $\rho(u_1, u_2)$. Where precision is required, we shall denote this by the symbol $\rho(y_1, y_2; x_1, x_2, \cdots, x_n)$ but where the context is clear, we shall use the shorter notation $\rho_{12.x}$. Hence

$$\rho(y_1, y_2; x_1, x_2, \cdots, x_n) \equiv \rho_{12.x} \equiv \rho[(y_1 - \mu_1), (y_2 - \mu_2)] \equiv \rho(u_1, u_2). \quad (9.53)$$

As an addition to the general properties of regression functions established previously we observe that

$$E(\mu_1 u_2) = 0 = E(u_1 \mu_2). \quad (9.54)$$

We can establish this result as follows:

$$E(\mu_1 u_2) = \iint \mu_1(x) u_2 f(x) \phi_2(y_2|x) \, dx \, dy_2$$
$$= \int \mu_1(x) f(x) [\int u_2 \phi_2(y_2|x) \, dy_2] \, dx = 0$$

and similarly for $E(u_1 \mu_2)$. Therefore, by techniques like those we have used several times already, we may show that if any two regressands y_1, y_2 are both estimated by the same regressors $x_1, x_2, \cdots, x_n$, the covariance of their regression functions equals the covariance between either regressand and the regression function of the other. Moreover, the covariance between the residuals u_1, u_2 equals the covariance of the regressands minus the covariance of the regression functions. In symbols:

$$\sigma_{\mu_1 \mu_2} = \sigma_{y_1 \mu_2} = \sigma_{\mu_1 y_2} \quad \text{and} \quad \sigma_{u_1 u_2} = \sigma_{y_1 y_2} - \sigma_{\mu_1 \mu_2}. \quad (9.55)$$

As a hint to the reader who may wish to supply the proof, we have

$$E(y_1 \mu_2) = E[(\mu_1 + u_1)\mu_2] = E(\mu_1 \mu_2),$$
$$E(\mu_1 y_2) = E[\mu_1(\mu_2 + u_2)] = E(\mu_1 \mu_2),$$
$$E(y_1 y_2) = E[(\mu_1 + u_1)(\mu_2 + u_2)] = E(\mu_1 \mu_2) + E(u_1 u_2),$$

and the rest of the proof involves merely the subtraction of products of means. We may now give a formula for the coefficient of partial correlation, namely

$$\rho_{12.x} = \frac{\sigma_{u_1 u_2}}{\sigma(u_1)\sigma(u_2)} = \frac{\sigma_{y_1 y_2} - \sigma_{\mu_1 \mu_2}}{\{[1 - \rho^2(\mu_1, y_1)]\sigma^2(y_1)[1 - \rho^2(\mu_2, y_2)]\sigma^2(y_2)\}^{1/2}} \qquad (9.56)$$

In practice, partial correlation is usually considered in terms of linear hypotheses. As the estimating function for y_1, let us put $m_1 = a_0 + a_1 X_1 + a_2 X_2 + \cdots + a_p X_p$, and as that for y_2 put $m_2 = b_0 + b_1 X_1 + b_2 X_2 + \cdots + b_p X_p$. The partial correlation will then be approximated (not necessarily very well) by the correlation between the residuals $y_1 - m_1$ and $y_2 - m_2$. Let us denote this by the symbol $\rho_{12.m}$ to bring out the fact that we are using linear hypotheses. That is,

$$\rho_{12.m} = \rho[(y_1 - m_1), (y_2 - m_2)]. \qquad (9.57)$$

Corresponding to (9.55), it is a straightforward matter to prove that

$$\sigma_{m_1 m_2} = \sigma_{y_1 m_2} = \sigma_{m_1 y_2} \quad \text{and} \quad \sigma_{z_1 z_2} = \sigma_{y_1 y_2} - \sigma_{m_1 m_2}, \qquad (9.58)$$

where $z_1 = y_1 - m_1$, $z_2 = y_2 - m_2$.

An interesting case arises when there is just one regressor; that is, $m_1 = a_0 + a_1 X_1$, $m_2 = b_0 + b_1 X_1$. To distinguish this special case from the general linear hypothesis, let us write the counterpart of $\rho_{12.m}$ as $\rho_{12.X_1}$. It turns out that this quantity is directly expressible in terms of simple correlations:

$$\rho_{12.X_1} = \frac{\rho(y_1, y_2) - \rho(y_1, X_1)\rho(y_2, X_1)}{[1 - \rho^2(y_1, X_1)]^{1/2} [1 - \rho^2(y_2, X_2)]^{1/2}}. \qquad (9.59)$$

This is established by substituting from (9.58) and (9.26) and canceling out the standard deviations. The corresponding sample estimate $\rho_{12.X_1}$ is given by

$$r_{12.X_1} = \frac{r(y_1, y_2) - r(y_1, X_1)r(y_2, X_1)}{[1 - r^2(y_1, X_1)]^{1/2} [1 - r^2(y_2, X_2)]^{1/2}}. \qquad (9.60)$$

While this estimate may be obtained formally by substituting sample estimates into (9.59), it actually represents the sample correlation between $y_1 - \hat{a}_0 - \hat{a}_1 X_1$ and $y_2 - \hat{b}_0 - \hat{b}_1 X_1$.

Typically, the partial correlation is smaller in magnitude than the simple correlation between two variables; it is mathematically possible, however, for the partial correlation to be much greater in magnitude than the simple correlation. The common situation is illustrated by Example 9–10 and the exceptional by Examples 9–11, 9–12.

EXAMPLE 9-10. $\rho(y_1, X_1) = \rho(y_2, X_1) = \rho(y_1, y_2) = 0.5$.

$$\rho_{12.X_1} = \frac{0.5 - 0.25}{[(1 - 0.25)(1 - 0.25)]^{1/2}} = \frac{0.25}{0.75} = \frac{1}{3}.$$

EXAMPLE 9-11. $\rho(y_1, X_1) = \rho(y_2, X_1) = 0.5$, $\rho(y_1, y_2) = 0$.

$$\rho_{12.X_1} = \frac{-0.25}{0.75} = -\frac{1}{3}.$$

EXAMPLE 9-12. $\rho(y_1, X_1) = \rho(y_2, X_1) = 0.5$, $\rho(y_1, y_2) = -0.5$.

$$\rho_{12.X_1} = \frac{-0.75}{0.75} = -1.$$

During the time of Karl Pearson (1857–1936), partial correlation attracted a great deal of attention in statistical literature, and a large number of algebraic properties were established. Many research workers apparently thought that partial correlation shed light on causal relationships, but such hopes were ill founded. Before computers were widely available, much use was made of partial correlation coefficients in solving the normal equations by working with formulas analogous to (9.60). These techniques, however, were of dubious computational efficiency even without machines. At present, very little practical use is made of partial correlation as an end in itself; but, at least in principle, it does have a permanent place in statistical theory because of its intimate association with tests of significance. We shall not devote space to showing this connection, because the significance tests do not make explicit use of partial correlations.

PROPERTIES OF SAMPLE ESTIMATES

Estimates of mean-square errors and other quantities bearing on the operating performance of the regression function involve the sampling variances and covariances of the empirical regression constants $\hat{a}_i$. Since these constants are determined by solving simultaneous equations, some simplifying assumptions are needed in order to make any headway with an elementary approach to the problem of sampling behavior. The classical model, which we adopt, consists of a two-fold assumption.

1. The hypothesized regression function m has the correct form; that is, $\mu(x,\alpha) = \alpha_0 + \alpha_1 X_1 + \cdots + \alpha_p X_p$.
2. The deviations from the true regression function are independent, identically distributed random variables over the entire range of the X_i.

As a consequence of the first assumption, the α's could be determined from any linearly independent set of $p + 1$ values of the regressors X_1, X_2, $\cdots$, X_p, together with the corresponding points on the true regression function. Hence the errors of the empirical regression coefficients are due solely to the deviations of the regressand from the regression function. Under the second assumption, these deviations have the same distribution irrespective of the values of the regressors and are mutually independent also. Hence we may obtain a restricted solution to the sampling problem, adequate for testing significance, by considering a sampling process that is random as far as the residuals from the regression function are concerned but which holds the *regressors fixed* at the same N values actually observed in the particular sample. By this scheme, the errors of the empirical regression coefficients become definite linear functions of the residuals and the requisite expected values are easily derived.

By hypothesis,

$$y_t \equiv \mu_t + u_t \equiv (\alpha_0 + \alpha_1 X_{1t} + \cdots + \alpha_p X_{pt}) + u_t. \tag{9.61}$$

If we substitute the latter expression in place of y_t in the empirical normal equations of (9.38) and transfer everything not containing u to the left side, we arrive at a set of equations identical in form to (9.38), but having the unknowns replaced by $\hat{a}_i - \alpha_i$ for $i = 0, 1, \cdots, p$ and, on the right side, y_t replaced by u_t. As a manipulative convenience, let us introduce the purely formal regressor $X_{0t} \equiv 1$, so that the system of equations will have the same structure throughout. The first row and column will now have the same form as the other rows and columns, since $N = \Sigma X_{0t}^2$, $\Sigma X_{it} = \Sigma X_{0t} X_{it}$, and $\Sigma u_t = \Sigma X_{0t} u_t$. Also put

$$S_{ij} = \Sigma X_{it} X_{jt} = S_{ji} \qquad \text{for } i = 0, 1, \cdots, p \quad \text{and} \quad j = 0, 1, \cdots, p;$$

and put

$$T_{iu} = \Sigma X_{it} u_t.$$

Then the simultaneous equations can be expressed in terms of the typical equation as

$$(\hat{a}_0 - \alpha_0)S_{i0} + (\hat{a}_1 - \alpha_1)S_{i1} + \cdots + (\hat{a}_p - \alpha_p)S_{ip} = T_{iu} \qquad (9.62)$$
$$\text{for } i = 0, 1, \cdots, p$$

Denote the determinant of all the S_{ij} as $|S|$ and the cofactor of the element S_{ij} as K_{ij}. Introduce the symbol S^{ij} to represent the quotient obtained when the cofactor is divided by the full determinant; that is,

$$S^{ij} = \frac{K_{ij}}{|S|}.$$

People familiar with matrix theory will recognize S^{ij} as the general element of the inverse matrix; the usual transposition is unnecessary because of symmetry. Solving Equation (9.62) by Cramer's rule for determinants, we may express the solution as

$$(\hat{a}_i - \alpha_i) = \sum_{k=0}^{p} S^{ki}T_{ku} = \sum_{k=0}^{p} S^{ik}T_{ku}. \qquad (9.63)$$

The problem now comes down to the behavior of the T_{ku}. The expected value of any T_{ku} is zero, because

$$E(T_{ku}) = E\left[\sum_t X_{kt}u_t \right] = \sum_t X_{kt}E(u_t) = 0. \qquad (9.64)$$

Therefore $\hat{a}_i$ is an unbiased estimate of α_i, since

$$E[(\hat{a}_i - \alpha_i)] = 0. \qquad (9.65)$$

Any cross-product term $T_{hu}T_{ku}$ is given by

$$T_{hu}T_{ku} = \sum_t X_{ht}X_{kt}u_t^2 + \quad \text{terms in } u_t u_q \qquad \text{for } q \neq t.$$

Therefore

$$E(T_{hu}T_{ku}) = \sum_t X_{ht}X_{kt}E(u_t^2) = \sigma^2(u)S_{hk} \qquad (9.66)$$

because $E(u_t^2) = \text{constant} = \sigma^2(u)$ for all t and $E(u_t u_q) = 0$ for all $q \neq t$. From (9.64) and (9.66) we may derive all of the required expected values. The main algebraic points to keep in mind are that a sum like

$\Sigma S_{ik} S^{kj}$ (where the summation is over k) vanishes if i and j are different, but the sum equals unity if i and j are the same; this is a basic property of cofactors. The manipulations are a bit tedious but straightforward, and the results are as follows:

$$\sigma^2(\hat{a}_i) \equiv E[\hat{a}_i - \alpha_i)^2] = \sigma^2(u)S^{ii}, \qquad (9.67)$$

$$\sigma_{\hat{a}_i \hat{a}_j} \equiv E[(\hat{a}_i - \alpha_i)(\hat{a}_j - \alpha_j)] = \sigma^2(u)S^{ij},$$

$$E(\Sigma[\hat{a}_0 - \alpha_0) + (\hat{a}_1 - \alpha_1)X_{1t} + \cdots + (\hat{a}_p - \alpha_p)X_{pt}]^2) = (p + 1)\sigma^2(u),$$

$$E[\Sigma(y_t - \hat{m}_t)^2] = [N - (p + 1)]\sigma^2(u),$$

where each summation is over t. An unbiased estimate of $\sigma^2(u)$ is thus given by

$$s^2(u) = \frac{\Sigma(y_t - \hat{m}_t)^2}{N - (p + 1)}, \qquad (9.68)$$

which is the same as (9.50).

While the statistic $s^2(u)$ is an unbiased estimate of the unexplained variance $\sigma^2(u)$ about the true regression function, it does not represent the error to be expected when the estimated regression $\hat{m}$ is applied to fresh data even assuming (as we must) the same conditions of sampling, that is, with prescribed values for the regressors. Not knowing the true regression function, we cannot determine this error for a specific $\hat{m}$; but we can estimate the mean-square error that would be obtained if a series of $\hat{m}$'s were derived from an indefinitely large number of independent samples of N observations of y and the errors of estimating y on new data were squared and averaged. This is the sense in which we refer to the error of application to new data. The error turns out to be a quadratic function of the regressors; therefore, we must focus attention on one particular combination of values $X_{1t}, X_{2t}, \cdots, X_{pt}$ where t will be held fixed, once chosen. Then we write

$$y_t - \hat{m}_t \equiv (y_t - \mu_t) - (\hat{m}_t - \mu_t) \qquad (9.69)$$

$$= u_t - [(\hat{a}_0 - \alpha_0) + (\hat{a}_1 - \alpha_1)X_{1t} + \cdots + (\hat{a}_p - \alpha_p)X_{pt}].$$

When we square this identity and take expected values, we obtain

$$E[(y_t - \hat{m}_t)^2] = \sigma^2(u) + \sigma^2(u) \sum_i \sum_j X_{it} X_{jt} S^{ij}, \qquad (9.70)$$

where i and j range independently from 0 to p (note inclusion of the dummy variable $X_0 \equiv 1$.) Equation (9.70), as we have said, applies to a particular combination of regressor values. It is interesting to see what happens when we sum this equation over t. We then find that the expected sum of squares of errors in estimating N new values of y is $(N + p + 1)\sigma^2(u)$, which shows that the unexplained sum of squares is, on the average, increased by the amount $(p + 1)\sigma^2(u)$.

EXAMPLE 9-13. The hypothesized regression function being $m = a_0 + a_1X_1 + a_2X_2$, a sample of 100 gave $\Sigma y_t^2 = 286.17$ and the normal equations were as follows:

$$100\hat{a}_0 + 40\hat{a}_1 + 70\hat{a}_2 = 168,$$
$$40\hat{a}_0 + 21\hat{a}_1 + 31\hat{a}_2 = 71,$$
$$70\hat{a}_0 + 31\hat{a}_1 + 51\hat{a}_2 = 120.$$

The solution is

$$\hat{a}_0 = 1.1, \qquad \hat{a}_1 = 0.4, \qquad \hat{a}_2 = 0.6;$$
$$\Sigma\hat{m}_t^2 = \hat{a}_0\Sigma y_t + \hat{a}_1\Sigma X_{1t}y_t + \hat{a}_2\Sigma X_{2t}y_t = 285.2;$$
$$\Sigma(y_t - \hat{m}_t)^2 = \Sigma y_t^2 - \Sigma\hat{m}_t^2 = 286.17 - 285.2 = 0.97.$$

Hence assuming that the true regression function is of the form of m, we have

$$s^2(u) = \frac{9.97}{100 - 3} = 0.01.$$

Let us estimate the mean-square error on new data at the point $X_1 = 0.4$, $X_2 = 0.7$. Here we find the values of S^{ij} to be

$$S^{00} = 1.1, \qquad S^{01} = 1.3 = S^{10}, \qquad S^{02} = -2.3 = S^{20},$$
$$S^{11} = 2.0, \qquad S^{12} = -3.0 = S^{21}, \qquad S^{22} = 5.0.$$

and remembering $X_0 = 1$ we get

$$\sum_i\sum_j X_{it}X_{jt}S^{ij} = (1)[(1)(1.1) + (0.4)(1.3) + (0.7)(-2.3)]$$

$$+ (0.4)[(1)(1.3) + (0.4)(2.0) + (0.7)(-3.0)]$$
$$+ (0.7)[(1)(-2.3) + (0.4)(-3.0) + (0.7)(5.0)]$$
$$= (1)(0.01) + (0.4)(0) + (0.7)(0) = 0.01.$$

Thus

$$\text{estimate of } E[(y_t - \hat{m}_t)^2] = s^2(u) + (0.01)s^2(u) = 0.0101.$$

The fact that the estimated regression constants incorporate linear functions of the residuals raises a question as to the extent to which a given sample result is due to chance. Partitioning the estimated regression function as

$$\hat{m}_t \equiv \mu_t + (\hat{m}_t - \mu_t) = \mu_t + \sum_{i=0}^{p} (\hat{a}_i - \alpha_i)X_{it}, \qquad (9.71)$$

squaring, summing on t, and taking expected values, we obtain from (9.67)

$$E(\Sigma \hat{m}_t^2) = \Sigma \mu_t^2 + (p + 1)\sigma^2(u), \qquad (9.72)$$

where the summations are over t. Thus we see that on the average, the empirical explained sum of squares is inflated by the amount $(p + 1)\sigma^2(u)$ which, it turns out, exactly equals the average increase in the unexplained sum of squares when empirical regression functions are applied to new data. Because of this inflation, we are not sure whether the explained sum of squares obtained from the sample reflects anything more than the effect of chance. Accordingly, significance tests are in order. There is a dual motive for testing significance. The principal one, of course, is to avoid self-delusion. The other is to avoid the penalty of increasing the unexplained sum of squares in future applications of $\hat{m}$ by retaining nonsignificant regressors. From (9.72) we see an obvious advantage of a large sample; whereas the inflation effect $(p + 1)\sigma^2(u)$ has a fixed magnitude for a given number of regressors, the significant contribution $\Sigma\mu_t^2$ increases with N and can thus be made arbitrarily large. On this basis, the actual distribution of the residuals becomes less important to the question of significance as the sample size increases.

In addition to the two assumptions already made, the standard significance tests available for regression problems require the further assumption that the residuals are normally distributed. The tests are derived from the general theory of linear and quadratic functions of normal variables. This subject is best treated from the matrix viewpoint. Here we shall merely state the results.

The significance of the entire regression function including $\hat{a}_0$ can be tested by the variance ratio

$$F = \frac{\Sigma \hat{m}_t^2}{(p + 1)s^2(u)} \qquad \begin{cases} \text{under the null hypothesis} \\ \text{that all } \alpha_i = 0. \end{cases} \qquad (9.73)$$

where F has $p + 1$ and $N - (p + 1)$ degrees of freedom. More often we are interested only in the significance of the variable portion of $\hat{m}$ exclusive of $\hat{a}_0$. For this the test is

$$F = \frac{\Sigma(\hat{m}_t - \bar{m})^2}{ps^2(u)} \qquad \begin{cases} \text{under the null hypothesis} \\ \qquad \alpha_1 = \cdots = \alpha_p = 0. \end{cases} \qquad (9.74)$$

where F has p and $N - (p + 1)$ degrees of freedom. The significance of an individual regression constant can be tested by the t-test. For the usual hypothesis that $\alpha_i = 0$ we set

$$t = \frac{\hat{a}_i}{[S^{ii}s^2(u)]^{1/2}} \qquad \begin{cases} \text{under the null hypothesis} \\ \qquad \alpha_i = 0. \end{cases} \qquad (9.75)$$

while for the more general hypothesis assigning a specific value, say α_i^*, to the parameter, we set

$$t = \frac{\hat{a}_i - \alpha_i^*}{[S^{ii}s^2(u)]^{1/2}} \qquad \begin{cases} \text{under the hypothesis} \\ \qquad \alpha_i = \alpha_i^*. \end{cases} \qquad (9.76)$$

In either case the variable t has $N - (p + 1)$ degrees of freedom, since this is the number of degrees of freedom in $s^2(u)$. Also, we may test the significance of the difference between any two regression constants $\hat{a}_i, \hat{a}_j$ derived from the same sample by computing t according to the formula

$$t = \frac{\hat{a}_i - \hat{a}_j}{[(S^{ii} - 2S^{ij} + S^{jj})s^2(u)]^{1/2}} \qquad \begin{cases} \text{under the hypothesis} \\ \qquad \alpha_i = \alpha_j. \end{cases} \qquad (9.77)$$

Despite the fact that we are testing the difference between two quantities, we still have only $N - (p + 1)$ degrees of freedom, because we have only one estimate of residual variance, namely $s^2(u)$. In all three cases, Equations (9.75) through (9.77), we should ordinarily use both tails of the t distribution, and if so, it would be somewhat more convenient to substitute the corresponding F-tests, thereby avoiding the extraction of square roots. The respective F's are merely the squares of the corresponding t's, and the degrees of freedom throughout are 1 and $N - (p + 1)$.

EXAMPLE 9–14. Assuming normally distributed residuals we apply test of significance to the data of Example 9–13 as follows. Here $s^2(u) = 0.01$ and $N - (p + 1) = 97$. For the entire regression function, $\Sigma \hat{m}_i^2 = 285.2$ and $p + 1 = 3$. Thus

$$F = \frac{285.2}{3(.01)} = 9507,$$

which with 3 and 97 degrees of freedom is significant very far beyond tabulated levels. For the explained sum of squares about the mean, we have $\Sigma(\hat{m}_t - \bar{m})^2 = 285.2 - (1.68)(168) = 2.96$, which gives

$$F = \frac{2.96}{2(.01)} = 148.$$

With 2 and 97 degrees of freedom, this value of F is significant greatly beyond the one percent level. For the regression constants the values of F are

$$\hat{a}_0: \quad F = \frac{(\hat{a}_0^2)}{S^{00}s^2(u)} = \frac{(1.1)^2}{(1.1)(.01)} = 110,$$

$$\hat{a}_1: \quad F = \frac{(\hat{a}_1)^2}{S^{11}s^2(u)} = \frac{(0.4)^2}{(2)(0.01)} = 8,$$

$$\hat{a}_2: \quad F = \frac{(\hat{a}_2)^2}{S^{22}s^2(u)} = \frac{(0.6)^2}{(5)(0.01)} = 7.2.$$

With 1 and 97 degrees of freedom, all of these are significant beyond the one percent level, and the value of F for the additive term $\hat{a}_0$ is significant far beyond one percent. To test the significance of the difference between $\hat{a}_1$ and $\hat{a}_2$ we have

$$F = \frac{(\hat{a}_1 - \hat{a}_2)^2}{(S^{11} - 2S^{12} + S^{22})s^2(u)} = \frac{(0.4 - 0.6)^2}{(2 + 6 + 5)(0.01)} = \frac{0.04}{0.13} < 1.$$

Since F is less than unity, the difference between $\hat{a}_1$ and $\hat{a}_2$ is not significant. Hence, unless we have cogent reasons to the contrary, we would do better in application to new data if we coalesced X_1 and X_2 by choosing as the revised regression function $m_t^* = a_0^* + a_1^* (X_{1t} + X_{2t})$. The normal equations are

$$100a_0^* + 110a_1^* = 168,$$

$$100a_0^* + 134a_1^* = 191;$$

the solution being

$$a_0^* = \frac{1502}{1300} = 1.16,$$

$$a_1^* = \frac{620}{1300} = 0.48.$$

To estimate the revised explained sum of squares, we must be careful in this particular case to guard against rounding errors. For this reason, we put

$$\Sigma m_t^{*2} = \frac{(1502)}{1300}(168) + \frac{(620)}{1300}(191) = \frac{370756}{1300} = 285.197.$$

Thus the residual sum of squares will be the same to two decimals as before (although mathematically it is greater to three decimals, since it is impossible to get a worse fit to the sample with $\hat{m}$ than m^*); hence, when we divide by 98 instead of 97, we get a somewhat smaller estimate of residual variance. We may explain this result by saying that the extra constant in $\hat{m}$ did not "pull its weight" in degrees of freedom.

A test of significance for the sample multiple correlation R can be derived from Equation (9.74) by substituting $\Sigma(\hat{m}_t - \bar{m})^2 = R^2\Sigma(y_t - \bar{y})^2$ and $\Sigma(y_t - \hat{m}_t)^2 = (1 - R^2)\Sigma(y_t - \bar{y})^2$. The result is

$$F = \frac{[N - (p + 1)]R^2}{p(1 - R^2)} \qquad \begin{cases} \text{under the null hypothesis} \\ \quad \rho(\mu, y) = 0, \end{cases} \qquad (9.78)$$

the degrees of freedom being p and $N - (p + 1)$. We need not use the corrected coefficient $\hat{R}$ because the F ratio already takes account of degrees of freedom. The simple correlation r can be tested by using (9.78) with $p = 1$. As an alternative (since in this case $F = t^2$) we may put

$$t = \frac{r(N - 2)^{1/2}}{(1 - r^2)^{1/2}} \qquad \begin{cases} \text{under the null hypothesis} \\ \quad \rho = 0 \end{cases} \qquad (9.79)$$

The latter equation can be obtained also by adapting Equation (9.75) when $p = 1$.

We notice that the assumption of normality does not enter the subject of regression until we consider tests of significance. At that point, the fact that the errors in the regression coefficients are linear functions of the residuals allows us to assume the normality of those sample estimates in large samples even if the residuals themselves are not normally distributed.

Appendix A:
Supplementary Exercises on Probability and Statistics

1. Among the digits 1, 2, 3, 4, 5 first one is chosen at random and then a second random choice is made among the remaining four digits. Find the probability that an even digit will be selected: (a) the first time, (b) the second time, (c) both times, (d) the first time and not the second time.

2. In problem 1 above, if both digits drawn are odd, what is the probability that their sum is 6? Do this solution using Bayes' theorem and identify here the events B_i and the event A in the statement of Bayes' theorem.

3. Two dice are thrown. Let A be the event that the sum of the faces is even, and let B be the event of at least one ace (i.e., a one spot.) List all 36 sample points, and assume that each one is equally probable. Find the probabilities of the events $A \cap B$, $A \cup B$, $A \cap B'$.

4. Four cards are drawn one by one from a full deck without jokers. What is the probability that each of the first and third will be an ace and each of the second and fourth will not be an ace? What is the probability of drawing three aces and one other card which is not an ace?

5. Given $f(x) = kx(1 - x)$ for $0 \leqslant x \leqslant 1$ with k so chosen to make $f(x)$ a density function,
   (a) Find the numerical value of k.
   (b) Find $F(x)$.
   (c) Find $E(x)$.
   (d) Find σ_x^2.
   (e) Find the mode.
   (f) Find the median.
   (g) Is there an axis of symmetry and if so, what is it?

375

(h) If 3 values of x are chosen at random, what is the probability that none exceeds $1/4$ in value?

(i) Compare the numerical values of

$$P\left[\frac{|x - \mu|}{\sigma} \geqslant \frac{\sqrt{20}}{4}\right]$$

as given (i) by the Tchebycheff inequality, and (ii) by direct integration.

6. Let a sample space consist of all points in the region between the curve $y = \sin x$ and the x-axis, from the origin to $x = \pi/2$. If all points are equally likely, find (a) the distribution of distances r from the y-axis, (b) the probability that $1 \leqslant r \leqslant 1.5$.

7. When a major oil company surveys a potential oil producing region, the probability density that the company will drill a new-field wildcat at coordinates (x,y) is given by

$$f(x,y) = \frac{1}{2\pi\sigma_x\sigma_y \sqrt{1 - \rho^2}} \exp\left\{-\frac{1}{2(1 - \rho^2)}\left[\left(\frac{x}{\sigma_x}\right)^2 - 2\rho\frac{x}{\sigma_x}\frac{y}{\sigma_y} + \left(\frac{y}{\sigma_y}\right)^2\right]\right\}.$$

Here x is positive to the east, and y positive to the north.

(a) Suppose you own a rectangular piece of land bounded by the lines $x = 1$, $x = 4$, $y = 2$, $y = 5$. Set up the integral that gives the probability that the wildcat will be drilled on your property. (Do not try to integrate.)

(b) Set up the integral that gives the probability that the wildcat will be drilled within a radius of R from the center $(0,0)$ of coordinates.

(c) Set up the integral which gives the probability density that the wildcat will have an east-west coordinate x irrespective of the value of the north-south coordinate y.

8. It may be shown that the marginal densities of $f(x,y)$ given in problem 7 are

$$f_1(x) = \frac{1}{\sqrt{2\pi}\sigma_x} \exp\left(-\frac{x^2}{2\sigma_x^2}\right) \quad \text{and} \quad f_2(y) = \frac{1}{\sqrt{2\pi}\,\sigma_y} \exp\left(-\frac{y^2}{2\sigma_y^2}\right).$$

(a) In the province of Mozambique, Portuguese East Africa, the Gulf Oil Corporation for economic reasons decided that the wildcat had to be drilled along the only existing road, which runs along the line $x = 1.5\sigma_x$. Then what is the probability density of drilling the wildcat at north-south coordinate y?

(b) In Saudi Arabia, the value of ρ is zero. Are x and y statistically independent or not? Give reason.

9. Given the density function $f(x)$, show that

$$\left.\frac{d^3 E(e^{\theta x})}{d\theta^3}\right|_{\theta=0} = E(x^3).$$

10. Given the joint probability density $f(x,y) = xe^{-x(1+y)}$ for $0 \leqslant x < \infty$, $0 \leqslant y < \infty$, find the moment generating function of u where $u = xy$.

11. Write the formula for (a) the number of permutations of n objects taken r at a time; (b) the number of combinations of n objects taken r at a time. Evaluate (a) and (b) for $n = 6$, $r = 4$.

12. From a group of n objects consisting of n_1 objects of type 1 and $n - n_1$ objects of type 2, a group of r objects is to be drawn at random, without replacement and with no regard to order. What is the probability that exactly r_1 objects of type 1 will be chosen? What is the range of r_1?

13. What is the probability of exactly three "aces" in 8 rolls of a die?

14. What is the probability that a bridge hand contains exactly 6 diamonds?

15. Write the formula for the binomial probability mass function $b(x;n,p)$.

16. Compute the moment generating function of the binomial probability mass function given in problem 15.

17. Write the formula for the Poisson probability mass function $p(x;\mu)$ where μ is the mean.

18. The total number of major business firms failing in the United States per year is found to fit a Poisson distribution with a mean of 12. What is the chance that no major business firm fails during a period of 3 months?

19. Write the formula for the Pascal probability mass function $f(k;r,p)$, where k is the number of trials where the rth success occurs.

20. Write the formula for the Pascal-geometric probability mass function $g(x)$ where x is the number of trials on which the first success occurs.

21. Personal-injury accidents in a certain factory are found to fit a Poisson distribution with a mean of 4 accidents per 8 hour shift.
    (a) Find the probability density that the next accident happens 20 minutes from now given that (i) an accident has just now happened, and (ii) the last previous accident happened 3 hours ago.
    (b) Find the median of the time between accidents.

22. An unbiased die is rolled 180 times. Find the probability that 20 or more "aces" occur.

23. In sampling from a normal distribution having a mean of μ and a variance of σ^2:
    (a) Prove that the sample mean $\bar{x} = (x_1 + x_2 + \cdots + x_n)/n$ is normally distributed with mean μ and variance σ^2/n.
    (b) Use this result to find the limits between which $\bar{x}$ will lie 92 percent of the time if $\mu = 20$, $\sigma = 4$, $n = 64$.

24. The exponential distribution has the probability density

$$f(x) = ae^{-ax} \quad \text{with} \quad a = \tfrac{1}{2} \qquad \text{for } 0 \leqslant x < \infty.$$

Find the probability density of the sample mean of n random observations. Find the expected value of the sample mean.

25. Two random draws x_1 and x_2 are taken from a uniform distribution with density function $f(x) = 1$ for $0 \leqslant x < 1$ and $f(x) = 0$ otherwise. Find the probability density of the quotient x_1/x_2.

26. Find the probability density of the t distribution with n degrees of freedom.

    HINT.

$$f(\chi^2) = \frac{1}{2^{n/2}\Gamma(n/2)} (\chi^2)^{n/2-1} e^{-\chi^2/2} \qquad \text{for } 0 \leqslant \chi^2 < \infty.$$

27. In a test to determine whether adding manganese strengthened a certain alloy, it was found that of 4 pieces to which manganese was added, 3 passed a certain test of strength, whereas of a control group of 4 pieces, 1 passed this same test. Is the effect of manganese significant at the 5 percent level?

28. Suppose, however, that as an intended part of the above test, breaking strengths were to be observed and that for the control group the distribution of breaking strength is known to be normal with mean 100, variance 25. The actual breaking strengths were:

$$\begin{array}{llll}
\text{Control group:} & 90, & 90, & 80, \quad 140 \\
\text{Manganese group:} & 110, & 110, & 130, \quad 90
\end{array}$$

Formulate a test variable which would utilize these data, or part of it, to investigate the hypothesis that manganese strengthens the alloy, and reevaluate the significance of the experimental observation.

29. In most applications of the F test, the appropriate critical region occupies the right portion of the distribution. With 2 and 4 degrees

of freedom, it is easy to show that under the null hypothesis, the probability of attaining or exceeding any particular value of F, say $F = A$, is given by

$$S(A) = \int_A^\infty f(F)dF = \frac{4}{(2 + A)^2}.$$

Two independent samples using this test variable (F with 2 and 4 degrees of freedom) gave $F = 4$ and $F = 6$ respectively. Determine whether or not the combined result is significant at the 5 percent level.

30. Two random samples from two binomial distributions gave the respective number of successes as $x_1 = 10$, $x_2 = 20$, where the numbers in the samples were respectively $n_1 = 50$, $n_2 = 75$. Are the distributions different at a 10 percent level of significance?

31. Given that $\sigma_1 = \sigma_2$ for a certain pair of normal populations x and y, test the hypothesis that $u_1 = u_2$ at the 5 percent significance level if observations yield

$$\bar{x} = 30, \qquad \bar{y} = 34, \qquad \sum_{i=1}^{n_1} (x_i - \bar{x})^2 = 800, \qquad \sum_{i=1}^{n_2} (y_i - \bar{y})^2 = 1320,$$

where $n_1 = 8$, $n_2 = 12$. Use the unbiased estimate of σ^2 given by

$$\frac{\Sigma(x_i - \bar{x})^2 + \Sigma(y_i - \bar{y})^2}{n_1 + n_2 - 2}.$$

32. According to Mendelian inheritance, offspring of a certain crossing should be colored red, black, or white in the ratios 9:3:4. If an experiment gave 72, 35, and 38 offspring in those categories, is the theory discredited at the 5 percent level of significance?

33. Five random samples of coal from each of two mines were analyzed for ash content (in percent) with the following results. Do the mean ash contents of the two sets of samples differ significantly?

Mine A: 18, 20, 22, 17, 23
Mine B: 17, 22, 15, 16, 20

[In working this problem, assume that the ash content (in percent) is normally distributed in each mine with equal variances.]

34. Perform an analysis of variance to determine whether there are significant differences between the means of processes A, B, C, where the following sets of three independent observations were made on each process.

A: 1, 2, 1;   B: 2, 4, 3;   C: 5, 5, 4

35. The theoretical probabilities associated with the various combinations of two attributes A, B are as follows:

Class	A_1	A_2	A_3
B_1	.20	.08	.04
B_2	.08	.20	.08
B_3	.04	.08	.20

A random sample exhibited the following frequency table.

Class	A_1	A_2	A_3
B_1	15	11	6
B_2	12	15	10
B_3	6	10	15

Is the theory tenable?

36. Find 90 percent confidence limits for the mean of a normal population if a sample of 17 observations gave $\bar{x} = 8$ and

$$\sum_{i=1}^{17} (x_i - \bar{x})^2 = 20.$$

37. Suppose that 5 men out of 100 and 25 women out of 10,000 are
color blind. A color-blind person is drawn at random. Assuming
males and females to be in equal numbers in the population, what
is the probability that this person is a male?

38. An insect lays eggs, the probability that n eggs are laid being

$$\frac{e^{-\lambda}\lambda^n}{n!}.$$

Each egg has probability p of developing, and develops independently of the other eggs. Show that the probability that exactly n
eggs develop is

$$\frac{e^{-\lambda p}(\lambda p)^n}{n!}.$$

39. The time between recordings of successive earth tremors on a seis-
mograph is exponentially distributed with a mean of 0.10 months.
Using the normal approximation, approximate to two decimal places
the probability of more than 20 tremors in a month.

40. Given $f(x) = 1/x^2$ for $1 < x < \infty$. Five observations are drawn at
random and ranked according to their magnitude from one to five.
Find the joint distribution of the second and fifth.

41. The probability of success for an individual event is 0.30. In 200

trials, what is the probability of the occurrence of either 56, 57, or 58 successes by using (a) the exact binomial (arithmetic not necessary); (b) the Poisson approximation (arithmetic not necessary); (c) the normal approximation (compute).

42. Let x and y be jointly distributed according to

$$f(x,y) = \frac{1}{2\pi\sigma_x\sigma_y} \exp\left[-\frac{1}{2}\left(\frac{x^2}{\sigma_x^2} + \frac{y^2}{\sigma_y^2}\right)\right].$$

Give the density functions of:

(a) $x/|y|$,   (b) x^2/y^2,   (c) $(x^2/\sigma_x^2) + (y^2/\sigma_y^2)$.

43. In a random sample of 13 observations from a normal population it is found that $\Sigma x = 52$, $\Sigma x^2 = 256$, and in another independent sample of 25 observations from another normal population the results are $\Sigma x = 30$, $\Sigma x^2 = 300$. Using the 10 percent level of significance, test the hypothesis that the populations sampled have equal variances. State the critical region involved in this test.

44. Given two normal populations with means $\mu_1 = 10$, $\mu_2 = 20$, and standard deviations $\sigma_1 = 2$, $\sigma_2 = 3$. A sample of 4 observations taken at random from the first population has a sample mean of 7, and a sample of 9 from the second population has a sample mean of 24. Is the original hypothesis concerning the populations borne out by the experiment?

45. Given the following table indicating the number of days in September (for years x, y, and z) for which the temperatures were above normal, normal, and below normal respectively. Can these three years be considered as comprising a homogeneous group, or is there probably a significant difference between them?

SEPTEMBER   STATION W

	Year		
	x	y	z
Above Normal	0	8	10
Normal	20	18	13
Below Normal	10	4	7
Total No. of Days	30	30	30

46. A motorist encounters four consecutive traffic lights, with each light equally likely to be red or green. Let k be the number of green lights

passed by the motorist before being stopped by a red light. What is the probability distribution of k?

47. A coin is tossed until for the first time the same result appears twice in succession. To every possible outcome requiring n tosses attribute the probability $1/2^n$. Find the probability of the following events: (a) the experiment ends before the sixth toss; (b) an even number of tosses is required.

48. Four cards are drawn one by one from a full deck without jokers. What is the probability that the second and fourth will be an ace and the first and third will not be an ace?

49. Of people with tuberculosis, 90 percent of the X-ray examinations detect the disease, but 10 percent go undetected. Of people free of tuberculosis, 99 percent of the X-ray tests are judged free of the disease, but 1 percent are diagnosed as showing tuberculosis. From a large population of which only 0.1 percent have tuberculosis, one person is selected at random, given a chest X-ray, and the radiologist reports the presence of tuberculosis. What is the probability that the person actually has tuberculosis?

50. If X and Y are random variables with joint probability density $f(x,y)$, then Professor Wiener (*Cybernetics*, 1948) defines the amount of information of X and Y as

$$\int_{-\infty}^{\infty} dx \int_{-\infty}^{\infty} dy\, f(x,y) \log_2 f(x,y).$$

Also, he defines the amount of information of X as

$$\int_{-\infty}^{\infty} dx\, f_1(x) \log_2 f_1(x),$$

where $f_1(x)$ is the marginal density function of X, and the amount of information of Y as

$$\int_{-\infty}^{\infty} dy\, f_2(y) \log_2 f_2(y),$$

where $f_2(y)$ is the marginal density function of Y. If X and Y are statistically independent, show that the amount of information is additive.

51. Let n independent observations $x_1, x_2, \cdots, x_n$ of a variable x having a density function $f(x)$ defined in the range $-\infty < x < \infty$ be taken at random. Denote the smallest of these n values by y. Find the distribution of y.

52. Given the probability density $f(x,y) = xe^{-x(y+1)}$ for $x \geq 0$ and $y \geq 0$. Find (a) the marginal density functions; and (b) the conditional density functions. (c) Determine whether x and y are independently distributed.

53. Consider a deck of cards consisting of the two, three, four, and five of each of the four suits. If 2 cards are drawn from this deck, and x and y denote the number of spades and hearts obtained respectively, find (a) the marginal distribution of x, and (b) the conditional distribution of y for $x = 1$.

54. In any sample of n independent observations from an infinite population with mean μ and variance σ^2, let x_i, x_j denote any pair of observations in the same sample and $\bar{x}$, the corresponding mean of the sample. Show that

$$E[(x_i - \bar{x})(x_j - \bar{x})] = \frac{-\sigma^2}{n} \qquad \text{for } i \neq j.$$

55. The moment generating function of the variable x is

$$M(\theta;x) = \cosh \theta = \frac{1}{2}(e^{\theta} + e^{-\theta}).$$

Find the mean and variance of x, and find the moment generating function of $\bar{x}$, the sample mean of n independent observations.

56. Independent discrete variables $x_1, x_2, \cdots, x_n$ all have the same distribution, consisting of two possible values, 0 and 1, with respective probabilities q and p, where $p + q = 1$. Find the moment generating function of $y = x_1 + x_2 + \cdots + x_n$ and state the distribution of y by inspection.

57. An engineer's report on the causes of failure of hot water heaters for home use revealed that 90 percent of the failures could be ascribed to one of three defects: Leaking Seams, Leaking Connections, or Pin-point Corrosion—the respective probabilities being 0.4, 0.3, 0.2. Neglecting the remote possibility of simultaneous defects and assuming independent trials, what is the probability that a random sample of five failures would contain two instances of leaking seams, two of leaking connections, one of pin-point corrosion, and none due to miscellaneous causes?

58. Personal-injury accidents in a certain factory are found to follow a Poisson distribution with a mean of 4 accidents per hour.
    (a) Find the median of the time between accidents.
    (b) Find the probability density of the time required for ten accidents to occur.

59. A sample of size n is drawn from a normal distribution with mean μ and variance σ^2:
    (a) Prove that the sample mean $\bar{x}$ is normally distributed with mean μ, variance σ^2/n.
    (b) Use this result to find the limits between which $\bar{x}$ will lie 92 percent of the time if $\mu = 20$, $\sigma = 4$, $n = 64$.

60. A retail hardware dealer kept count of the defective one-inch wood screws purchased from a certain manufacturer and found that one in fifty screws come throuth without slots. Using the normal approximation to the binomial, find the probability that in a random sample of 2500 there will be 64 or more without slots.

61. (a) Independent variables x_1 and x_2 have the Poisson distribution with respective parameters $\mu_1 = 0.27$, $\mu_2 = 0.32$. Determine the probability that $y \geq 4$, where $y = x_1 + x_2$.
    (b) The variable x has the Poisson distribution with parameter μ. Find the mean and variance of the variable u defined as $u = (x - \mu)/\sqrt{\mu}$. If $\bar{u}$ denotes the sample mean of n independent observations from the population of u, find the mean and variance of the variable y, where $y = \bar{u}\sqrt{n}$.

62. If the random variable x has probability density

$$f(x) = 3x^2 e^{-x^3} \qquad \text{for } 0 \leq x < \infty$$

find the probability density of the random variable y where $y = x^3$.

63. Independent random variables x and y are distributed as chi-square with $2a$ and $2b$ degrees of freedom respectively. Making the change of variables

$$u = x + y, \qquad v = \frac{x}{x + y}$$

derive the distribution of v.

64. The joint distribution of x,y is

$$f(x,y) = \tfrac{1}{8}(x^2 - y^2)e^{-x} \qquad \text{for } 0 \leq x < \infty, \text{ and } -x \leq y \leq x.$$

Derive the joint and marginal distributions of u,v where

$$u = \frac{x + y}{2}; \qquad v = \frac{x - y}{2}.$$

Recall that

$$J = \begin{vmatrix} \dfrac{\partial x}{\partial u} & \dfrac{\partial y}{\partial u} \\[2ex] \dfrac{\partial x}{\partial v} & \dfrac{\partial y}{\partial v} \end{vmatrix}$$

65. Given the joint density function $f(x,y) = e^{-(x+y)}$ for $x \geq 0$, $y \geq 0$. Prove that the distribution of $z = e^{-(x+y)}$ is given by $h(z) = -\log z$. What is the range of z here?

66. In not more than two pages summarize the theory of hypothesis testing from the viewpoint of probability. Build your essay around these main topics: Simple hypotheses, significance levels, critical regions, two types of errors, power functions, working hypotheses, null hypotheses, randomization. Omit the discussion of specific testing procedures.

67. Given the density function

$$f(x) = \frac{\theta}{(x\theta + 1)^2} \qquad \text{for } 0 \leq x < \infty.$$

Let the null hypothesis be that $\theta = 1$, and let the critical region for the alternative hypothesis $\theta < 1$ be the interval $r \leq x < \infty$.
   (a) Determine r so that the probability of an error of the first kind is $\alpha = 0.10$.
   (b) Given $\alpha = 0.10$, find the probability of an error of the second kind as a function of θ.
   (c) Would a random draw of $x = 8.8$ be significant?

68. Two independent tests were made using a continuous test variable which has a density function $f(x)$ for admissible region $0 \leq x < \infty$. The measure of significance $S(u)$ is given by

$$S(u) = \int_u^\infty f(x)\, dx.$$

If the value of S is 0.1 for the first test and p for the second, how small must p be in order that the combined result be significant at the 5 percent level. Use the table of chi-square.

69. Two independent measures x_1, x_2 are available for the estimation of a certain parameter μ. In repeated trials, the mean values of both measures are equal to μ, but since their variances σ_1^2, σ_2^2 are unequal, the measures differ in precision. Treating p as an arbitrary constant, show that the statistic

$$y = px_1 + (1 - p)x_2$$

is an unbiased estimate of μ and then determine p such that the variance of y is as small as possible, thus yielding a linear estimate of maximum precision. Express the final equation for y in simplified form.

70. Data on the durability of cast iron, heat-treated and not heat-treated are summarized in the following table. Test to see whether the variability of the durability is affected by heat-treating at a 10 percent level of significance. Let the area outside the acceptance region be divided equally between the lower and upper extremes of the distribution.

Quantity	Heat-Treated	Not Heat-Treated
n	201	101
$\bar{x}$	165	182
$\Sigma(x - \bar{x})^2$	2400	1100

71. Aircraft company A observed that 3 test pilots were lost out of 200 runs, while aircraft company B manufacturing the same jet observed that 7 test pilots were lost out of 250 runs. Using the normal approximation, is it reasonable to suppose that the two percentages represent a common level of test pilots lost, at a 5 percent level of significance and therefore you, a test pilot, would be equally willing to work for either company? Would you be more worried about an error of the first kind or of the second kind, or equally worried, or not worried at all? Why?

72. Prove that in order to obtain the lower critical limit for an F distribution, we may take the reciprocal of the upper critical limit of the "F" with interchanged degrees of freedom. Find the lower critical limit for a 5 percent level of significance for F with 7 and 10 degrees of freedom.

73. In a large sample of size n from a binomial population with known probability p of success on an individual trial, denote the observed number of successes by 0_1, the expected number of successes by E_1, the observed number of failures by 0_2, and the expected number of failures by E_2. Show that the statistic

$$\sum_{i=1}^{2} \frac{(0_i - E_i)^2}{E_i}$$

is distributed approximately as chi-square with one degree of freedom. State the generalization of this theorem to the case of sampling from a known multinomial population, and discuss how it may be used to test goodness to fit.

74. A factory has two machines that make roller bearings, and because of the small standard deviation (which means that the physical minimum of zero length is a very large negative value in standardized measure), the output of each machine can be considered normally distributed. Independent samples of 5 from each machine were measured to the nearest hundredth of an inch, with the following results:

Machine I	1.52	1.51	1.48	1.49	1.50
Machine II	1.52	1.49	1.50	1.53	1.51

Having first shown that the variances can be considered equal, test whether the means of the two machines are significantly different. In both steps use 5 percent levels of significance and state the appropriate critical regions.

75. Assuming a normal population, propose an exact method of testing whether the value of s^2 obtained from a small sample is consistent with a stated theoretical value of σ^2. Using $\alpha = 0.1$, $\sigma^2 = 4$, carry out this test for the following sample: 4,5,5,7,7,8.

76. Prove that the covariance between the regression function and the residual is zero and that the variance of the regression function equals the covariance between the regression function and the regressand.

77. If

$$f(x,y) = \frac{1}{2\Gamma(n + 1)} x^{n+3} y^n e^{-xy-x} \qquad \text{for } 0 \leqslant x < \infty, 0 \leqslant y < \infty,$$

then

$$f_1(x) = \tfrac{1}{2} x^2 e^{-x} \qquad 0 \leqslant x < \infty;$$

$$f_2(y) = \frac{(n + 3)(n + 2)(n + 1)y^n}{2(1 + y)^{n+4}} \qquad 0 \leqslant y < \infty.$$

Find both regression functions.

78. Let x be a random variable on the range $0 \leqslant x < \infty$. Let

$$P(x \geqslant a) = (1 + a/k)^{-k} \qquad \text{for any } a \geqslant 0.$$

(a) Find the cumulative distribution function $F(x)$ and the probability density function $f(x)$.

(b) If $k = 2$ and three observations are drawn at random, what is the probability that one will lie between 0 and 2, another between 3 and 5, and the other will be greater than 6?

79. If x is a random variable with probability density function $f(x)$ for $-\infty < x < \infty$, and

$$y = \int_{-\infty}^{x} f(t)\, dt,$$

what is the distribution of: (a) y; (b) v, where $v = y^2$?

80. A mass producer of men's shirts finds the average number of defects (e.g., button missing, improper stitching, snag in the cloth) is one per shirt. In 10,000 shirts, what is the probability that 500 or more of them will be free of defects?

81. A random sample of n observations are made from an unknown continuous probability distribution. Prove that the probability that an additional $(n + 1)$st observation is greater in magnitude than any of the n already drawn is $1/(n + 1)$.

82. The variable x has the Cauchy distribution

$$f(x) = \frac{1}{\pi} \frac{1}{1 + x^2} \qquad \text{for } -\infty < x < \infty.$$

Find the distribution of $y = \tan^{-1}x$. Using the distribution of x, find the moment generating function of y, and verify your result by using the distribution of y.

83. Suppose a sample of 10 observations from a normal distribution gave $\bar{x} = 20$ and $\Sigma(x - \bar{x})^2 = 144$. Find 90 percent confidence limits for the mean μ of the population, symmetric about $\bar{x}$.

84. A company has 5 sales districts, and total sales for a year in each district is known to be a Poisson variable with mean μ_i, $i = 1, 2, \cdots, 5$. Tabulated below are the values of μ_i and also the 1955 sales for each of the districts. Would you conclude that total company sales were significantly larger than normal for 1955?

District	1	2	3	4	5
μ	12	10	15	4	20
1955 sales	14	9	20	7	30

85. In a sample of size n, the variable y is predicted from the formal variable x_0 (defined as $x_0 = 1$) and the bona fide variables x_1, x_2, x_3. Denote the sample multiple correlation by R, the sample correlation of y and x_1 by r_{y1}, the sample partial correlation of y and x_2 with respect to the regressor x_1 by $r_{y2.1}$, and the sample partial correlation of y and x_3 with respect to the regressors x_1 and x_2 by $r_{y3.12}$. Prove that

$$1 - R^2 = [1 - r_{y1}^2][1 - r_{y2.1}^2][1 - r_{y3.12}^2].$$

HINT. Utilizing the fact that R is unaffected by a linear change of variables replace x_1, x_2, x_3 by their residuals $x_1 - \hat{x}_1$, $x_2 - \hat{x}_2$, $x_3 - \hat{x}_3$ where each $\hat{x}_i$ is the linear regression in terms of the previous variables.

86. A man wishes to unlock his door in the dark, and has a ring of m keys. The right key can only be found by trying it in the lock. He can select successive keys at random, or by rotation key by key around the ring.
   (a) Find the probability of success at the nth trial.
   (b) Show, thereby, that rotation is more efficient than random selection.

87. The time between successive accidents on the Pennsylvania turnpike is exponentially distributed with a mean of 0.10 months. Using the normal approximation, approximate to two decimal places the probability of more than 20 accidents in a month.

88. The variables x, y are jointly distributed thus:

$$f(x,y) = \frac{|x|}{2\pi y^2} \exp\left[-\frac{x^2}{2}\left(1 + \frac{1}{y^2}\right)\right] \qquad \text{for } -\infty < x < \infty, \ -\infty < y < \infty.$$

Find the distribution of $v = x/y$. Choose $u = x$.

89. Given $f(x) = 1/x^2$ for $1 < x < \infty$. Five observations are drawn at random and ranked according to their magnitude. Find the joint distribution of the largest and smallest.

90. Two random samples from normal populations gave the results:

$$\bar{x} = 150, \qquad s_x = 14, \qquad n_x = 10,$$

$$\bar{y} = 170, \qquad s_y = 19, \qquad n_y = 12.$$

Are the populations significantly different? Do not estimate any parameters.

91. If the number of automobiles passing a point on a rural highway is a random variable with a Poisson distribution having a mean of 10 per hour, find 98 percent confidence limits for the number passing the point in any eight-hour period.

92. Using chi-square, test the hypothesis that the respective populations represented by the following independent samples have the same distribution. Is chi-square an exact test in this connection? Why?

Observed Frequencies

Category	I	II	III	IV	V
Sample A	7	8	20	12	3
Sample B	3	12	20	8	7

93. The yield rate, y, of a chemical process was observed at various temperatures, t, and the following pairs recorded:

y	5	6	8	9	12
t	4	5	7	13	21

   (a) Find the slope of the regression line of y on t.
   (b) Test the hypothesis that the true slope is 0.15.

94. A, B, and C play a game in which the probabilities of their winning are equal. In playing three games, find the probability that one or the other of the players will win all the games.

95. If x has the cumulative distribution function $F(x) = x/2$, over the range $0 < x < 2$, and $y = \sin x$, find the distribution on y. Find the mean and the variance of y.

96. A pays a dollars to toss three coins, and receives b dollars for each head that appears. Set up the moment generating function for A's total winnings in n tosses of three coins.

97. (a) A point is taken at random in the interval $0 < x < 1$, the probability of the point falling in the interval of length dx being dx. Let u be the value of x so obtained. Now let a point be taken at random on the interval $u < x < 1$ and let v be the value of x found. Find the joint density function of u and v and also the marginal densities $g(u)$ and $h(v)$.
   (b) Find the probability that if k points are taken at random in this same interval $0 < x < 1$ all but one of them will fall in the interval $1/k < x < 1$.

98. Assume x and y are independently distributed as $f(x)$ and $g(y)$. If $f(x) = 3x^2$ for $0 < x < 1$ and $g(y) = 4/y^3$ for $2 < y < 4$, find the density of $x^2 + y^2$.

99. Three points are taken at random on a circle. Show that the probability that they lie on the same semicircle is $3/4$. Assume the limit of elementary intervals of arc are equally probable.

100. Let r be the total number of dots obtained in n throws of a true

die considering only the faces 2, 4, 6, and let s be the total considering only the faces 1, 3, 5. Set up the moment generating function of r and s, and find the covariance between r and s. What is the moment generating function of $r + s$?

101. Two subway trains A,B arrive at random between 8 A.M. ($t = 0$) and 9 A.M. ($t = 1$). Assume that the arrival times t_A and t_B of A and B respectively are independent, each with uniform density in the interval $0 \leqslant t \leqslant 1$. Then $\tau = \min{(t_A, t_B)}$ is the arrival time of the earlier train. Find the probability density function of τ.

102. In a certain manufacturing process let the fractional output, that is, the number of finished pieces divided by the starting quantity be denoted by x. The probability distribution of the fractional output is represented very closely by a curve of the form

$$f(x) = kx^{a-1}(1 - x)^{b-1} \qquad \text{for } 0 \leqslant x \leqslant 1.$$

Suppose a customer orders an amount A and will accept $(1.1)A$. The maximum operating efficiency will be obtained when the finished pieces lie in the acceptable range. Let e_1 be the proportionate efficiency rating if the number of finished pieces falls short of A and let e_2 be the efficiency rating if the finished quantity exceeds $(1.1)A$. These efficiency ratings will be fractions which depend upon the long-term importance of the two events they represent. Assume that e_1 and e_2 are constant.

If S is the (unknown) starting quantity, set up an expression for the overall efficiency E and show that it is a maximum when the value of S is so chosen that

$$\frac{A}{S} = \frac{1 - c}{1 - (1.1)c},$$

where

$$c = \left[\frac{(1.1)^a(1 - e_2)}{1 - e_1} \right]^{1/(b-1)}.$$

103. What is the probability that two randomly chosen numbers between 0 and 1 will have a sum no greater than 1 and a product no greater than $2/9$?

104. The random variable x has the distribution function

$$F(x) = \begin{cases} \frac{1}{4} e^x & \text{for } -\infty < x < 0 \\ \frac{1}{2} & \text{for } 0 \leqslant x < 1 \\ 1 - \frac{1}{2} e^{-(x-1)} & \text{for } 1 \leqslant x < \infty \end{cases}$$

(a) What is the probability of each of the events: $x < 0$, $x \leqslant 0$, $x = 0$, $x \geqslant 0$, $x = 1$, $|x| < 1$, $0 < x < 1$?

(b) Find the moment generating function, mean, and variance.

105. Find the mode and median of the exponential distribution.

106. Given a density $f(x)$ with $\mu = 10$, $\sigma = 2$, $\mu_3 = {}^1/_{10}$ and $\mu_4 = 4$. A sample of five observations are drawn x_1, x_2, x_3, x_4, and x_5 at random from this distribution; and three statistics y_1, y_2, and y_3 are computed where

$$y_1 = x_1 + 2x_2 + x_3, \qquad y_2 = x_2 + 2x_3 + x_4, \qquad y_3 = x_3 + 2x_4 + x_5.$$

Find the mean, standard deviation, skewness, and flatness of the distribution of $y_1 + y_2 + y_3$.

107. If the covariance is $\rho\sigma^2$ between each two individuals in a sample of n where σ^2 is the variance of the random variable in the population, show that the variance of the mean of the sample is $(\sigma^2 / n) + [1 - (1/n)]\rho\sigma^2$ and hence that the mean of the sample does not converge stochastically to the mean of the population unless $\rho = 0$.

108. Suppose a college entrance examination has $2k$ questions on it. Let x_1, x_2, $\cdots$, x_{2k} be the variables indicating scores on the questions. Suppose the distribution function $f(x_1, x_2, \cdots, x_{2k})$ for a population of entrance candidates is such that the variance of each x is σ^2, and the correlation between each pair of questions has a positive value ρ. Let T_1 be the total score obtained by adding up the scores on the odd questions, and T_2 that for the even ones. Show that the correlation coefficient between T_1 and T_2 is $\rho k/(1 - \rho + \rho k)$, which can be made as near to unity as we please by making the examinations sufficiently long.

109. The random variables x and y have identical continuous distributions. Two observations are drawn at random, one from x and one from y. Denoting their values by x' and y' respectively prove that $P(x' \leqslant y') = {}^1/_2$ regardless of the form of the distribution function.

110. The random variable x has density function $f(x) = cx^{-\alpha}$ for $1 \leqslant x < \infty$, and $f(x) = 0$ otherwise, where $\alpha > 1$ is a given constant and c is a constant depending on α. Compute the value of c and the distribution function $F(x)$, the mode, the median, the mean, and the variance.

111. Let x be a discrete random variable having the probability mass function $f(-1) = f(1) = {}^1/_2$. For $\epsilon = 0.9, 0.99, 1$, and 2, compute

exact values of $P\{|x| > \epsilon\}$, and compare these values with those obtained from the upper bound $1/\epsilon^2$ provided by Chebyshev's inequality.

112. Let x, y, and z be independent random variables whose respective standard deviations are 12, 5, and 9. Find numerical values for the covariance and the coefficient of correlation of the two random variables $u = x + y$ and $v = x - z$.

113. From an urn containing a white balls and b black ones, a certain number of balls, k, is drawn, and they are laid aside, their color unnoted. Then one more ball is drawn. Find the probability that it is a white ball.

114. If x is standard normal and $y = x^2$, we have shown that the distribution of y is given by

$$f(y) = \frac{1}{\sqrt{2\pi}} y^{-1/2} e^{-y/2} \qquad \text{for } 0 \leqslant y < \infty.$$

(a) Find the mean and variance of y and check your answers from the facts that $E(x^2) = 1$, $E(x^4) = 3$.
(b) Using these results find the mean and variance of chi-square with n degrees of freedom by direct appeal to the definition of chi-square, together with basic theorems concerning the mean and variance of a linear function of independent variables.

115. The variable z is defined by the equation $z = x/y$. Find the expected value of z if x and y are independent and

$$f_1(x) = \frac{1}{(1 + x^2)^{3/2}} \qquad \text{for } 0 \leqslant x < \infty,$$

$$f_2(y) = ye^{-y} \qquad \text{for } 0 \leqslant y < \infty.$$

116. If x_1 and x_2 are random observations of the variable x where $f(x) = 1$ for $0 \leqslant x \leqslant 1$ and $y = x_1 + x_2$, we have previously proved that

$$g(y) = y \qquad \text{for } 0 \leqslant y \leqslant 1,$$
$$g(y) = 2 - y \qquad \text{for } 1 \leqslant y \leqslant 2.$$

From the distribution of y, it can then be shown that

$$M(\theta;y) = \frac{(e^\theta - 1)^2}{\theta^2}.$$

Arrive at this same moment generating function without making use of the distribution of y.

117. Making an initial substitution $u^2 = 2y$, reduce the following integral to a form that can be evaluated as a beta function, and show that it is equal to unity.

$$\int_0^\infty \frac{6du}{(2 + u^2)^{5/2}}.$$

118. The independent variables x,y are distributed as follows:

$$f(x) = \frac{2}{\sqrt{2\pi}} e^{-x^2/2} \qquad \text{for } 0 \le x < \infty,$$

$$g(y) = 2y^3 e^{-y^2} \qquad \text{for } 0 \le y < \infty.$$

Taking $u = x/y$, $v = y$, show that the distribution of u is

$$h(u) = \frac{6}{(2 + u^2)^{5/2}} \qquad \text{for } 0 \le u < \infty.$$

To receive credit on this problem, the student must carry out the integrations and perform the necessary manipulations.

119. Find the expected value of $(x + y)^2$ where x and y are independent standard normal variables.

120. A sample of n items is drawn without replacement from a lot of N items, of which Np are defective. What is the expected value of the number of defective items in the sample?

121. Over the right-triangular region bounded by the lines $y = 0$, $y = x$, $x = 1$ the joint density function of x and y is

$$f(x,y) = 24y(1 - x),$$

and the density is zero outside this region. Find the marginal density of x, the conditional density of y, and the regression function of y on x.

122. Find the mean and standard deviation of $y = 1/x$ where

$$f(x) = \tfrac{1}{2} x^2 e^{-x} \qquad \text{for } 0 \le x < \infty.$$

123. Given the joint density function $f(x,y) = \tfrac{5}{2} (1 - y)$ defined for the area lying in the first quadrant and bounded by $x = 0$, $y = 0$, and $y = 1 - x^2$.
    (a) Find the marginal densities.
    (b) Find the conditional densities.
    (c) Find the two regression curves.
    (d) Calculate ρ.

124. The variable x has the Cauchy distribution:

$$f(x) = \frac{1}{\pi} \frac{1}{1 + x^2} \qquad \text{for } -\infty < x < \infty.$$

Find the distribution of y if $y = 1 - x^3$.

125. Let a random sample $x_1, x_2, \cdots, x_n$ be drawn from a rectangular distribution with density function

$$f(x) = \begin{cases} 1/b & \text{for } 0 \leqslant x \leqslant b, \\ 0 & \text{otherwise.} \end{cases}$$

Find the distribution function of $y = \max (x_i)$, and hence show that

$$E\left[\left(1 + \frac{1}{n}\right)y\right] = b.$$

126. Given $f(x) = Cx^2 e^{-x}$ for $0 < x < \infty$.
   (a) Find $M(\theta;x)$.
   (b) Find μ_x and σ_x.
   (c) If a sample of size n is drawn at random from the given distribution and the average $(\bar{x})$ of this sample is computed, this average will have a distribution. Derive $\mu_{\bar{x}}$ and $\sigma_{\bar{x}}$.

127. Independent random variables x and y are distributed as chi-square with $2a$ and $2b$ degrees of freedon respectively. Making the change of variables $u = x + y$, $v = x/(x + y)$, derive the distribution of v.

128. The random variables x,y are jointly distributed as

$$f(x,y) = \frac{|x|}{2\pi y^2} \exp\left\{-\left[\frac{x^2}{2}\left(1 + \frac{1}{y^2}\right)\right]\right\} \qquad \text{for } -\infty < x < \infty, \ -\infty < y < \infty.$$

Find the distribution of $v = x/y$. Choose $u = x$.

129. Independent variables x_1, x_2, x_3 have the Poisson distribution with respective parameters $\mu_1 = 0.27$, $\mu_2 = 0.32$, $\mu_3 = 0.41$. Determine the probability that $y \geqslant 4$, where $y = x_1 + x_2 + x_3$.

130. According to Mendelian theory, the probability of a certain trait is 0.25. In a sample of 192 independent observations, 51 instances of this trait were observed. Using the normal approximation, test the significance of this result.

131. The independent variables x_1, x_2 are normally distributed with respective means 3, -2 and variances 4, 3. Write down the distribution of the function $y = 1 + x_1 + 2x_2$.

132. (a) Find $E[(y/x^2) + (x/y)]$ if x and y are independent with

$$f(x) = 12x^2(1 - x) \qquad \text{for } 0 < x < 1,$$

$$g(y) = 2y \qquad \text{for } 0 < y < 1.$$

(b) Find the moment generating function of the probability density (with $a > b > 0$).

$$f(t) = \frac{ab}{a - b}(e^{-bt} - e^{-at}) \qquad \text{for } 0 \leqslant t < \infty.$$

Thus complete the sentence: "The random variable t is the sum of two independent ——— variables."

133. Let x_1, x_2, x_3 be the coordinates of a point in a set of three-dimensional rectangular coordinates, and let $f(x_1, x_2, x_3)$ be a probability density function defined in this space. If new coordinates r, ϕ_1, ϕ_2 are introduced, the transformation equations being $x_1 = r \cos \phi_1$, $x_2 = r \sin \phi_1 \sin \phi_2$, $x_3 = r \sin \phi_1 \cos \phi_2$, what is the transformed density function? What is the transformed density function in case $f(x_1, x_2, x_3)$ depends only on $x_1^2 + x_2^2 + x_3^2$?

134. Two unbiased coins and an unbiased die are tossed simultaneously. Let x be the number of heads showing on the two coins and y be the number of spots on the upturned face of the die. Describe a sample space for this experiment, derive the joint probability function of x and y, directly compute $E(xy)$, and show by actual computation that the covariance of x and y is 0.

135. The joint density function of x and y has the form

$$f(x,y) = \begin{cases} cx(x - y) & \text{for } 0 < x < 2, \ -x < y < x, \\ 0 & \text{otherwise.} \end{cases}$$

Evaluate the constant c, and find the marginal density functions of x and y.

136. The independent random variables x and y have exponential density functions with parameters $\lambda = 2$ and $\lambda = 1$ respectively. Find the probability that $x + y > 2$.

137. Compute the covariance and the coefficient of correlation of two random variables having the joint density function $f(x,y) = .8xy$ for $0 < x < 1$, $0 < y < x$, and $f(x,y) = 0$ otherwise.

138. Two urns identical in appearance contain respectively three white and two black balls; two white and five black balls. One urn is selected and a ball taken from it. What is the probability that this ball is white?

139. A window is constructed in the form of a rectangle surmounted

by a semicircle. The height of the rectangle is x and this is a number that is equally likely to be any number from 0 to 1. The width y is the side on which the semicircle is constructed; y is equally likely to be between two and four. What is the probability that the area of the window will be greater than $2(\pi + 2)$?

140. What is the probability that the birthdays of six people will fall in exactly two calendar months? (Assume equal probabilities for the 12 months.)

141. Three urns contain respectively one white and two black balls; three white and one black ball; two white and three black balls. One ball is taken from each urn. What is the probability that among the balls drawn there are two white and one black?

142. What is the probability of throwing nine heads twice in five throws of 10 true coins?

143. In firing a rifle at a target from a given distance, suppose the probability of hitting the bull's-eye is 0.4. What is the smallest number of shots which must be fired in order to have a probability of at least 0.9 of hitting the bull's-eye at least once?

144. Let the proportions of families in a given state having 0, 1, 2, 3, $\cdots$ children be $p_0, p_1, p_2, p_3, \cdots$ ($\Sigma p_i = 1$). Find the proportion of all families having exactly j sons. (Assume that the probability of having sons or daughters is equal.)

145. Let the radius of the base of a circular cone be x, and the height of the cone be y. If all values of x from 0 to 1 are equally likely, and all values of y from 0 to 2 are equally likely, what is the probability that the volume of the cone be less than $\pi/3$.

146. Suppose that the probability that n items are made per day in a factory is $e^{-\lambda}\lambda^n/n!$ for $n = 0, 1, 2, \cdots$. Each item has probability p of working correctly. Show that the probability that exactly x working items are made per day is $e^{-\lambda p}(\lambda p)^x/x!$.

147. Two observations x_1 and x_2 of the variable x are taken at random. Given that

$$f(x) = \frac{1}{\sqrt{\pi}} x^{-1/2} e^{-x} \qquad \text{for } 0 \leqslant x < \infty,$$

derive the density of $y = x_1 + x_2$.

148. The variables x, y are jointly distributed as follows:

$$f(x,y) = 120x(y - x)(1 - y) \qquad \text{for } 0 \leqslant x \leqslant y, 0 \leqslant y \leqslant 1$$

(a) Find the marginal density of y.
(b) Find the conditional density of x for a fixed value of y.
(c) Derive the density of the quotient $u = x/y$.

149. The cost of operating a hydroelectric plant for a specific period consists of a fixed charge of K_1 dollars independent of the amount of electricity generated and a cost of K_2 dollars per kilowatt hour for generation. All of the power generated can be sold at a price which is K_3 dollars per kilowatt hour. The possible amount of generation is related to the area's rainfall, which has a certain probability distribution thus permitting the computation of the probability density function of kilowatt hours generated as $f(x) = \alpha^2 x e^{-\alpha x}$ for $0 < x < \infty$. What is the expected value of the profit?

150. In reference to problem 149, by finding the probability density function of the profit, determine the probability of going into bankruptcy at the end of the period if only K_4 dollars are available as reserves at the beginning of the period with no borrowing capacity. $K_4 < K_1$.

151. In a building with n lamps lighted the same time per day, we replace a bulb immediately when it burns out. The costs for replacing one bulb consists of purchasing cost a_1, the cost a_2 of changing the bulb (i.e., taking out the old and putting a new one in the lamp) and all the costs a_3 for bringing the bulb, ladder, and any other necessary things to the required place. Another alternative is to change the bulbs at the same time periodically with time period T, but still change a bulb when it burns out. When all the bulbs are to be changed together, the cost a_1 and a_2 are the same per bulb, but instead of a_3, we now have a cost a_4 for bringing all the bulbs and other materials to the place. We want to determine T such that the average cost per unit of time the bulbs are lit becomes a minimum. Assume the lifetime of a bulb is a random variable exponentially distributed with mean m.

152. The volume v of shipping from a given warehouse per week is subject to random fluctuations and has the probability density $f(v) = v^2(1 - v)$ for $0 < v < 1$, where the volume is measured in thousands of pieces. The cost per piece is subject to random variations and we shall assume this to be a uniform distribution of the form $g(c) = 10$ for 25 cents $< c <$ 35 cents. What is the probability that the cost of operating the shipping room will exceed 275 dollars for a given week?

153. A manufacturer on a particular item has a fixed cost of $50,000. He has a cost of $1.00 per piece for manufacturing, including raw

material. He has other variable costs which decrease with volume v and can be represented by $15/v$ where v is expressed in millions of pieces produced. From previous experience with the sales department, he determined the probability density for the sales of the form $f(v) = v^3 e^{-v}/6$ for $0 < v < \infty$ where v is again expressed in millions of pieces. What is the lowest price he can sell the item for in order that the expected value of the profit is \$100,000.

154. Automobiles stopping for gasoline at a service station represent events in a Poisson process with density $\lambda = 20$ per hour. What is the probability that the time interval between successive automobiles will be (a) longer than 3 minutes, (b) longer than 6 minutes, and (c) between 2 and 4 minutes?

155. The probability density of a certain sales volume v during a given period is $f(v) = (^3/_{2500})v^2(10 - v)$ for $0 < v < 10$ and the units are in hundreds of pieces. The cost of the raw material is \$3.00 per item with an added cost of \$2.00 per item for manufacturing cost. The raw material is only made up as orders are received, and raw material that is not used will be sold for scrap at \$1.00 per piece. Find an expression for the number of items of raw material we should order at the beginning of the period as a function of the mark-up price m per item, under the restriction that the expected value of the profit will be a maximum. The raw material must be on hand at the beginning of the period for fulfillment of orders.

156. A team for a certain game of skill is made up of two players; an individual player can score 0, 1, or 2 points, and the score for the team is the sum of the scores of the two players. Assuming the scores of the two players independently distributed, find the probability distribution of scores for the team represented by the following information. What laws of probability are involved in the solution of this problem?

| | Probability | |
Score	Player A	Player B
0	0.5	0.2
1	0.3	0.3
2	0.2	0.5

157. In matching pennies three people A, B, C flip one coin each at a given signal and then compare results. If one coin shows a different side from the other two, the "odd" player wins; if all three are alike the toss is repeated until one is different. A "match" is

defined as any situation with one coin different from the other two. List the distinct possibilities. Assuming fair play with unbiased coins, show that each player has an equal chance of winning. Find the probability that in a series of 6 such matches, C would lose at least 5 times.

158. In a series of n matches as described in Problem 157, C loses every time. What quantity would be needed in order to determine, by Bayes' Theorem, the probability that A and B are in collusion (i.e., by a prearranged system, always showing opposite faces of the coin)? Find this probability as a function of said quantity. Explain what the probability means in this connection.

159. Find the density of $y = 1/x$ where $f(x) = 320x/(x + 2)^6$ for $0 \leqslant x < \infty$.

160. Multiple choice: Choose the best answer.
   (A) The distribution of the waiting time for the emission of 1000 α-particles is: (a) Poisson, (b) hypergeometric, (c) beta, (d) gamma, (e) exponential.
   (B) What distribution applies to sampling from a finite population without replacement? (a) binomial, (b) Poisson, (c) geometric, (d) hypergeometric, (e) multinomial.
   (C) If x is a nonnegative variable with a skewed distribution, then the distribution of which of the following tends to be more nearly symmetrical than the distribution of x? (a) $x^{1/2}$, (b) x^2, (c) x^3, (d) $-x$, (e) e^x.
   (D) The distribution of the sum of two independent variables drawn from the same uniform distribution is: (a) uniform, (b) triangular, (c) normal, (d) gamma, (e) hypergeometric.
   (E) If x is a variable with density function $f(x)$, then $y = |x|$ has the density function: (a) $f(y) + f(-y)$, (b) $f(y)$, (c) $f(-y)$, (d) $f'(y)$, (e) $|f(y)|$.
   (F) The sum of 10 independent variables all governed by the same exponential distribution is a/an: (a) exponential variable, (b) Poisson variable, (c) normal variable, (d) gamma variable, (e) beta variable.
   (G) If x is a gamma variable with $r = 3$, $\lambda = 1$, then the probability is 0.90 that x will exceed: (a) 2.000, (b) 1.102, (c) 5.322, (d) 0.436, (e) 1.414.
   (H) The probability transformation exists for: (a) only the normal distribution; (b) only for equally likely distributions; (c) only for independent events of the classical theory; (d) for all events of the classical theory; (e) for all distributions.

(I) The mean of the discrete variable x distributed as $f(x) = (1 - \alpha)\alpha^x$, $(0 < \alpha < 1$, $x = 0, 1, 2, \cdots)$, is: (a) 0, (b) αx, (c) $\alpha/(1 - \alpha)$, (d) $(1 - \alpha)/\alpha$, (e) $\alpha - x$.

(J) If x is a standard normal variable, then the probability of $|x|$ exceeding 1.4 is: (a) 0.0808, (b) 0.1615, (c) 0.8385, (d) 0.9192, (e) 0.1497.

(K) If x and y are two random variables with density functions $f(x)$ and $g(y)$ respectively, then the product xy: (a) has no distribution; (b) has density function $f(x)g(y)$, (c) is a parameter of the joint distribution; (d) is a random variable; (e) has density function $(1/2)f(x)g(y)$.

(L) A continuous random variable with a bona fide distribution: (a) fluctuates from sample to sample; (b) is parametric; (c) takes a fixed value depending on the distribution; (d) is a normal deviate; (e) is approximately normal except under sampling conditions.

(M) The joint density function of independent random variables can be constructed by taking the product of their respective (a) distribution functions; (b) marginal density functions; (c) nth partial derivatives; (d) Jacobians; (e) normal density functions.

(N) The value of a continuous distribution function at the point x gives the probability that: (a) the random variable lies between x and $x + dx$; (b) the random variable is equal to $x + dx$; (c) the random variable is greater than x; (d) the random variable is less than x; (e) the random variable is equal to $x \, dx$.

161. Given $f(x) = 1/x^2$ for $1 < x < \infty$. Five observations are drawn at random and ranked according to their magnitude from one to five. Find the distribution of the smallest one.

162. From a company of 100 soldiers, 3 soldiers are to be chosen for a suicide mission. The captain puts 100 tickets into a hat, three of the tickets being marked with an X. The 100 soldiers are lined up and each in turn picks a ticket at random from the hat. Are the chances of drawing an X equal for all the soldiers, and if not what is the best place in the line for a soldier to be?

163. Given $f(x) = 2x$ for $0 < x < 1$ and $g(y) = 1/2$ for $1 < y < 3$. Find the density function of $u = x/y$.

164. The volume of sales of a particular item is equally likely to be anywhere between one and two million pieces, the cost to make the first million is $100,000 or 10 cents a piece. The cost per piece

of the excess x over one million is

$$C = 2 \cdot 10^{-13}(x - 5 \cdot 10^5)^2 + 0.05 \qquad \text{for } 0 \le x \le 1,000,000.$$

If the selling price is 15 cents and the scrap value for those not sold is 5 cents, how many should be made to maximize the expected profit?

165. Multiple choice: Choose the best answer.
(A) If the random variable x has the density function $f(x)$ and the independent random variable y has the density function $g(y)$ then the sum $x + y$: (a) has density function $f(x) + g(y)$, (b) has density function $\frac{1}{2}[f(x) + g(y)]$; (c) has density function $\int f(z - y)g(y)dy$; (d) has no density function; (e) has density function $g(x + y)f(x + y)$.
(B) The distribution function of a discrete random variable is not: (a) a step function; (b) a monotonic function; (c) a continuous function; (d) a discontinuous function; (e) a distribution function in the ordinary sense.
(C) A typical example of a Poisson variable is: (a) any exponential variable; (b) chi-square variable; (c) normal approximation; (d) the number of α-particles emitted per unit time; (e) the time for the emission of n α-particles.
(D) The density function may be found from the distribution function by: (a) integration; (b) summation; (c) use of the Jacobian; (d) change of variables; (e) differentiation.
(E) An example of a monotonic increasing function is: (a) x^2 for all values of x; (b) the normal density function; (c) distribution function; (d) $\sin x$; (e) any statistical transformation.
(F) The sum of two independent normally distributed variables is distributed: (a) normally, (b) as chi-square, (c) exponentially, (d) geometrically, (e) hypergeometrically.
(G) A normal random variable has: (a) a fixed value; (b) a continuous distribution function; (c) a finite range; (d) $\mu = 0$, $\sigma = 1$; (e) no finite distribution.
(H) The probability transformation is defined by: (a) $y = f(x)$; (b) the null hypothesis; (c) $y = F(x)$; (d) the relative frequency theory; (e) the classical theory.
(I) If $y = x^2$ for all values of x, if the density function of x is $g(x)$, and if the density function of y is $f(y)$, then:

(a) $g(y) = \dfrac{f(\sqrt{y}) + f(-\sqrt{y})}{2\sqrt{y}}$;   (b) $f(y) = \dfrac{g(\sqrt{y}) + g(-\sqrt{y})}{2\sqrt{y}}$;

(c) $f(y) = g(x)\left|\dfrac{dx}{dy}\right|$;   (d) $g(x) = f(x^2)$;   (e) $g(y) = 1$

(J) The distribution of a random variable irrespective of the values of any other random variables is the: (a) conditional distribution; (b) marginal distribution; (c) normal distribution; (d) joint distribution; (e) inverse distribution.

(K) Which of the following does not permit the simplification of one-parameter tabulation? (a) the gamma distribution; (b) the chi-square distribution; (c) the beta distribution; (d) the exponential distribution; (e) the Poisson distribution.

(L) What distribution is characterized by a single parameter? (a) binomial, (b) hypergeometric, (c) normal, (d) beta, (e) geometric.

(M) If x is a standard normal varible, the probability of x^2 exceeding 1 is: (a) 0.2420, (b) 0.8413, (c) 0.1587, (d) 0.3173, (e) 0.6827.

(N) For a beta distribution, an interchange of parameters yields: (a) a concentration in the neighborhood of the mode; (b) an inversion of the distribution; (c) the normal approximation; (d) the mirror image of the initial density function; (e) a negative density function.

166. Three urns contain respectively one white, and two black balls; three white and one black ball; two white and three black balls. One ball is taken from each urn. What is the probability that among the balls drawn there are two white and one black?

167. Suppose the probability of hitting the bull's-eye on a single shot is 0.4. What is the smallest number of shots that must be fired in order to have a probability of at least 0.9 of hitting the bull's-eye at least once?

168. A machine sorts turkey eggs and chicken eggs by weight x. If the density function of the chicken egg population is

$$f(x) = \begin{cases} -2 + x & \text{for } 2 \leqslant x \leqslant 3, \\ 4 - x & \text{for } 3 \leqslant x \leqslant 4, \\ 0 & \text{otherwise;} \end{cases}$$

and the density function of the turkey egg population is

$$g(x) = \begin{cases} 1 & \text{for } 3.5 \leqslant x \leqslant 4.5, \\ 0 & \text{otherwise.} \end{cases}$$

Find the critical region for $\alpha = 0.08$. (Let the type I error be the machine throwing a chicken egg into the turkey egg box.) What is the value of β?

169. Let x be a random variable on the range $0 \leqslant x < \infty$. Let

$$P(x \geqslant a) = (1 + a/k)^{-k}$$

for any $a \geq 0$. Find the distribution function and the density function of x.

170. What is the probability of obtaining a given sum s of points with n dice?

171. The position of a random point with coordinates (x,y) is equally probable inside a square with side 1 and whose center coincides with the origin. Determine the probability density function of the random variable $u = xy$.

172. The average height of men 50 years ago was 170 cm (= μ) with a standard deviation of 6.7 cm (= σ). Assume that today the standard deviation is still 6.7 cm. It has been claimed that today the average height of men is 172 cm or less. After examining the heights of 50 randomly selected men you found a sample mean of 173.5 (= $\bar{x}$). Does this sample entitle you to deny the statement that μ is 172 or less?

173. Derive the least-square normal equations for fitting a modified exponential, $y = c + ae^{bx}$, to a set of n points, and indicate why these equations would be difficult to solve.

174. Derive the normal equations for fitting the polynomial

$$y = a_0 + a_1x + a_2x^2 + \cdots + a_kx^k$$

to the points (x_i, y_i), $i = 1, 2, \cdots, n$, in the sense of the principle of least squares.

Appendix B:
Mathematics

**DIFFERENTIATION OF AN INTEGRAL
WITH RESPECT TO A PARAMETER**

Often we work with an integral of the form

$$\phi(\alpha) = \int_{a(\alpha)}^{b(\alpha)} f(x,\alpha)\, dx.$$

This is an integral over x whose lower limit, upper limit, and integrand are all functions of the parameter α. If we desire to obtain the derivative $d\phi(\alpha)/d\alpha$, we could first integrate with respect to x and then differentiate with respect to α. However, it is usually more efficient to use instead the following formula directly:

$$\frac{d\phi(\alpha)}{d\alpha} = \int_{a}^{b} \frac{\partial}{\partial\alpha}\,[f(x,\alpha)]\, dx + f(b,\alpha)\frac{db}{d\alpha} - f(a,\alpha)\frac{da}{d\alpha}.$$

If we apply this formula to

$$\phi(\alpha) = \int_{\alpha}^{\alpha^2} (1 + \alpha x)\, dx,$$

then

$$\frac{d\phi(\alpha)}{d\alpha} = \int_{\alpha}^{\alpha^2} x\, dx + (1 + \alpha^3)2\alpha - (1 + \alpha^2)(1)$$

$$= \frac{\alpha^4}{2} - \frac{\alpha^2}{2} + 2\alpha + 2\alpha^4 - 1 - \alpha^2 = \frac{5\alpha^4}{2} - \frac{3\alpha^2}{2} + 2\alpha - 1.$$

STIELTJES INTEGRAL

If $F(x)$ is a distribution function of a random variable x where $a < x \leq b$, and $\alpha(x)$ is a continuous function, then the expectation of $\alpha(x)$ is

$$E[\alpha(x)] = \int_a^b \alpha(x)\, dF(x).$$

This integral is a Stieltjes integral, which we now define. First, partition the given interval $a < x \leq b$ into n half-open subintervals $x_{i-1} < x \leq x_i$ by choosing $n + 1$ points $x_0, x_1, \cdots, x_n$ such that $a = x_0 < x_1 < \cdots < x_n = b$. Then choose n points $x_1', x_2', \cdots, x_n'$ such that $x_{i-1} < x_i' < x_i$ for $i = 1, 2, \cdots, n$. The Stieltjes integral of $\alpha(x)$, with respect to $F(x)$ over $a < x \leq b$, is then defined by

$$\int_{a+0}^b \alpha(x)\, dF(x) = \lim_{\delta_n \to 0} \sum_{i=1}^n \alpha(x_i')[F(x_i) - F(x_{i-1})],$$

where the limit is taken over all partitions of the interval $a < x \leq b$ as the maximum length $\delta_n = \max(x_i - x_{i-1})$ for $i = 1, 2, \cdots, n$ of the subintervals tends to 0. (Of course, this requires that n approach ∞).

One can readily see that the Stieltjes integral is a generalization of the ordinary Riemann integral studied in calculus. To do this, replace $F(x)$ by the special function $F(x) = x$, and note that the foregoing equation becomes

$$\int_a^b \alpha(x)\, dx = \lim_{\delta_n \to 0} \sum_{i=1}^n \alpha(x_i')(x_i - x_{i-1}).$$

Let the symbol $c - 0$ imply that c is included in the interval of integration, and the symbol $c + 0$ that c is excluded. Then, if $\phi(c) \neq 0$ and if the distribution function $F(x)$ has a jump at $x = c$, then

$$\int_{c-0}^{c+0} \alpha(x)\, dF(x) = \alpha(c)[F(c + 0) - F(c - 0)].$$

This means that the Stieltjes integral, extended over an interval that reduces to a single point, can yield a result different from zero. In particular if the value of the distribution function $F(x)$ changes only at the discrete points $c_1, c_2, c_3, \cdots$, then we have

$$\int_a^b \alpha(x)\, dF(x) = \sum_{n=1}^\infty \alpha(c_n)[F(c_n + 0) - F(c_n - 0)],$$

which shows that the Stieltjes integral reduces to a discrete summation in such a case. Because the probability mass function is given by the jumps in the distribution function, i.e., because the probability mass function is

$$p(c_n) = F(c_n + 0) - F(c_n - 0),$$

we see that the Stieltjes integral becomes the familar expression

$$\int_a^b \alpha(x)\, dF(x) = \sum_{n=1}^{\infty} \alpha(c_n) p(c_n),$$

which holds in the case of a discrete random variable.

In the case of a continuous random variable with density $f(x)$, the Stieltjes integral reduces to the familar expression

$$\int_a^b \alpha(x)\, dF(x) = \int_a^b \alpha(x) f(x)\, dx,$$

where $dF(x) = f(x)\, dx$. Engineers like to introduce the notion of probability density for a discrete random variable by setting

$$f(x) = \sum_{n=1}^{\infty} p(c_n)\, \delta(x - c_n),$$

where $p(c_n)$ is the probability mass function and $\delta(x)$ is the Dirac delta function. Then the Stieltjes integral (whose claim to fame is that it is good for both discrete and continuous random variables) can be disregarded, and instead the Riemann-looking integral

$$\int_a^b \alpha(x) f(x)\, dx$$

can be used, as it now holds for both continuous and discrete random variables. For example, in the discrete case, the Riemann-looking integral becomes

$$\int_a^b \alpha(x) \sum_{n=1}^{\infty} p(c_n)\, \delta(x - c_n)\, dx = \sum_{n=1}^{\infty} p(c_n) \int_a^b \alpha(x)\, \delta(x - c_n)\, dx$$

$$= \sum_{n=1}^{\infty} p(c_n) \alpha(c_n),$$

which is the familar expression. In conclusion, one has a choice to use either (1) separate expressions in the case of discrete and continuous variables, (2) the same Stieltjes integral for both, or (3) the same Riemann-looking integral for both but with the use of the Dirac delta function.

A USEFUL EQUATION (BY PAUL SMITH, THE UNIVERSITY OF KANSAS)

Given the expression:

$$1 + 1 = 2 \tag{1}$$

It is never in good taste in engineering to express a sum directly. Therefore, it is necessary to simplify this equation. A knowledge of mathematics gives us:

$$1 = \ln e \tag{2}$$

$$1 = \sin^2 x + \cos^2 x. \tag{3}$$

Therefore equation (1) becomes:

$$\ln e + \sin^2 x + \cos^2 x = 2. \tag{4}$$

But

$$2 = 2 \sum_{n=1}^{\infty} \frac{1}{2^n}. \tag{5}$$

Therefore it follows:

$$\ln e + \sin^2 x + \cos^2 x = 2 \sum_{n=1}^{\infty} \frac{1}{2^n}. \tag{6}$$

This equation may be further simplified by use of the expression:

$$\cos^2 x + \sqrt{\frac{(1 - \cos 2x)^2}{4}} = 1. \tag{7}$$

Equation (6) then reduces to:

$$\ln e + \sin^2 x + \cos^2 x = 2 \sum_{n=1}^{\infty} \frac{\cos^2 x + \sqrt{\dfrac{(1 - \cos 2x)^2}{4}}}{2^n} \tag{8}$$

It is intuitively obvious to the most casual observer that equation (8) is much simpler and more easily understood than equation (1). In most elementary courses in calculus this topic in mathematics is not usually included, and therefore is presented here.

Chronological Historical Bibliography

Girolamo Cardano (1501–1576), *De Ludo Aleae*, published posthumously in 1663, 15 folio pages.

Johann Kepler (1571–1630), *De Stella Nova in Pede Serpentarii*, 1606.

Galileo Galilei (1564–1642), *Considerazione sopra il Giuco dei Dadi*, publication date unknown.

Blaise Pascal (1623–1662), Three letters of Pascal to Fermat written in 1654, published in *Varia Opera Mathematica D. Petri de Fermat*. Tolosae, 1679, pp. 179–188.

Christiaan Huygens (1629–1695), *De Ratiociniis in Ludo Aleae*, 1657.

John Graunt (1620–1674), *Natural and Political Observations made upon the Bills of Mortality*, London, 1662.

Gottfried Wilhelm Leibniz (1646–1716), *Dissertatio de Arte Combinatoria*, 1666.

Sir Isaac Newton (1642–1727), on the Binomial Theorem, Letter of June 13, 1676.

John Wallis (1616–1703), *A Discourse of Combinations, Alternations, and Aliquot Parts*, London, 1685.

Jakob Bernoulli (1654–1705), *Journal des Scavans*, 1685; *Acta Eruditorium*, 1690; *Ars Conjectandi*, published posthumously in 1713, 306 pages.

Edmund Halley (1656–1742), *An estimate of the degrees of the mortality of mankind*, *Philosophical Transactions*, London, 1693.

Pierre Remond de Montmort (1678–1719), *Essai d'Analyse sur les Jeux de Hazards*, First Edition 1708, 189 pages; Second Edition, 1714, 414 pages.

Johann Bernoulli (1667–1748), Letter to Montmort, 1714.

Nicolas Bernoulli (1687–1759), *Specimina Artis conjectandi, ad quaestiones Juris applicatae*, Basle, 1709.

Abraham de Moivre (1667–1754), *De Mensura Sortis, seu, de Probabilitate Eventuum in Ludis a Casu Fortuito Pendentibus*, *Philosophical Transactions*, volume 27, pp. 213–264, 1711; *The Doctrine of Chances, or a Method of Calculating the Probabilities of Events in Play*, 1718, 189 pages.

Daniel Bernoulli (1700–1782), *Specimen Theoriae Novae de Mensura Sortis, Commentarii Acad. Petrop.*, volume 5, 1730, pp. 175–192.

George Louis Leclerc de Buffon (1707–1788), Acad. Sci. Paris, 1733; *Essai d'Arithmetique Morale*, 1777, 103 pages.

Thomas Simpson (1710–1761), *The Nature and Laws of Chance*, London, 1740.

Leonhard Euler (1707–1783), *Calcul de la Probabilite dans le Jeu de Recontre*, Berlin, 1751.

Jean Le Rond d'Alembert (1717–1783), *Croix ou Pile, Encyclopedie ou Dictionnaire Raisonne*, 1754.

Thomas Bayes (1702–1761), *An Essay toward solving a Problem in the Doctrine of Chances, Philosophical Transactions*, volume 53, pages 370–418, 1763; *A Demonstration of the Second Rule in the Essay towards the Solution of a Problem in the Doctrine of Chances, Philosophical Transactions*, volume 54, 1764, pages 296–325.

Joseph Louis Lagrange (1736–1813), *Memoire sur l'utilite de la methode de prendre le milieu entre les resultats de plusieurs observations*, 1770.

Johann Bernoulli (1744–1807), *Sur les suites ou sequences dans la loterie de Genes*, Berlin, 1771.

Johann Heinrich Lambert (1728–1777), *Examen d'une espece de Superstition ramenee au calcul des probabilites*, Berlin, 1771.

Pierre Simon de Laplace (1749–1827), *Recherches sur l'integration des Equations differentielles aux differences finies, et sur leur usage dans la theorie des hasards*, 1773; *Theorie analytique des Probabilities*, 1812.

Karl Friedrich Gauss (1777–1855), *Theoria motus corporum coelestium in sectionibus conicis solem ambientium*, 1809.

Simeon Denis Poisson (1781–1840), *Recherches sur la probabilite des jugements en matiere criminelle et en matiere civile*, 1837.

John Venn (1834–1923), *The Logic of Chance*, 1866; *Symbolic Logic*, 1881.

Pafnuti Lvovich Chebyshev (1821–1894), *On two theorems concerning probability*, Zap. Akad. Nauk., 1887.

Bibliography

Allen, A. O., *Probability, Statistics, and Queuing Theory with Computer Science Applications*, Academic Press, New York, 1978.

Arley, N., *On the Theory of Stochastic Processes and Their Application to the Theory of Cosmic Radiation*, Gads Forlag, Copenhagen, 1943.

Ash, R. B., *Basic Probability Theory*, Wiley, New York, 1970.

Bharucha-Reid, A. T., *Elements of the Theory of Markov Processes and Their Applications*, McGraw-Hill, New York, 1960.

Bush, R. R., and Mosteller, F., *Stochastic Models for Learning*, Wiley, New York, 1955.

Bush, R. R., and Estes, W. K., eds., *Studies in Mathematical Learning Theory*, Stanford University Press, Stanford, 1959.

Clarke, A. B., and Disney, R. L., *Probability and Random Processes for Engineers and Scientists*, Wiley, 1970.

Cramer, H., *Mathematical Methods of Statistics*, Princeton University Press, Princeton, 1946.

David, F. N., *Games, Gods, and Gambling*, Macmillan, New York, 1962.

Drake, A. W., *Fundamentals of Applied Probability Theory*, McGraw-Hill, New York, 1967.

Dwass, M., *First Steps in Probability*, McGraw-Hill, New York, 1967.

Dwass, M., *Probability Theory and Applications*, Benjamin, New York, 1970.

Feller, W., *An Introduction to Probability Theory and Its Applications*, 2 vols., Wiley, New York, 1957 and 1966.

Fisher, R. A., *Statistical Methods for Research Workers*, 11th ed., Oliver & Boyd, Edinburgh, 1950.

Fry, T. C., *Probability and Its Engineering Uses*, Van Nostrand, Princeton, 1968.

Gnedenko, B., *The Theory of Probability*, MIR Publishers, Moscow, 1969.

Gnedenko, B. V., and Kolmogorov, A. N., *Limit Distributions for Sums of Independent Random Variables* (translated from the Russian by K. L. Chung), Addison-Wesley, Reading, Mass., 1954.

Herdan, G., *The Advanced Theory of Language as Choice and Chance*, Springer-Verlag, New York, 1966.

Hoel, P. G., Port, S. C., and Stone, C. J., *Introduction to Probability Theory*, Houghton Mifflin, Boston, 1971.

Hogg, R. V., and Tanis, E. A., *Probability and Statistical Inference*, Macmillan, New York, 1977.

Johnson, N. L., and Kotz, S., *Distributions in Statistics: Discrete Distributions*, Houghton Mifflin, Boston, 1969.

Johnson, N. L., and Kotz, S., *Distributions in Statistics: Continuous Univariate Distributions*, vols. 1, 2, Houghton Mifflin, Boston, 1970.

Karlin, S., *A First Course in Stochastic Processes*, Academic Press, New York, 1966.

Karlin, S., and Taylor, H. M., *A First Course in Stochastic Processes*, Academic Press, New York, 1975.

Kemeny, G., and Snell, J. L., *Finite Markov Chains*, Van Nostrand-Reinhold, New York, 1960.

Kemeny, J. G., Snell, J. L., and Knapp, A. W., *Denumerable Markov Chains*, Van Nostrand, Princeton, 1966.

Kendall, M. G., and Stuart, A., *The Advanced Theory of Statistics*, Macmillan, New York, 1980.

Kolmogorov, A. N., *Foundations of the Theory of Probability* (translation of the original), Chelsea, New York, 1933, 1950.

Krickeberg, K., *Probability Theory*, Addison-Wesley, Reading, Mass., 1965.

Larson, H. J., *Introduction to Probability Theory and Statistical Inference*, Wiley, New York, 1974.

Lass, H., and Gottlieb, P., *Probability and Statistics*, Addison-Wesley, Reading, Mass., 1971.

Lieberman, G. J., and Owen, D. B., *Tables of the Hypergeometric Distribution*, Stanford University Press, Stanford, 1961.

Loeve, M., *Probability Theory*, Van Nostrand, Princeton, 1963.

Miller, I., and Freund, J. E., *Probability and Statistics for Engineers*, Prentice-Hall, Englewood Cliffs, 1977.

Mises, R. von, *Probability, Statistics, and Truth*, 2nd rev. English ed. (prepared by H. Geiringer), Macmillan, New York, 1957.

Molina, E. C., *Poisson's Exponential Binomial Limit*, Van Nostrand, Princeton, 1942.

Mood, A. M., and Graybill, F. A., *Introduction to the Theory of Statistics*, McGraw-Hill, New York, 1963.

Mosteller, F., Rourke, R. E. K., and Thomas, G. B., *Probability with Statistical Applications*, Addison-Wesley, Reading, Mass., 1961.

Olkin, E., Gleser, L. J., and Derman, C., *Probability Models and Applications*, Macmillan, New York, 1980.

Parzen, E., *Modern Probability Theory and Its Applications*, Wiley, New York, 1960.

Parzen, E., *Stochastic Processes*, Holden-Day, Oakland, 1962.

Pearson, K., *Tables of the Incomplete Beta-Function*, University Press, Cambridge, 1934.

Prabhu, N. U., *Stochastic Processes: Basic Theory and Its Applications*, Macmillan, New York, 1965.

Rasch, G., *Probabilistic Models for Some Intelligence and Attainment Tests*, Studies *in Mathematical Psychology I*, Nielson and Lydiche, Copenhagen, 1960.

Riordan, J., *Stochastic Service Systems*, Wiley, New York, 1962.

Robinson, E. A., *An Introduction to Infinitely Many Variates*, Macmillan, New York, 1959.

Robinson, E. A., *Random Wavelets and Cybernetic Systems*, Macmillan, New York, 1962.

Robinson, E. A., *Least Squares Regression Analysis in Terms of Linear Algebra*, IHRDC Press, Boston, 1981.

Robinson, E. A., *Statistical Reasoning and Decision Making*, IHRDC Press, Boston, 1981.

Rutherford, E., Chadwick, J., and Ellis, C. D., *Radiations from Radioactive Substances*, Cambridge University Press, Cambridge, 1920.

Rutherford, E., and Geiger, H., The probability variations in the distribution of a particle, *Philosophical Magazine*, 20 (1910): 698–707.

Savage, L. J., *The Foundations of Statistics*, Wiley, New York, 1954.

Shooman, M. L., *Probabilistic Reliability: An Engineering Approach*, McGraw-Hill, New York, 1968.

Syski, R., *Introduction to Congestion Theory in Telephone Systems*, Oliver & Boyd, Edinburgh, 1960.

Trivedi, K. S., *Probability and Statistics*, Prentice-Hall, Englewood Cliffs, 1982.

Whitworth, W. A., *Choice and Chance*, 5th ed., Macmillan, New York, 1948.

Whitworth, W. A., *DCC Exercises in Choice and Chance*, Macmillan, New York, 1959.

Williamson, E., and Bretherton, M. H., *Tables of the Negative Binomial Distribution*, Wiley, New York, 1963.

Index